I0110043

BISON
BOOKS

BOOKS BY PAUL A. JOHNSGARD PUBLISHED BY
THE UNIVERSITY OF NEBRASKA PRESS

Waterfowl: Their Biology and Natural History (1968)

Grouse and Quails of North America (1973)

North American Game Birds of Upland and Shoreline (1975)

The Bird Decoy: An American Art Form (1976)

The Plovers, Sandpipers and Snipes of the World (1981)

The Grouse of the World (1983)

The Platte: Channels in Time (1984; 2nd ed. 2008)

Those of the Gray Wind: The Sandhill Cranes (1986; 2nd ed. 2017)

Diving Birds of North America (1987)

This Fragile Land: A Natural History of the Nebraska Sandhills (1995)

Crane Music: A Natural History of American Cranes (1997)

The Nature of Nebraska: Ecology and Biodiversity (2001)

Lewis and Clark on the Great Plains: A Natural History (2003)

The Niobrara: A River Running through Time (2007)

Sandhill and Whooping Cranes: Ancient Voices over the America's Wetlands (2011)

Seasons of the Tallgrass Prairie: A Nebraska Year (2014)

PAUL A. JOHNSGARD

WILDLIFE of NEBRASKA

A Natural History

University of Nebraska Press | LINCOLN

© 2020 by the Board of Regents
of the University of Nebraska

Acknowledgments for the use of copyrighted
material appear on page xix, which constitutes
an extension of the copyright page.

All rights reserved. Manufactured
in the United States of America. ∞
Library of Congress Control Number: 2020007888

Designed and set in New Baskerville ITC Pro by L. Auten.

Contents

Illustrations

Acknowledgments

I can't begin to thank all the people who have helped me since I first began wandering over the prairies, sandhills, and wetlands of Nebraska nearly 60 years ago and who have enriched my life and knowledge of this wonderful state. I also thank my many academic colleagues, the staffs of the 17 book publishers who have kindly tolerated my writing, and especially my family, who never complained about the countless hours I have spent writing, rewriting, illustrating, and proofing a seemingly constant and endless series of writing projects. I especially thank the University of Nebraska Press for publishing 18 of my books.

Although the majority of the text in this book is new, while writing it I have often relied, out of a mixture of convenience and the lethargy of advancing age, on some of my older publications. I have extracted and modified parts of the narrative profiles of several species from a companion book on Wyoming wildlife (Johnsgard 2019) and a few from another much earlier survey of Great Plains birds (Johnsgard 1979a). The descriptions of Nebraska's major wildlife viewing areas (chapter 6) are partly derived from my recent *Naturalist's Guide to the Great Plains* (Johnsgard 2018c). Among the people who read this manuscript, I especially thank Prof. Russell Benedict, who critically reviewed the mammal accounts. I also thank Cris Trautner, who let me adapt three articles I wrote for Lincoln's now sadly vanished monthly progressive newspaper, *Prairie Fire* ("Secrets of the very long dead: Ashfall Fossil Beds State Historical Park," "Nebraska: Where the West begins and the East peters out," and "Secrets of the most sincerely dead: Agate Fossil Beds National Monument"). Unless otherwise indicated, the line illustrations and maps are all mine and are in my copyright. The splendid cover was done at my request by Dr. Allison Johnson.

Abbreviations and Symbols

Di: Dispersed (occurs over ca. half of Nebraska)

HL: Highly localized (reported in one or two counties)

Lo: Local (reported in maximum of ten counties)

SD: Slightly dispersed (occurs over ca. one-quarter of Nebraska)

Ub: Ubiquitous (statewide)

Wi: Widespread (occurs over ca. three-quarters of Nebraska)

Cen: Central Nebraska

Ea: Eastern Nebraska

NC: North-central Nebraska

NE: Northeastern Nebraska

No: Northern Nebraska

NW: Northwestern Nebraska

Pan: Panhandle

PR: Pine Ridge

RB: Rainwater Basin

SC: South-central Nebraska

SE: Southeastern Nebraska

SH: Sandhills

So: Southern Nebraska

SW: Southwestern Nebraska

UP: Upland Prairie

We: Western Nebraska

WH: Wildcat Hills

?: Information inadequate for geographic attribution

Ab: Abundant

Co: Common

Oc: Occasional

Ra: Rare

Un: Uncommon

VR: Very rare

?: Abundance status uncertain

OTHER ABUNDANCE/CONSERVATION CATEGORIES

Accidental: A bird far outside its usual range, with very few documented Nebraska records

CC: "At-risk" species of conservation concern in Nebraska (Panella 2010)

CC1: "Watch list" bird species of continental conservation significance (Rich et al. 2004)

Extinct: No longer alive anywhere in the world

Extirpated: Eliminated from Nebraska; still present elsewhere

Hypothetical: Reports of Nebraska presence are not adequately documented

NT: Considered near threatened by Birdlife International

T1-NNLP-18: Tier 1 species (of highest concern) (Schneider et al. 2018)

T2-NNLP-18: Tier 2 species (of secondary concern) (Schneider et al. 2018)

Vagrant: A species far outside its usual range; very few documented Nebraska records

Vul: Considered vulnerable by Birdlife International Seasons

Sp: Spring (March–May)

Su: Summer (June and July)

Fa: Fall (August–November)

Wi: Winter (December–February)

Mig: Spring and fall migrant

Res: Permanent resident (breeding assumed or proven)

RMig: Rare migrant

SpMig: Spring migrant

Vis: Visitor (nonbreeder; present during various seasons)

VRMig: Very rare migrant

VRRes: Very rare resident

VRSuRes: Very rare summer resident

VRWiMig: Very rare winter migrant

WiMig: Wintering migrant

?: Seasonal status uncertain

* Note that the status of some species might include more than one category, depending on regional environments; short-term weather variations; racial, sexual, and/or individual differences in migration tendencies; and other variables.

HABITATS*

FoUD: Forest, upland deciduous (70 spp.)

FoWC: Forest, western coniferous (40 spp.)

GrAll: All grasslands (59 spp.)

GrMG: Mixed-grass prairie (9 spp.)

GrSG: Shortgrass prairie (15 spp.)

GrSH: Sandhills prairie (16 spp.)

GrTG: Tallgrass prairie (14 spp.)

HuCL: Human-modified, cultivated lands (12 spp.)

HuFR: Human-modified, farmsteads, ranches (25 spp.)

HuPG: Human-modified, parks, gardens, bird feeders (47 spp.)

HuTC: Human-modified, towns, cities, villages (12 spp.)

RiDe: Riverine, deciduous woody shoreline (63 spp.)

RiHe: Riverine, herbaceous shoreline (9 spp.)

RiLo: Riverine, lotic (flowing-water) habitats (7 spp.)

RiSS: Riverine, shrub shoreline (15 spp.)

RiWe: Riverine and wetland habitats generally (33 spp.)

RoBL: Rocky, badlands (14 spp.)

RoCl: Rocky, cliffs (8 spp.)

RoUc: Rocky, outcrops (16 spp.)

SaSG: Sage-shortgrass (21 spp.)

TeAll: Diverse terrestrial habitats (10 spp.)

WeDW Deep-water wetlands (lakes, reservoirs) (50 spp.)

WeSW: Shallow-water wetlands (permanent/seasonal wetlands) (131 spp.)

WoJu: Woodlands, juniper (9 spp.)

WoUp: Woodlands, general upland (35 spp.)

* The number of species in parentheses indicates the total number classified in the text as associated with each habitat type in Nebraska.

Wildlife of Nebraska

CHAPTER 1

Introduction to Nebraska and Its Biological Environment

THE GEOGRAPHY AND BIOGEOGRAPHY OF NEBRASKA

The surface topography of Nebraska is primarily that of a slightly inclined plane, sloping down from west to east at an average gradient of about 9 feet per mile. The state's elevations range from more than 5,000 feet in the western Panhandle region to about 825 feet in the extreme southeast. Precipitation likewise increases from the northwest to the southeast from about 15 to 33 inches of total annual precipitation and occurs mostly during spring and summer. The two largest river valleys in the state are the Missouri Valley and the Platte Valley, which both tend to be broad and very gradual, whereas the more recently formed Niobrara Valley in the northern part of the state is deeper and narrower, the shorelines often lined with steep bluffs. Eroded hills and sedimentary escarpments are also typical of the Pine Ridge area in the northern Panhandle, the Wildcat Hills in Scotts Bluff and Banner Counties, and the upper portions of the North Platte Valley (map 1).

The native vegetation of Nebraska once consisted of about 95 percent perennial grasslands (map 2). Major grassland components are the tallgrass bluestem prairie of the eastern third of the state, the mixed-grass prairie lying immediately to the west and merging with the tallgrass prairies, and the Sandhills prairie of north-central Nebraska. Minor components include the sand sage prairie of southwestern Nebraska, the Kansas mixed-grass prairie of southernmost Nebraska, the Dakota prairie of the Pine Ridge, and the arid shortgrass and mixed-grass mosaic of the Panhandle's high plains (Kaul, Sutherland, and Rolfsmeier 2012).

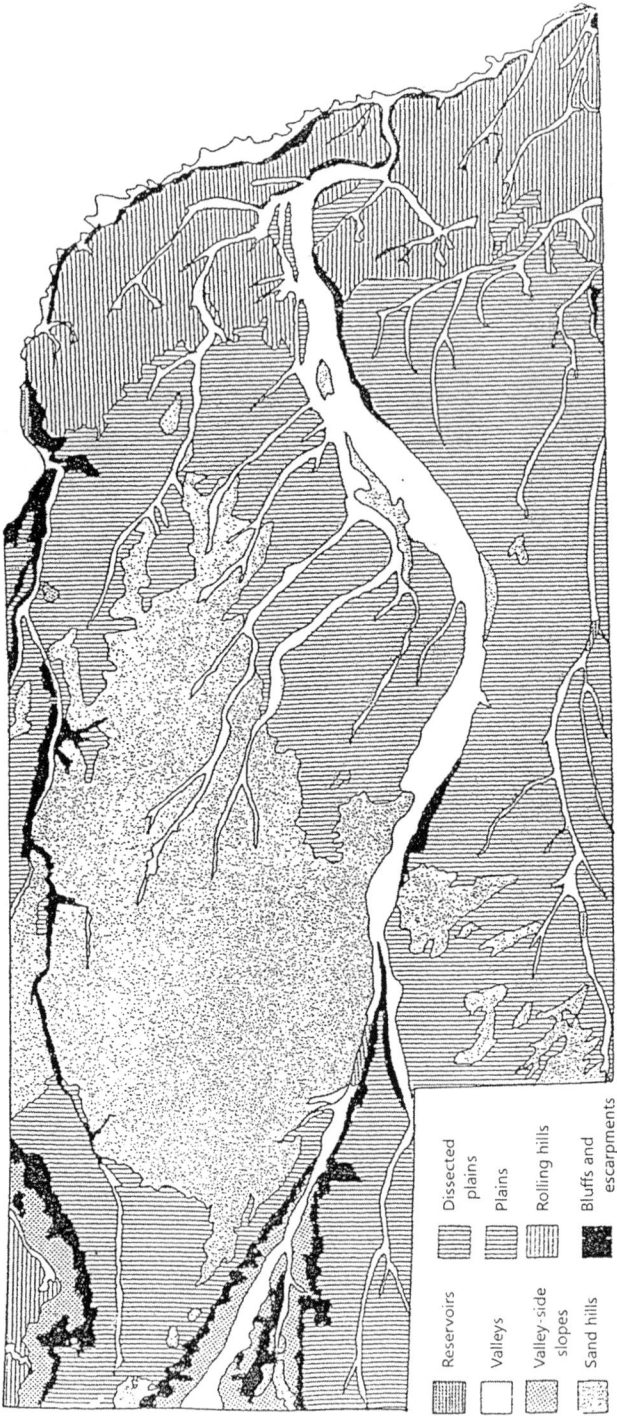

MAP 1. Nebraska landforms. After a map by the Conservation and Survey Division, University of Nebraska.

Reservoirs

Valleys

Valley-side
slopes

Sand hills

Dissected
plains

Plains

Rolling hills

Bluffs and
escarpments

Grass-dominated communities

Mixed prairie

Mixed-Tallgrass transition zone

Tallgrass Bluestem prairie

Kansas mixed prairie

Shortgrass prairie

Dakota prairie

Sandsage prairie

Sandhills prairie

Communities dominated by deciduous or coniferous trees

Rocky Mountain forest

Rocky Mountain–deciduous transition zone

Eastern deciduous forest

Floodplain forest and prairie

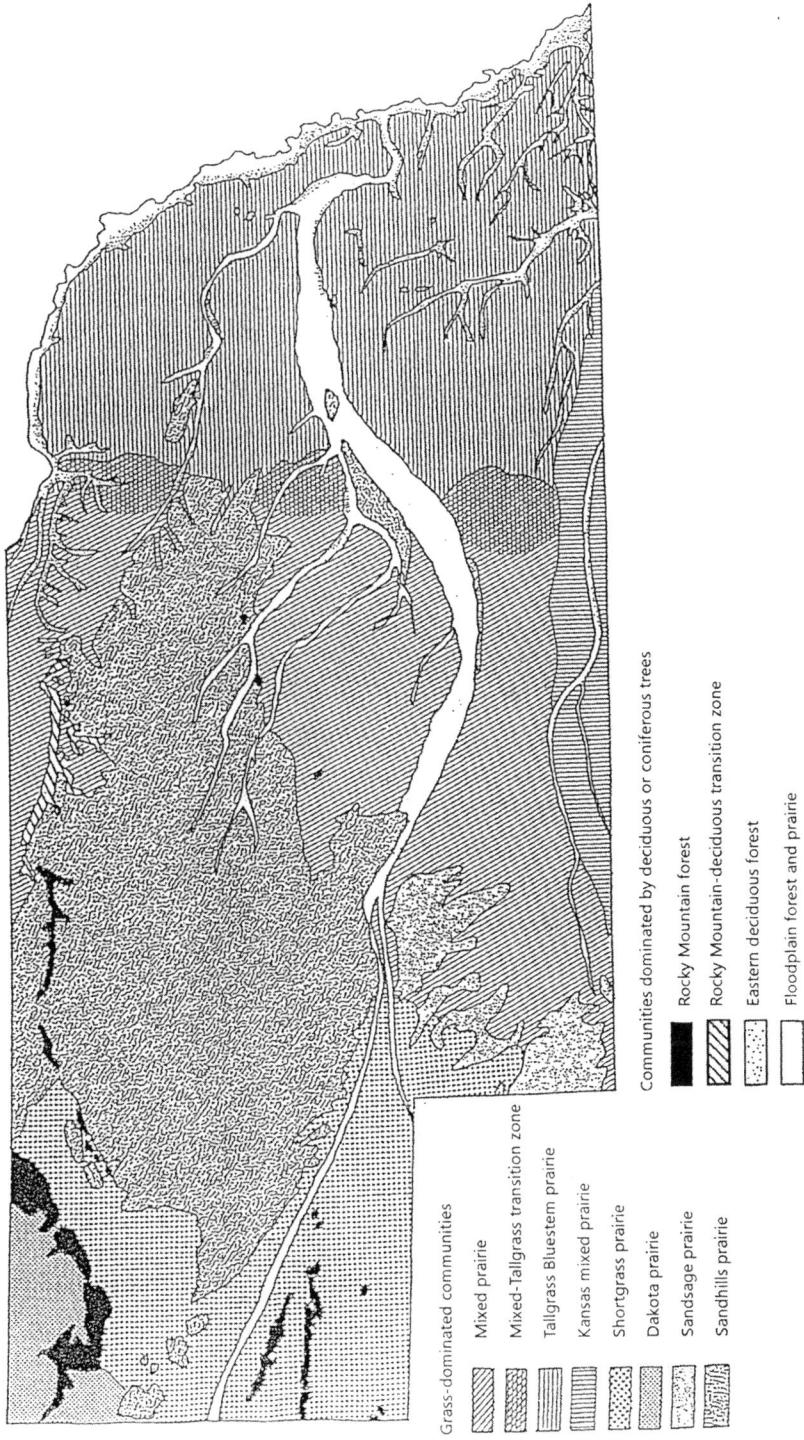

MAP 2. Native vegetation of Nebraska. After a map by Robert Kaul.

3

The wetter eastern portions of the state that still support remnants of tallgrass prairie ("true prairie") contain up to 50 or more species of perennial grasses (e.g., *Andropogon, Panicum, Sorghastrum*), some of which grow to more than 5 feet tall, and a diverse vascular plant flora that might contain 400 species in an area of a few hundred acres (Johnsgard 2018). Along the well-watered central Platte Valley, plant diversity is similar to that of tallgrass prairie. In *A Guide to the Natural History of the Central Platte Valley of Nebraska* (2007a), I listed 73 species of grasses and sedges, 217 forbs, 27 shrubs and woody vines, and 21 trees as representing typical central Platte Valley flora, totaling 338 species.

The significantly drier central portions of the state support mixed-grass prairies (e.g., *Stipa, Aristida, Schizachyrium*) on loess soils that are dominated by lower-stature plants and a typical plant diversity of 164–239 species (Johnsgard 2001). More than a fifth of the state, or 19,000 square miles, consists of the Sandhills and their unique prairie, also dominated by perennial grasses. These plants often have sand-related adaptations, such as long roots that are able to reach groundwater or stems and leaves that can resist the effects of blowing sand, that allow them to survive on the largest region of stabilized sand dunes in North America (Johnsgard 1995). Typical Sandhills grasses (e.g., *Sporobolus, Eragrostis, Redfieldia*) are usually no more than about 3 feet tall, and individual plants also tend to be more scattered than are the densely growing and sod-forming grasses of tallgrass prairie and the bunchgrasses of mixed-grass prairie. In *This Fragile Land: A Natural History of the Nebraska Sandhills* (1995), I listed 32 species of grasses and sedges, 73 forbs and small shrubs, and 11 large shrubs, vines, and small trees, totaling 116 species, as representative of typical Sandhills grassland flora. The Sandhills vertebrate fauna includes 35 mammals, 312 birds (including 34 accidentals), 20 reptiles, and 8 amphibians (Bleed and Flowerday 1988).

The westernmost part of the state, its geographic "Panhandle," primarily consists of shortgrass ("steppe") communities, with a relatively small number of arid- and often alkaline-adapted grasses that are usually only 1–2 feet tall, such as grama grasses (*Bouteloua*) and buffalo grass (*Buchloe*) (Johnsgard 2005). These plant communities often exist as a mixture of short perennial grasses, many annual weedy forbs, and a few drought-adapted shrubs. The so-called sage-steppe shrubs include several species of sage (*Artemisia* spp.) and similarly drought-tolerant rabbitbrush (*Chrysothamnus* spp.). On more alkaline soils, other highly salt-tolerant shrubs are present, such as greasewood (*Sarcobatus vermiculatus*) and saltbush (*Atriplex* spp.). Arid, shrub-dominated scrublands and variably permanent, sometimes alkaline wetlands probably historically constituted up to 2 percent of the state's surface area.

Perhaps up to about 3 percent of Nebraska was once covered by forests, about 25 percent of which were of coniferous trees and 75 percent of deciduous trees. Forests in Nebraska were historically confined to the state's larger river valleys and floodplains in eastern Nebraska and mostly consisted of deciduous trees, such as riverine and floodplain cottonwoods (*Populus* spp.), ashes (*Fraxinus* spp.), hackberries (*Celtis*), and elms (*Ulmus* spp.), with oaks (*Quercus* spp.) and hickories (*Carya* spp.) dominating the uplands. In contrast, the upper Niobrara Valley, the Pine Ridge uplands, and the Wildcat Hills of western Nebraska are all dominated by coniferous trees, mainly fairly tall ponderosa pine (*Pinus ponderosa*) and shorter junipers (*Juniperus* spp.), the latter often forming rather densely spaced woodlands with little herbaceous growth below.

The present-day floodplain forests and adjacent woodlands still range from pines and junipers in the west to a diversity of oaks, cottonwoods, hickories, and woody shrubs in the east, with significant recent losses of American elms to disease and the present-day loss of ashes to invasive beetles. Following the twentieth-century suppression of uncontrolled prairie fires, which once prevented woody vegetation from maturing and reproducing, many Nebraska streams that once lacked large trees have since developed mature corridor forests often dominated by cottonwoods, ashes, and elms on their floodplains. On the other hand, logging, river channeling, dam construction, and associated reservoir impoundments have

destroyed most of the old-growth floodplain forests that historically occupied the richly vegetated bottomlands along all the larger
Nebraskan rivers, especially the Missouri (map 3).

The rivers and smaller streams of Nebraska typically support
dense shrubby and small tree riparian shoreline vegetation such
as willows (*Salix*) and dogwoods (*Cornus*). There are more than
20 named rivers in Nebraska (map 4), and their collective lengths
exceed 4,000 miles. In sequence of diminishing average water dis-
charges, the ten largest of these are the Missouri, Niobrara, Platte,
Loup, Elkhorn, Big Blue, Republican, Big Nemaha, Little Blue,
and Dismal. There are also many smaller rivers, creeks, and irrigation canals. The state's streams collectively have an estimated total
length of 23,600 miles of land-water interface (Zuerline 1983), a
remarkable total (nearly equal to the circumference of the earth
at the equator) that seems improbably high. Some of the state's
major creeks have essentially disappeared as a result of irrigation
extractions, such as Pumpkin Creek in Banner and Morrill Counties.

Nearly all the larger surface waters of Nebraska are the result
of twentieth-century impoundments. The largest state reservoir
is Lake McConaughy, in the central Platte Valley, impounding up
to 35,700 acres. Excluding Lewis and Clark Lake, a 31,000-acre
reservoir on the Missouri River that is shared with South Dakota,
Harlan County Lake, on the Republican River, is second largest
and covers 13,250 acres. Other large Nebraska impoundments are
Calamus Reservoir, on the North Loup River, extending over 5,100
acres, and Johnson Lake, an irrigation canal diversion in Gosper
County that impounds up to about 2,500 acres.

Besides its reservoirs, Nebraska also has thousands of small permanent to semipermanent wetlands (lakes, marshes, wet meadows,
etc.), especially in the Sandhills region (map 5). All of Nebraska's
natural "lakes" are associated with the Sandhills, where the water
table lies very close to the land surface, although most of them are
shallow enough to be better described as marshes. Sandhills wetlands range in size up to about 2,000 acres, but most of the wetlands
are smaller than 100 acres. In a survey of nearly 1,700 lake-sized

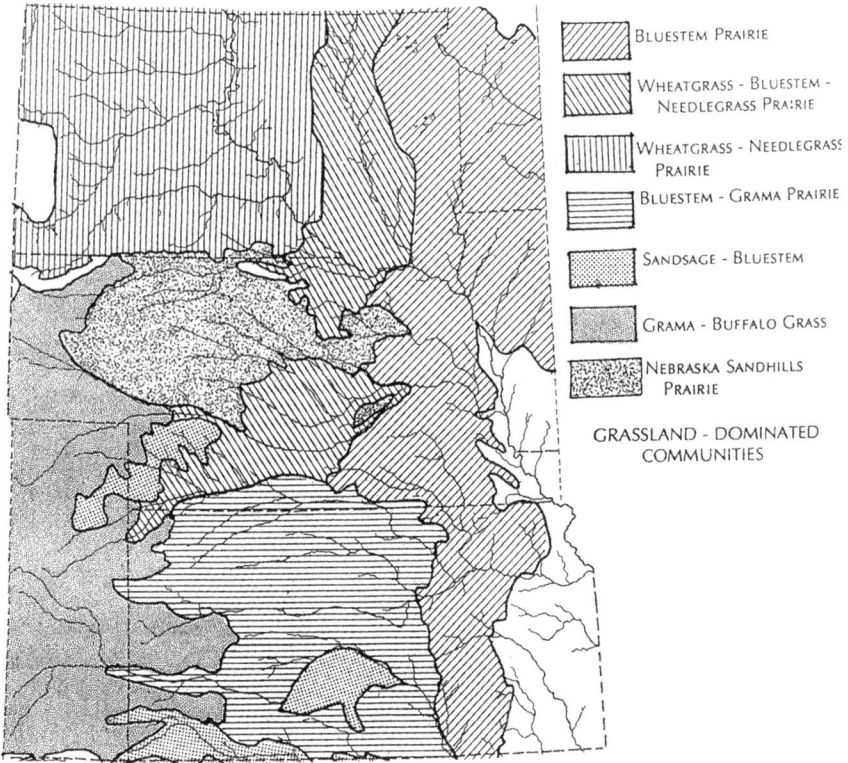

Legend:

- BLUESTEM PRAIRIE
- WHEATGRASS - BLUESTEM - NEEDLEGRASS PRAIRIE
- WHEATGRASS - NEEDLEGRASS PRAIRIE
- BLUESTEM - GRAMA PRAIRIE
- SANDSAGE - BLUESTEM
- GRAMA - BUFFALO GRASS
- NEBRASKA SANDHILLS PRAIRIE

GRASSLAND - DOMINATED COMMUNITIES

MAP 3. Native grassland–dominated areas of Nebraska and central Great Plains states. Unshaded areas are tree-dominated areas. From P. A. Johnsgard. 1979. *Birds of the Great Plains: Breeding Species and Their Distribution.* Lincoln: University of Nebraska Press. See also 2009 Supplement and Revised Maps, https://digitalcommons.unl.edu /bioscibirdsgreatplains/1/.

wetlands in 13 Sandhills counties, the average water area was 60 acres, and the average depth was about 3 feet (McCarraher 1977). These wetlands are often of very high foraging and nesting value for many species of waterfowl (Oberholser and McAtee 1920), as well as for shorebirds, and as habitats for other aquatic animals and plants (Novacek 1989).

Additionally, there are temporary seasonal wetlands (ephemeral ponds and shallow marshes, which are colloquially called "lagoons" in Nebraska, and playas farther south) that are too transitory to mea-

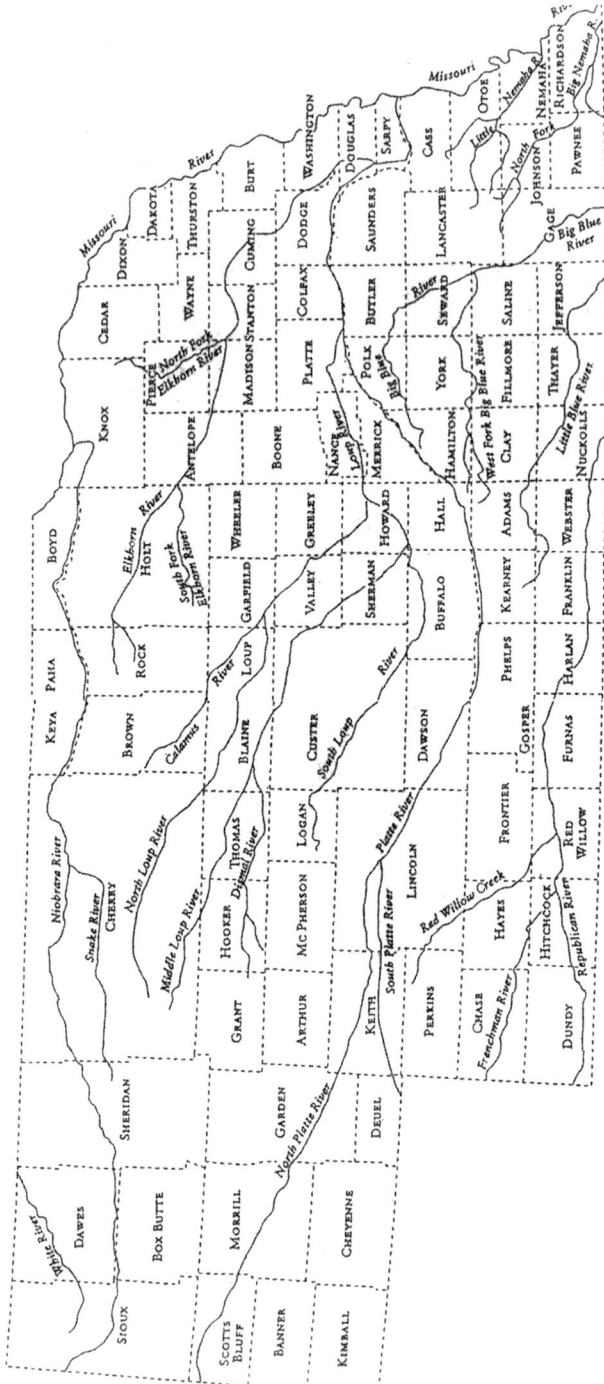

MAP 4. Nebraska rivers and counties. Courtesy of the University of Nebraska Conservation and Survey Division.

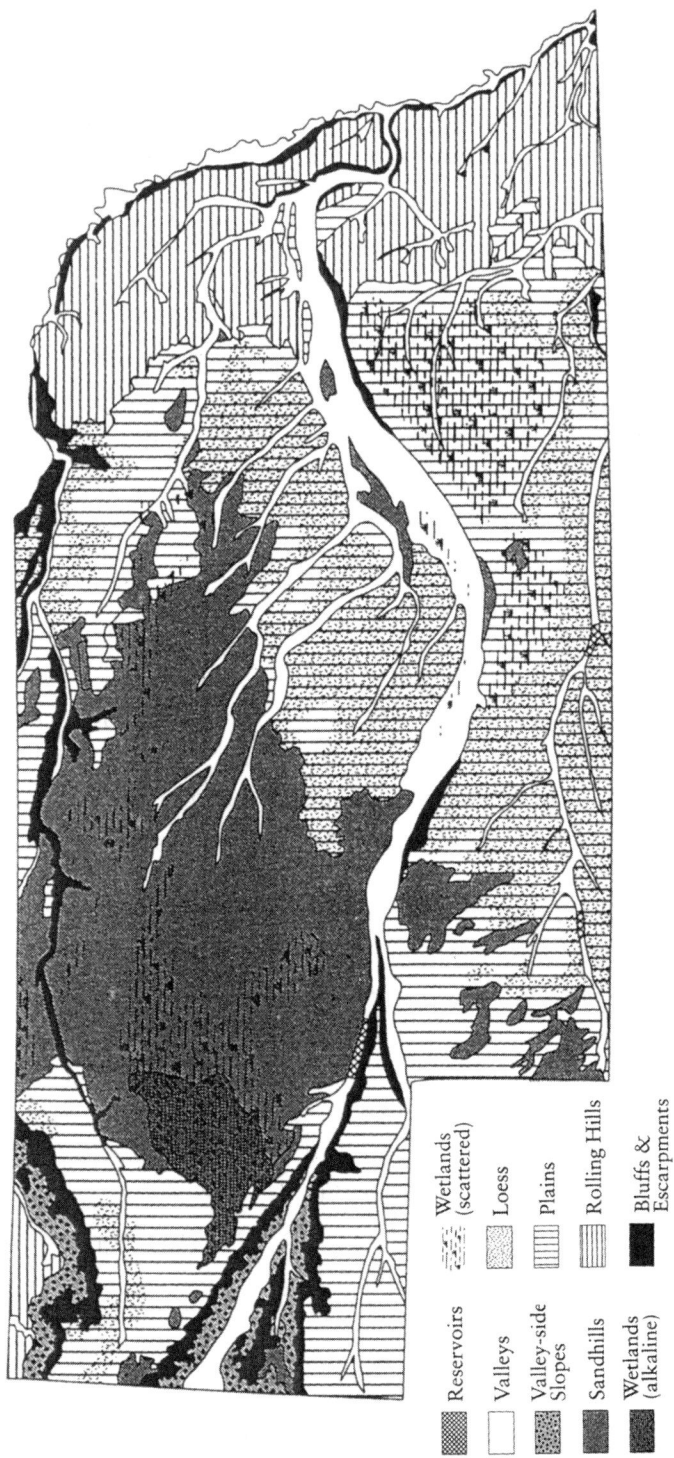

MAP 5. Nebraska topography and wetlands. Adapted from a map by the University of Nebraska Conservation and Survey Division.

Reservoirs

Valleys

Valley-side Slopes

Sandhills

Wetlands (alkaline)

Wetlands (scattered)

Loess

Plains

Rolling Hills

Bluffs & Escarpments

sure and count accurately but are mostly found in central Nebraska where clay soil deposits prevent rapid subsurface drainage. In spite of all the wetland drainage that has occurred during the past century in Nebraska, out of an estimated original 100,000 acres there might still be as much as about 20,000 acres of temporary and permanent wetlands left, and they are among the most biologically productive of all North American ecosystems (LaGrange 2005). In Nebraska, wetland ecosystems provide habitats for at least 231 species of birds, 35 species of mammals, 30 species of reptiles and amphibians, and more than 40 species of fish, as well as countless invertebrates and plants. They are the most valuable of all our state's habitats from the standpoint of their biological diversity and productivity relative to their area (Johnsgard 2012a).

NEBRASKA AS A BIOLOGICAL TRANSITION AND GENETIC SUTURE ZONE

After its recognition as a state in 1867 and the gradual development of the modern state of Nebraska, 93 counties were established (map 4). By the mid-1800s the westward-bound immigrants following the North Platte River upstream knew they had finally entered the American West as they approached Chimney Rock, the most easterly of the famous monoliths that travelers would encounter along the Oregon Trail. This iconic landmark, located at longitude 103.2° west, confirmed that, in the vernacular of the day, the immigrants had finally "seen the elephant" and were perhaps at least halfway to Oregon Territory. A general awareness that Nebraska represents a transition zone between East and West was formalized by the state legislature in 1963 in adopting our official state motto as "Welcome to Nebraska, Where the West Begins."

There is some biological evidence for this rather precise geographic definition. In 1887 Charles Bessey, botany professor at the University of Nebraska, reported finding a "meeting place" of eastern and western floras in western Rock County's Niobrara Valley, near the mouth of Long Pine Creek, at longitude 99.8° west. And in an analysis of the zoogeography of more than 200 species of breeding birds in the Great Plains (Johnsgard 1979a), I concluded that the 100th meridian represents an objective division line between the state's eastern and western bird faunas. It also closely conforms to the middle of several hybrid zones that exist in several of the occasionally interbreeding species of western and eastern Great Plains birds, such as the flickers, orioles, buntings, and grosbeaks, as well as the distributional transition zones of several other eastern and western counterpart animals (Johnsgard 2015c).

It is convenient to use conspicuous typical plants or plant communities in judging biogeographic classifications as relatively stationary evidence of definable and climatically based geographic units. For example, we think of characteristic shortgrass plains species such as buffalo grass (*Buchloe dactyloides*) and grama grasses (*Bouteloua* spp.) as indicators of the American High Plains. Some Rocky Mountain coniferous forest species such as ponderosa pine (*Pinus ponderosa*) and limber pine (*Pinus flexilis*) also serve as reliably iconic species typical of the Rocky Mountain West. Similarly, characteristic plants of the tallgrass prairies such as big bluestem (*Andropogon gerardii*) and Indiangrass (*Sorghastrum nutans*) help define the central Great Plains. Lastly, Nebraska's floodplain trees (elms, maples, cottonwoods, and ashes) are members of tree groups that are otherwise largely associated with the vast hardwood forests of eastern North America.

As an example of these biogeographic indicators, in eastern Nebraska there are six native oak species, only two of which extend west beyond the relatively moist Missouri Valley of extreme southeastern Nebraska. The red oak (*Quercus rubra*) ranges north along this well-timbered valley to Dakota County and west along the Platte Valley to Saunders County. However, the much more drought- and fire-tolerant bur oak (*Quercus macrocarpa*) occurs west along the Niobrara Valley to Dawes County and along the Platte Valley ranges west to Custer County. Among the maples, the moisture-dependent silver maple (*Acer saccharum*) is widespread in riverine forests of eastern North America but extends across only the eastern third of Nebraska.

In contrast, the smaller and relatively arid-adapted Rocky Mountain maple (*Acer glabrum*) barely enters the Panhandle from the west by eking out a marginal existence in the shady canyons of Sioux County. A typical Rocky Mountain conifer, the arid- and timberline-adapted limber pine barely reaches southwestern Nebraska in Kimball County. Big sagebrush (*Artemisia tridentata*) probably once occupied a million or more square miles in western North America and still dominates nearly half of Wyoming, but it occurs locally only in northwestern Nebraska's dry grasslands of Sioux, Dawes, and Sheridan Counties. However, ponderosa pine, the highly adaptable and most widespread of the major western American conifers, penetrates east along the Niobrara Valley to Keya Paha and Rock Counties, its easternmost range limit, and that terminus point was one of the reasons Charles Bessey selected Rock County as the state's primary east–west floral transition zone, where species reflective of both eastern and western floras occur in close proximity.

It is impossible to define precisely the meeting point of East and West over the broad geographic, climatic, and biological gradients that exist in Nebraska or across the broader Great Plains. However, Kaul, Kantak, and Churchill (1988) documented the importance of the central Niobrara Valley as a major transition and contact zone for many closely related eastern and western plants and animals. The transitional influence of both eastern and western biotic communities in the Pine Ridge–Niobrara Valley region is reflected in its relatively high diversity of breeding birds; Mollhoff (2016) reported that Sioux, Sheridan, and Cherry Counties were among the four counties having the highest degree of avian species diversity during both of Nebraska's breeding bird atlas surveys.

DISPERSAL CORRIDORS AND HISTORIC BIOGEOGRAPHIC CHANGES

While expanding their ranges, plants and animals often travel via convenient natural corridors along which they can survive, reproduce, and disseminate progeny into new areas of dispersal. In eastern Nebraska, the primary north–south corridor for many woodland species has been the Missouri Valley. However, there are many characteristic Missouri Valley plant species that probably have been unable to adapt and expand into the drier forests somewhat farther northwest, including trees such as the American hazel (*Corylus americana*), bitternut (*Carya cordiformis*) and shagbark (*C. ovata*) hickories, black cherry (*Prunus serotina*), ironwood (*Ostrya virginiana*), pawpaw (*Asimima trilobata*), and rock elm (*Ulmus thomasii*).

Some of many woodland herbaceous forbs with similarly restricted lower Missouri Valley distributions in southeastern Nebraska include the Indian-pipe (*Monotropa uniflora*), pale touch-me-not (*Impatiens pallida*), showy orchid (*Galearis spectabilis*), willow aster (*Aster praealtus*), Turk's-cap lily (*Lilium canadensis*), and wild columbine (*Aquilegia canadensis*) (Kaul, Sutherland, and Rolfsmeier 2012).

Similarly, breeding by some eastern- or southeastern-oriented birds such as the Acadian flycatcher, Kentucky, prothonotary, and yellow-throated warblers, Louisiana waterthrush, summer tanager, and tufted titmouse is still mostly limited in Nebraska to the mature, relatively moist forests of southeastern Nebraska's Missouri Valley.

Some other birds, such as the eastern whip-poor-will, chuck-will's-widow, and pileated woodpecker, are examples of eastern and southeastern deciduous forest-breeding species that have been moving very slowly northward along this narrow Missouri Valley corridor during historic times. However, the pileated woodpecker has recently been seen north to Dakota County, and the chuck-will's-widow has similarly been observed as far north as Knox County, suggesting future breeding range expansions.

Nebraska has several major river valleys that cross the state from east to west, most notably, the Platte and Niobrara, and that offer opportunities for east–west riverine corridor movements. The barred owl has been moving westward at a notable rate in recent decades and has been seen as far west along the Platte Valley as North Platte. The Mississippi kite, red-bellied woodpecker, and northern cardinal all seem likely to reach and colonize eastern Wyoming before long, as the wood duck, indigo bunting, and blue grosbeak already have. Silvia (1995) described the similar influence of the central Platte River as an important east–west dispersal corridor for small mammals. Mammals that are apparently expanding westwardly along wooded east–west waterways include the Virginia opossum, northern myotis, evening bat, eastern red bat, woodchuck, white-footed mouse, and gray fox. In contrast, the porcupine has expanded eastwardly recently along wooded rivers.

Considering a few woodland bird species, the American redstart and black-and-white warblers, ovenbird, and red-breasted nuthatch are all largely restricted along the Niobrara Valley to this narrow east–west wooded corridor. The absence of fairly cool and shady deciduous canopies might prevent the three warblers from extending much farther west in the valley than Cherry County, whereas the eastern limits of coniferous forest might prevent the nuthatch from any future range expansion farther east than Keya Paha County. Other songbird species that are only loosely dependent on a well-wooded corridor and can also use woodland edge for breeding are much more widespread and more common, such as the house wren, gray catbird, and brown thrasher.

East–west transition zones also present possibilities of interspecies hybridization. Many interesting biological situations arise in locations where two closely related western- and eastern-oriented species meet and variably interact, producing overlapping ecological niches and potentially initiating interspecies competition for the same environmental resources. For very closely related eastern and western species, hybridization might not be harmful and conceivably might even be beneficial if the two species are able to exchange their most desirable genes. However, the more distantly related two hybridizing species are, the less likely their two gene complexes can interact successfully, and the more probable that the resulting hybrids will be ill adapted to the environment and unable to breed successfully.

As an example of east–west hybridization between plants, the Rocky Mountain juniper (*Juniperus scopulorum*) extends from Nebraska's western border counties east to Cherry County. Over part of its Nebraska range this shrubland-associated juniper is in contact with the more widespread and less drought-tolerant eastern red-cedar (*Juniperus virginianus*), and the two freely hybridize. Similar examples occur in some Nebraska birds. The red-shafted (western) and yellow-shafted (eastern) populations of the northern flicker have likewise come into broad contact across much of Nebraska and elsewhere in the Great Plains, probably during post-Pleistocene times. There they have thoroughly interbred, producing a wide spectrum of intergrades and blurring their genetic differences to the point that both populations are now classified as a single species (Short 1965).

Several other woodland- and forest-adapted species pairs of songbirds that currently have both eastern and western relatives apparently also had continuous ancestral distributions across central North America that were probably split during Pleistocene times. When tundra and grasslands expanded south over the Great Plains, they replaced forests and separated those populations that took refuge and survived in eastern hardwood forests from their counterparts that similarly adapted to life in western woodlands, initiating the speciation process.

As late or post-Pleistocene wooded riparian corridors across the plains developed, many of these species' ranges expanded back across the Great Plains, slowly reducing the geographic distances separating them. Over the past few centuries, some of their expanded ranges have come into geographic and ecological contact along these riverine corridors, and the once-separated and now genetically more distinct populations now variably interact with their ancestral relatives.

Four such species pairs with interacting western and eastern counterpart populations are, respectively, the Bullock's and Baltimore orioles, the lazuli and indigo buntings, the black-headed and rose-breasted grosbeaks, and the spotted and eastern towhees. In areas where these species pairs geographically overlap, the presence of mixed matings and occasional hybrids provides an insight into the speed and mechanisms of speciation (Sibley and West 1959; Sibley and Short 1959, 1964; West 1962; Rising 1983; Emlen, Rising, and Thompson 1975).

Depending on the incidence of hybridization and the relative reproductive fitness of hybrid offspring, it is possible to judge the degree of evolutionary divergence between the parental populations. If the hybrids are unable or less likely to reproduce, putting them at a genetic disadvantage with nonhybrids, the two parental populations are best considered as genetically isolated species. Such is the case with all four of the bird species pairs just mentioned. Some similarly coexisting species will probably ultimately evolve sufficient genetically based differences that will reduce or prevent crossbreeding, which seems to be happening with the eastern Baltimore and western Bullock's orioles in the Great Plains.

GLOBAL WARMING AND CHANGING CLIMATES

During the current trend of global warming and associated vegetational changes, plant and animal ranges will continue to change, and extinctions of some species might occur. As many as 22 mammalian species (28 percent of the state's mammalian fauna) might be currently shifting their distributional patterns in Nebraska (Benedict, Genoways, and Freeman 2000). In contrast to western and northern range expansions among woodland species, many herbivorous mammals that require large tracts of prairie have undergone contracting ranges.

Mammals that have moved north into southern Nebraska from more southern climates include the hispid cotton rat (Genoways and Schlitter 1967) and the nine-banded armadillo (Benedict, Genoways, and Freeman 2000). Many southern bird species have also expanded north into Nebraska during recent decades and now breed here regularly, such as the white-faced ibis, cattle egret, Mississippi kite, white-winged dove, scissor-tailed flycatcher, and great-tailed grackle (Mollhoff 2016; Johnsgard 2018a). Some clearly arid-adapted southwestern bird species that appear to be moving north into western Nebraska are the ash-throated flycatcher, the lesser goldfinch, and the western race (*Polioptila caerulea amoenis-*

sima) of the blue-gray gnatcatcher (Mollhoff 2016). Other species that are associated with the American tropics or subtropics are also now appearing in small but increasing numbers, including the fulvous whistling-duck, glossy ibis, Neotropic cormorant, and Inca dove.

Contrastingly, several arctic or boreal-oriented birds that once regularly overwintered in Nebraska, such as the Bohemian waxwing and snow bunting, now increasingly winter farther north and only reach Nebraska during very cold winters (Johnsgard 2015a). As would be expected from global warming trends, no Nebraska birds have exhibited apparent southward wintering or breeding expansions. Contrarily, a few mammals, such as the masked shrew, meadow vole, and least weasel, have apparently expanded southwardly (Benedict, Genoways, and Freeman 2000). Among southeastern Nebraska mammals, little or no recent range expansion is apparent in the southern flying squirrel and eastern chipmunk, possibly because of having specific niche requirements not found elsewhere to the north or west (Benedict, Genoways, and Freeman 2000).

In the central Great Plains, we can anticipate that a progressively warmer and drier climate will develop unless effective international controls on carbon consumption are adopted. In Nebraska, we can also expect to experience more extreme weather events, recurrent droughts, and reduced water availability through excessive draw-downs of our aquifers and declining surface waters. Thus, in the near future, homeowners should consider planting buffalo grass and drought-tolerant fescues rather than the water-hungry Kentucky bluegrass and yuccas rather than ornamental roses, or just give up hope and settle for rocks. When planting trees, drought-tolerant species, such as bur oaks, are far better choices than more thirsty trees, such as maples. Nebraskans should also not be surprised to soon encounter armadillos digging up their backyards and gardens and someday might begin seeing both black vultures and turkey vultures soaring ominously overhead as a somber harbinger of hotter and drier days ahead.

CHAPTER 2

Mammals

To walk today on the land that was the home of the last great aquatic reptiles from the Late Cretaceous period in western North America (see table 1), simply wade in the Niobrara River anywhere from Meadville downstream. Its bedrock base is comprised of a hard layer of Pierre shale, which is as much as 700 feet thick not far north in South Dakota. The shale was formed from sediments deposited between about 146 and 70 million years ago by the Western Interior Seaway, a shallow sea that once extended from what is now southern Manitoba to New Mexico. It impounded all of what was destined to become Nebraska until the final stages of the Cretaceous period, which ended abruptly 65 million years ago.

In that warm Cretaceous sea, whale-sized mosasaurs and plesiosaurs (see map 6 for sketches of representative examples) roamed about in search of smaller prey, and above it flew a strange array of winged pterosaurs, birdlike flying reptiles that preyed on fish and invertebrates and even scooped up plankton. By far the largest of these was *Quetzalcoatlus*, a probable fish or carrion eater with a wingspread of nearly 40 feet. Most pterosaurs were much smaller, and probably many were invertebrate eaters. Less than 100 million years ago, a loon-sized flightless bird, *Hesperornis*, also swam about in the sea, diving for fishes. A tern-like bird, *Ichthyornis*, lived along the seashore, perhaps tearing up carrion with its sharply toothed jaws, a reptilian trait it shared with *Hesperornis* and with the much more famous and more ancient "ancestral bird," *Archaeopteryx*. This important fossil represents the single most important anatomical evidence connecting birds and reptiles and sets the apparent time of divergence of these two major vertebrate groups at no less than 150 million years.

On tropical Jurassic-age shorelines and dry uplands great populations of dinosaurs spread over the land and occupied swamps. Although most of them were herbivores, consuming the abundant terrestrial vegetation, some were bipedal carnivores with the general body form of the fearsome *Tyrannosaurus*, but they were much smaller and probably far more agile. One such predatory group, the troödonts, also had large brains and deadly claws on their forelegs. They additionally had bird-like skeletons, were perhaps warm-blooded, and possibly were even feather-covered. Other highly agile species, such as *Ornithomimus*, had toothless, bird-like beaks and probably preyed on arthropods or other lizards. Still other bird-like predators, the oviraptorids, certainly had feathers but were flightless. By mid-Jurassic times (about 150 million years ago), birds had completed the transition to achieving flight, probably gradually replacing pterosaurs as masters of the air.

The giant predaceous tyrannosaurids of the Cretaceous period, such as *Tyrannosaurus* and *Albertosaurus*, still ruled over North America and Asia until less than 100 million years ago. Their prey included diverse browsing herbivores, such as the oddly crested hadrosaurs, the variously horned and often ornately shielded ceratopsians such as *Triceratops*. There were also heavily armored ankylosaur dinosaurs that were covered with heavy protective dorsal plates, such as *Niobrarasaurus* (which was confusingly named not for Nebraska's Niobrara River but rather for the Niobrara chalk beds in Kansas). Some ankylosaurs also had club-like tails and heavy spikes protruding from their shoulders, often resembling a modern science-fiction creation more than a once-living animal.

All of these great beasts, and indeed all the dinosaurian fauna, were doomed to die in an eye-blink of time about 65.5 million years ago, when a giant meteorite crashed into the earth just north of the present-day Yucatan Peninsula, producing the gigantic but now-submerged Chicxulub crater, and suddenly changed the world's climate for months, years, and possibly centuries. Its shock waves, associated earthquakes, and volcanic dust had lethal planet-wide effects. It caused the catastrophic deaths of much of the world's terrestrial plant life and nearly all the plant-eating reptiles and their carnivorous predators and ended the planetary reptilian dominion. The survivors, including a few groups of reptiles, birds, and mammals that somehow had been spared, began the slow process of recolonizing the earth and initiating the so-called Age of Mammals, or Cenozoic era (map 6).

By the start of the Cenozoic era, the land that would eventually become Nebraska was indeed land, although perhaps still a bit soggy. The vast inland sea that once covered much of what is now the Great Plains during the latter parts of the Cretaceous period had essentially disappeared. This drying out was the result of both general land uplift and the accumulation of vast amounts of sands and gravels in the form of water-carried sediments from streams originating from mountains that were rising much farther west.

As quickly as these new mountains arose they began to erode and deposit their dislodged materials over the lower lands to the east. Wind-carried volcanic ash deposits, originating from these western mountains, also helped to eliminate the seaway during a prolonged mountain-building period (the Laramide orogeny) that extended from 75 to 45 million years ago (Maher, Engelmann, and Shuster 2003).

The world's climate then was tropical, and the vegetation was replete with palms, tree ferns, cycads, and other tropical forest flora. Early forms of flowering plants attracted insects for more efficient pollination, and the evolution of fruits and berries facilitated better seed dissemination by browsing vertebrates. Grasses were then also evolving, and as world climates became drier and cooler, the advantages of the quickly maturing life cycle strategy of grasses and other herbaceous plants over the slow-growing and late-maturing life cycles of woody plants would eventually win out over much of the world's land masses.

Just as plants were evolving into a diversity of herbaceous species during the Eocene epoch, there was a similar trend toward the

GEOLOGIC TIMETABLE

Era	Period	Epoch	Years before Present	Major Events (*particularly in Nebraska*)
CENOZOIC	Quarternary	*Holocene*	11,000–0	Postglacial period to present
		Pleistocene	1.6–0.01 MYA*	Several glaciations; Sandhills form; also widespread loess, till, and alluvial deposits
	Tertiary	*Pliocene*	5.2–1.6 MYA	Grasslands spreading; cooler and drier
		Miocene	23.3–5.2 MYA	Ashfall Beds (10 MYA); Agate Fossil Beds (20 MYA)
		Oligocene	35–23.3 MYA	Mountain building in the West; early grasslands and grazing mammals on plains
		Eocene	60–35 MYA	Major modern mammal and bird groups appear
		Paleocene	65–60 MYA	Modern plants appear; Nebraska emerging
MESOZOIC	Cretaceous		135–65 MYA	Last of dinosaurs; Nebraska mostly submerged
	Jurassic		197–135 MYA	Peak of dinosaurs; early birds and mammals
	Triassic		225–197 MYA	Early dinosaurs appear
PALEOZOIC	Permian		280–225 MYA	Cooler and drier; many extinctions worldwide
	Carboniferous		345–280 MYA	Early reptiles appear
	Devonian		405–345 MYA	Seed plants appear
	Silurian		425–405 MYA	First land plants and early amphibians appear
	Ordovician		500–425 MYA	Early fishes appear
	Cambrian		570–500 MYA	Abundant marine life; many invertebrates

*MYA = millions of years ago.

Source. P. A. Johnsgard. 2001. *The Nature of Nebraska: Ecology and Biodiversity.* Lincoln: University of Nebraska Press.

evolution of a herbivorous diet for many mammals. This trend was aided by the evolution of broad, flattened teeth capable of harvesting leaves and stems, crushing and chewing plant materials, and developing digestive tracts able to effectively digest starches, sugars, and other carbohydrates. Although primitive mammals had evolved fairly early in the Mesozoic era, they had remained small and inconspicuous during the 155-million-year reign of the dinosaurs, probably by using their large brains, warm-blooded high metabolic rates, and superior visual and olfactory abilities to avoid being eaten by snakes, lizards, and huge predatory birds. Although many Nebraskan fossils provide physical evidence of the Western Interior Seaway's aquatic vertebrates, such as mosasaurs and plesiosaurs, scarcely any fossil evidence of Eocene-era tropical forests and their inhabitants has been found in Nebraska.

NEBRASKA'S MAMMALS, 58–24 MILLION YEARS AGO

The only fragmentary evidence of land mammals in Nebraska that date from the earlier part of the Eocene epoch (about 58 million years ago) can be found in Knox County, northeastern Nebraska, where a few scattered fossils indicate the presence of crocodiles and, more importantly, the remains of the earliest known presumed progenitor of modern horses, the terrier-sized *Hyracotherium*, or "dawn horse," which had four front toes and three hind toes, had low-crowned generalized teeth, and was probably a forest-dwelling browser.

MAP 6. (*opposite*) Geological time scale relative to Nebraska geography. The time scale is depicted as one mile of I-80 mileage per million years of time, totaling 450 million years (Ordovician period to the end of the Pleistocene epoch). A few important fossil-bearing Cenozoic strata of northwestern Nebraska are listed on the right. Some fossil reconstructions are also shown; many of them also existed before or after the time frame for which they are identified. Note: mybp = millions of years before present. Adapted from P. A. Johnsgard. 2001. *The Nature of Nebraska: Ecology and Biodiversity* (Lincoln: University of Nebraska Press).

1. Coral (Paleozoic)
2. Trilobite (Paleozoic)
3. Early jawless fish
4–6. Early jawed fish
7. Lobe-finned fish
8. Early amphibian
9. Early lungfish
10. Early reptile
11. Dimetrodon (Permian)
12. Spinosaurus (Cretaceous)
13. Ichthyosaurus (Jurassic)
14. Plesiosaurus (Cretaceous)
15. Mosasaurus (Cretaceous)
16. Early crocodile
17. Velociraptor (Cretaceous)
18. Stegosaurus (Jurassic)
19. Pterodactylus (Jurassic)
20. Allosaurus (Jurassic)
21. Triceratops (Cretaceous)
22. Titanothere (Eocene)
23. Dawn horse (Eocene)
24. Early camel
25. Mastodon (Pliocene)
26. Pliohippus (Pliocene)
27. Mammoth (Pleistocene)

Millions of Years Before Present

1 mile = 1 million years

Paleocene: 60–65 mybp
Eocene: 35–60 mybp
Oligocene: 23.3–35 mybp
Miocene: 5.2–24.3 mybp
Pliocene: 1.6–5.2 mybp
Pleistocene: 1.6–0.01 mybp

Ordovician
Silurian
Devonian
Carboniferous
Permian
Triassic
Jurassic
Cretaceous
Tertiary
Quaternary

PALEOZOIC ERA
MESOZOIC ERA
CENOZOIC ERA

Omaha
Lincoln
I-80
450
425
405
345
280
222
197
135
65
20
Sidney
Potter
Kimball
Bushnell
I-80

Thirty million years later, a deposit of mudstone, sandstone, and volcanic ash that dates from the late Eocene (ending about 37 million years ago) and the early parts of the Oligocene epoch (extending from 37 to 24 million years ago) is present along the White River's badland topography of Dawes County. This deposit is known as the Chadron formation, a thick layer of silts and clays that are the source of some of the most important early mammal fossils in all of North America. They include a diversity of now primarily tropical mammalian types, such as primates, marsupials, and dermopterans (the "flying lemurs" that are now found only in southeastern Asia). There are also a variety of reptilian fossils that have living descendants, including alligators, turtles, snakes, and lizards.

Most of the Chadronian mammals were small but included many hooved mammals, such as *Mesothippus* ("middle horse"). This three-toed horse, the size of a large dog, provides an intermediate step in the long lineage of North American horses and was slightly bigger than its tiny presumed progenitor *Hyracotherium*. There was also a small, primitive rhino, *Toxotherium*, and a rabbit-sized ruminant (cud-chewing) herbivore, *Leptomeryx*, that had longer hind legs than forelegs, as occur in modern rabbits.

One abundant group of fossil mammals of the Chadronian deposits differs from the other hooved mammals in their members' sheer size, the titanotheres (family Brontotheridae). These rather rhinolike grazing mammals had paired bony horns behind their nostrils, might have weighed as much as 2.5 tons, and ranged up to 8 feet in height. However, their horns and massive size didn't prevent them from going extinct about 30 million years ago, when they were apparently unable to adapt to the more fibrous plant life associated with gradually drying and cooling world-wide climates that developed during the next 20 million years.

North of the White River badlands in northeastern Sioux County is Toadstool Geological Park. Above its Chadronian deposits there are volcanic siltstones of the Lower Oligocene epoch that are part of the fossil-rich Brule formation (34–29 million years ago). Small horses and miniature camels were abundant during that time, and all the major groups of the dog family were present in North America by the Oligocene epoch. There were other predators, namely, primitive dog-like species with large jaws and enlarged molars. These were members of a family of predators called hyaenodonts, which had up to three rows of flesh-slicing molars. They probably were efficient predators but nevertheless became extinct by about mid-Miocene times.

All the major groups of the dog family were present by the Oligocene epoch, but ancestors of the New World cats, such as the cougar, lynx, and bobcat, arrived in North America from Asia in late Miocene times (8.5–8 million years ago) via the Bering land bridge. Also present during late Miocene times was *Barbourofelis*, a lion-sized ferocious predator and member of a group called the false saber-toothed cats. It was named in honor of Edwin Barbour, third director of the Nebraska State Museum.

Elsewhere in northern Sioux County are sandstone deposits known as the Arikaree group (late Oligocene and early Miocene epochs, 29–19 million years ago), as well as a grayish volcanic ash layer that is also visible on the upper parts of Nebraska's three most iconic landmarks: Chimney Rock, Scotts Bluff, and Courthouse Rock. The fossils in these strata include various grazing mammals such as oreodonts, which somewhat resembled sheep and had tall, crowned teeth adapted for eating fibrous foods in a dry and sandy environment.

In beds of the same age there were also similarly drought-adapted camels (*Stenomylus*) no more than 3 feet tall. Later camels evolved into forms as large as their modern-day relatives; they eventually migrated west from their North American homeland into Eurasia about 5–4 million years ago. This outward migration left North America with no camel-like mammals, but some modern descendants such as llamas and guanacos still survive in South America. South America also was the source of a few unusual mammals that ventured north out of the tropics via a Central American corridor that existed during late Pliocene or early Pleistocene times, adding opossums and armadillos to our present-day mammalian diversity.

For further nontechnical coverage of this period in Nebraska's complex early mammalian history, see the summary by Voorhies in the special issue of *Nebraskaland Magazine* (1994); it was the primary information source for this brief survey. Prothero and Emry (2004) provided a more technical review of the fossils reflecting this important phase of North American mammalian evolution.

NEBRASKA'S MAMMALS, 23–22 MILLION YEARS AGO

Agate Fossil Beds National Monument, in Nebraska's northwestern Panhandle, is situated among arid grassy plains and ancient bluffs, where the still-tiny Niobrara River cuts a meandering thin blue line through an otherwise mostly bleak grassy landscape. The late nineteenth-century human history of this mostly neglected region is a saga of betrayed indigenous tribes, hopeful immigrants, and hardy ranchers. It was also a period of bitter professional competition for fossils among paleontologists, who were then just discovering the wealth of early mammal fossils hidden in the eroding badlands strata of western Nebraska and the Dakotas.

Not long after James Cook, an ex-army volunteer scout from the Indian Wars, moved into western Nebraska, he acquired a large ranch about 20 miles south of Harrison, at Agate Springs. In his spare time, Cook was fond of collecting rocks and Native American artifacts that he found on his ranch. He shipped one of his prize finds, a fossil leg bone that he had picked up in 1885, to Professor Edwin Barbour at the State Museum. Cook encouraged Barbour to visit Agate Springs, resulting in the start of a long series of investigations by paleontologists from Nebraska's fledgling University Museum. Representatives of several major eastern museums, such as the Carnegie Museum in Pittsburgh and Yale University's Peabody Museum of Natural History, soon followed. In 1904 paleontologists from the Carnegie Museum excavated the famous "bone bed," which exposed an amazing number and array of mammalian fossils that have since been exhibited in museums around the world.

Although this was also a period when discoveries of remarkable dinosaur fossils in Wyoming and South Dakota were occurring, no significant dinosaur fossils have been found in Nebraska, since what is now Nebraska was inundated during much of the Mesozoic era. The only major Mesozoic vertebrate fossils so far found in the state have been from aquatic reptiles such as plesiosaurs, mosasaurs, and turtles that had been living in the Western Interior Seaway, which by late Cretaceous times (less than 100 million years ago) was slowly retreating south and merging into the Gulf of Mexico.

The last dinosaurs became extinct 65 million years ago, more than 40 million years before the Agate beds were deposited. Agate's fossil beds are about 23–22 million years old in strata deposited during early Miocene times. North America's western plains were then undergoing a gradually drying climate, with the Rocky Mountains and other mountains still rising to the west, accompanied by periodic massive volcanic eruptions. As the growing mountains increasingly intercepted moisture originating in the Pacific Ocean, the climate of the plains to the east was slowly changing from the rich temperate forests of the Oligocene epoch to a drier, more savanna-like vegetation in the early Miocene epoch.

In the Agate region, a proto–Niobrara River was then a well-developed river, bringing in sand and silt sediments from the nascent Rocky Mountains. Seasonally dry winds deposited volcanic ash over the river sediments of the adjacent valley, forming a layer of materials that would later be geologically defined as the Harrison Formation (see map 6).

At some point, a prolonged drought descended on the region, drying up the wetlands and reducing the river to no more than a trickle. Water-dependent mammals and other animals perished in the thousands, if not tens of thousands, many dying where they had taken refuge in the last surviving waterholes. During the drought, their remains were buried in sediments that filled drying ponds, waterholes, and abandoned stream channels. The bones of a few species of large mammals, such as rhinos, accumulated in huge random aggregations, extending over a large area within the Harrison Formation. Later Miocene sediments at Agate also show evidence of severe and prolonged droughts, during which countless additional skeletal remains of rhinos, camels, and other mammals accumulated and gradually fossilized.

The myriad of fossilized mammals that has been found at Agate is mostly comprised of more than six hundred examples of a tiny rhino (*Monoceras*) with small paired horns near its nose, but there are also many gazelle-like camels (*Stenomylus*) and a few predators. One of the larger predators slightly resembled a greyhound and is called a beardog (*Daphaenodon*), but it is not related to either the dog or bear group. Paleontologists have also found the bones of a small oreodont (*Merychyus*), the beardog's probable major prey species, in beardog dens. Oreodonts were mostly rather sheep-like and paired-toed herbivores with short legs and rounded faces and were then widespread in North America.

In addition to these variably familiar mammalian groups, there were examples of several strange and now long extinct mammal families. One of the common forms (*Dinohyus*) resembled a giant pig and was a member of a large group of generalized omnivores called entelodonts. *Dinohyus* was as large as a modern bull, had powerful paired hooves and massive jaws, and was perhaps the most important predator-scavenger of the time. There were also nearly twenty examples of a horse-sized herbivore (*Moropus*) belonging to a now long extinct family of browsers called chalicotheres. *Moropus* had long forelimbs tipped with three long claw-like hooves and short strong hind limbs. It probably walked in a sloping, hyena-like posture and might also have stood erect, using its claws to strip vegetation from trees.

Among the most remarkable hooved mammals found was an antelope (*Syndyoceras*) that superficially resembled the modern pronghorn. Like the pronghorn, it had a pair of bony horns near its ears, but it also had another Y-shaped horn located just behind its nose. There was also a primitive colonial beaver (*Paleocastor*) that dug strange corkscrew-shaped dryland burrows rather than building aquatic lodges in the manner of modern beavers. The odd helical shape of the burrows, resembling a spiral staircase, might have made vertical climbing easier.

Over millions of years, this great bone bed became covered by several hundred feet of sands and silts that were dropped by streams and carried in by winds, eventually burying the bones but preserving the skeletons. Later erosion cut away and reshaped the landscape, carving out two large buttes where harder caprock prevented the softer fossil-rich sediments below from being carried away downstream. These two adjacent buttes are known, west to east, as Carnegie and University Hills, which together contain the major fossil beds and arguably represent the two most famous and scientifically valuable hills in the state of Nebraska (Johnsgard 2014a ["Secrets"]).

NEBRASKA'S MAMMALS, 12–10 MILLION YEARS AGO

In contrast to Agate Fossil Beds of Nebraska's dry and heavily eroded western Panhandle, Ashfall Fossil Beds is located in Antelope County, northeastern Nebraska, within the well-watered valley of the broad and spring-fed lower Niobrara River, most of which consists of fertile farmland long devoted to raising corn. It is just west of the westernmost influences of two Pleistocene glaciers that strongly shaped the modern topography and river drainage patterns of eastern Nebraska and left glacial till as far west as adjacent western Pierce County.

These early Pleistocene glaciers, sequentially the Nebraskan and Kansan, made eastern Nebraska more fertile by the wealth of glacial till and river-carried sediments associated with materials they brought in from much farther north. These glaciers also carried in large boulders of red (Sioux) quartzite south from eastern South Dakota and Minnesota. The boulders were deposited haphazardly as the glacial ice melted on moraine-shaped hillsides, making them impossible to plough. However, they provide modern ecologists and botanists with a few still-virgin prairies and occasional relict plants and animals that were isolated and left behind on steep shady slopes and valleys as the ice sheets retreated northward more than ten thousand years ago.

One day in the summer of 1971, University of Nebraska paleontologist Mike Voorhies and his wife, Jane, were walking along a streambed tributary of Verdigre Creek, gathering data for a planned geological map. Mike knew the area well, having grown up in the small town of Orchard, 8 miles away. Walking along the streambed ravine, he noticed an exposed layer of ash about a foot in thickness and partway up the face of a steep ravine.

This pale grayish ash layer is part of the widespread evidence of a volcanic explosion that had occurred nearly 12 million years ago, during late Miocene times. The ash originated in what is now southwestern Idaho, where a gigantic "hotspot" of magma once erupted. Because its overlying North American tectonic plate has drifted slowly to the west during the past 12 million years, the hotspot that produced the Ashfall eruption is currently located under Yellowstone National Park, where it is responsible for all of that region's thermal features and earthquake-prone landscape.

As Mike walked along the ravine, he noticed a small fossilized jaw and teeth. After some careful excavation, the entire skull of a baby rhinoceros about a foot in length slowly emerged. Further excavation exposed some of the neck vertebrae, leading him to believe the entire skeleton might be present. Returning the next day, he gradually uncovered the baby rhino's entire skeleton, along with three more skeletons, including one of an adult rhinoceros.

In spite of his excitement, Mike didn't have the time and equipment to continue digging, but he recognized this as the prize find of a lifetime. Under his direction, in 1977 a museum crew began to remove the layer of more recent sandstone deposited above the volcanic deposits. From an exposed area of about 200 square feet the group collected several more rhino skeletons. With the support of a *National Geographic* grant, Mike brought a group of students back in 1978, and they began large-scale excavations over an area of about 6,000 square feet.

By 1979 some two hundred skeletons of various mammals had been found among the volcanic matrix, in one part of which twelve rhino skeletons were clustered into an area not much bigger than the size of an average living room. The site had once been a watering hole that had gradually filled with ash. Lung failure, caused by inhaling the volcanic dust, eventually killed all the unlucky animals that had huddled together in it. The smallest rhinos evidently died soonest, as their remains are largely situated at the lower level of the volcanic layer, with the middle-sized and larger animals being sequentially located above.

Besides the rhinos, which were robust, short-legged animals belonging to a genus (*Teleoceros*) known as the barrel-bodied rhinoceros, there also were five genera of horses, three camel genera, and two members of the dog family. There was also a small saber-toothed deer closely resembling the modern musk deer (*Mochus*) of Asia whose distinctive protruding canine teeth were used by males for fighting.

Most of the mammal fossils found so far are of species that had previously been discovered, but the Ashfall site is unique in the number and completeness of specimens and in the extremely high details of bone preservation. The tiny particles of volcanic dust served as a perfect casting material, preserving tiny body parts, such as the tiny middle ear bones of larger mammals and the windpipes of cranes. Other remarkable finds include an unborn rhino calf inside the skeleton of its mother, a bird skeleton with small pebbles in its gizzard that had once served for grinding food, and fossil grass seeds from the throat of a rhino.

More than fifty other species of plants and animals have also since been identified from the site, including smaller mammals, turtles, and a few birds. Although so far no elephants have been found at Ashfall, they have been found in nearby Niobrara Valley excavations, and the fossil history of elephant fossils in Nebraska is perhaps the most continuous and complete of any state.

In 2009 a huge enclosed building, the Hubbard Rhino Barn, was constructed. Inside, a walkway allows visitors a close view of the excavated skeletons and of scientists continuing the excavation work (Johnsgard 2014b ["Secrets"]). No doubt the prize fossils are the numerous intact skeletons of barrel-chested rhinos (*Teleoceros*), most of which lay crouched with their legs tucked under them or lying on their sides. More than a dozen were found tending young as they died. The clustered grouping suggests that this rhinoceros was much more social than are the present-day species of rhinos, which are solitary animals. Evidently, like elephants, these rhinos formed herds of adult females and calves, accompanied by single adult males. Of about one hundred skeletons, only seven were adult males, with adult females outnumbering them by a ratio of more than six to one. The staggered ages of the young indicate that they were born seasonally rather than throughout the year.

Rhinos, camels, and wild horses now only occur in the Old World. This geographic pattern shows how the distributions of many mammals and birds have shifted over the past 12 million years and how evolution has molded them during the intervening time span. For example, three of the five species that were found at Ashfall had feet with three well-developed toes that were probably useful for walking in soft terrain or for dodging predators. One species (*Protohippus*) also had three toes, but two of them were reduced in size and perhaps of no functional value. Another species, the large, single-toed *Pliohippus*, closely approached the modern horse, having broad hooves that were best adapted to running fast over hard surfaces but probably had limited cornering abilities.

In contrast, the toes of camels have evolved to allow walking over relatively soft substrates. Modern camels' feet have two toes and undivided soles, helping to maximize each foot's surface area and to help spread their substantial body weight. Other than the rhinos and horses, camels were the most common large mammals found at Ashfall, with several dozen individual being found. Most of them are of a type called *Procamelus*, which, as the name implies, is believed to have been ancestral to the true Old World camels.

Some of the more notable avian finds were the skeletons of an extinct crane species that, at least in terms of its skeletal features, is nearly identical to the present-day African crowned cranes. The two living species of crowned cranes are believed to be the most "primitive" of all fifteen living cranes, with some unique features, such as long hind toes that allow them to perch in trees. In contrast to the more advanced cranes, the windpipes of crowned cranes extend directly from the gullet to the lungs rather than forming a loop that penetrates the keel of the breastbone. This loop greatly extends the length of the windpipe, enhancing the birds' vocal resonating abilities, and probably increased the volume of their calls. The discovery of crowned cranes at Ashfall also proves that relatively primitive cranes once ranged over North America millions of years before the sandhill crane and whooping crane evolved.

Several other species of fossil birds have also been found at Ashfall, including a hawk (*Apatosagittarius*) with convergent similarities to the modern African secretary bird. Another major recent discovery is a primitive vulture-like hawk, *Archigyps voorhiesi*, which seems to connect anatomically the typical hawks and eagles of North America with the Old World vultures. It was named in honor of Mike Voorhies, whose contributions to Nebraska paleontology are unparalleled (Voorhies 1981, 1994a, 1994b, 1994c, 1994d, 1994e).

More than 10 million years of evolution have occurred since the Ashfall event, including multiple periods of glaciation on the northern plains, bringing in some entirely new animal and plant populations and eliminating others. Postglacial climates have further shaped biological distributions over the northern plains (Hoffmann and Jones 1970; Frey 1992), and the appearance of humans on the Great Plains less than fifteen thousand years ago marked the start of even greater changes in mammalian distribution and abundance. The climate-caused changes have been relatively slow compared with the rate and extent of the biological changes occurring during the past 15 centuries caused by the influence of humans.

About 50 miles to the northeast of Agate Fossil Beds is a site of fossil deposits about ten thousand years old: the Hudson Meng Bison Bone Bed near Crawford. There, at a very early Native American site, the remains of up to six hundred bison can also be seen (Agenbroad 1978). The bison were of a species transitional between the modern bison and a directly ancestral species, *Bison antiquus*. Together with Nebraska's Agate and Ashfall Fossil Beds, the Hudson Meng site provides visitors with a stunning portrait of mammalian evolution in

North America spanning more than 20 million years and physically associating modern humans with our mammalian brethren.

SELECTED SPECIES PROFILES

The following profile summaries include 60 percent of Nebraska's native mammals. Of the state's 89 native species of mammals (Genoways et al. 2008), several groups, such as shrews, moles, and pocket gophers, are not likely to be seen by the average person during an entire lifetime, and the bats are usually seen only as flitting shadows in the moonlight. Rodents, the largest group of Nebraska's mammals, comprise 30 percent of the total and are nearly all nocturnal and inconspicuous. Humans generally regard rodents with disgust, as they include some major pest species, such as the house mouse (*Mus musculus*) and Norway rat (*Rattus norvegicus*). Both of these were introduced from Europe and are not described in this book. However, rodents represent some of Nebraska's most interesting mammals in terms of their behavior and ecology.

Of the 89 Nebraska mammal species listed in the synopsis (chapter 5), the 53 that have been selected to describe here in some detail are intended to be a sampling of the major mammalian groups. They include representatives of nearly all the genera that have well-represented Nebraska species, most of the very common or important species, and a few of the rare or endangered ones. They were also chosen in part to illustrate the amazing diversity and fascinating life histories of some of our nearest mammalian relatives.

Measurements and weights given in the species profiles are mostly those of Jones, Armstrong, and Choate (1985) or Clark and Stromberg (1987). Much of the life history information that is not specifically attributed to its source was derived from Clark and Stromberg (1987) or from Armstrong, Fitzpatrick, and Meaney (2011). See the list of abbreviations and symbols in the front matter for the meanings of the code letters that follow the species name. They provide shorthand clues to its range and abundance in Nebraska.

Family Didelphidae (Opossums)

VIRGINIA OPOSSUM. *DIDELPHIS VIRGINIANA*. DI, CO

Identification: This is North America's only marsupial, but the female's unique external brooding pouch is unlikely to be visible at a distance, even with young inside. However, the long naked tail is conspicuous, as are the large black ears and long pink nose. Opossums are about the size of house cats, weighing from about 4 to 14 pounds. They are as likely to be seen in a tree as on the ground and are active both diurnally and nocturnally. They don't hibernate during winter but might retreat to a sheltered location during freezing weather. Nebraska is near the northern edge of their range, and the Virginia opossum is not only the most northern opossum species but also the only marsupial that stores fat seasonally. Length 643–900 mm (25–35 in.), tail 250–440 mm (9.8–17.3 in.), weight 1.9–2.8 kg (4.2–6.2 lb.).

Voice: Opossums hiss when threatened. Both sexes also utter clicking sounds in aggressive situations, and the female uses clicking sounds when communicating to offspring. These sounds stimulate ambulatory babies to follow her, either clinging to her back or belly or running along beside her.

Status: Opossums are widespread in Nebraska and seem to be one of the commonest of medium-sized mammals in cities, judging from the number of run-over carcasses that can be seen along city streets. They have poor emergency responses to oncoming vehicles and become dazzled by headlights into an often-fatal "freezing" stance. Together with the badger, opossum populations did not significantly change between the 1940s and 2000 (Landholt and Genoways 2000).

Habitats and Ecology: Opossums are nearly omnivorous in their diets; their foods range from carrion to live animals, they often eat grain such as corn in the winter, and they gradually shift to insects, other invertebrates, birds' eggs, small mammals, and plant materials during summer. During fall and early winter they also consume fruits and berries.

Breeding Biology: Females become sexually mature during their first year and in Wyoming have two litters per year, one in late January or February and another in May or June. The gestation period is about 13 days, after which 4–23 embryo-like young emerge in rapid succession, either singly or in groups. Their forelegs are developed well enough for them to climb up and into the female's marsupium, where they try to find one of the approximately 13 teats, although some of these are nonfunctional. The average number of pouched young is about 8.5. As the young grow they are able to climb well, and their prehensile tail helps them maneuver in trees, but they can't hang upside-down from a branch by their tail alone for more than a few seconds, as I have personally observed. Opossums are short-lived in the wild, with few living more than two years. Their tendency to "play possum" by becoming immobile and feigning death when threatened is of no survival value in the modern car-dominated world. Presidents and politicians might be able to survive indefinitely in the twenty-first century by dissembling, but not opossums.

Selected References: Clark and Stromberg 1987; Armstrong, Fitzpatrick, and Meaney 2011; Buskirk 2016.

Family Leporidae (Hares and Rabbits)

WHITE-TAILED JACKRABBIT. *LEPUS TOWNSENDII.* DI (NO), UN

BLACK-TAILED JACKRABBIT. *LEPUS CALIFORNICUS.* WI (NO), RA

Identification: The white-tailed jackrabbit is the largest rabbit in North America; I have sometimes thought I had flushed a white-tailed deer fawn when I was suddenly startled by a bounding white-tailed jackrabbit. Nebraska's two species of jackrabbits differ most obviously in their tail color, the black-tailed distinctive in being black on the tail's upper half rather than entirely white. They also differ in having relatively longer (albeit actually slightly shorter) ears. The white-tailed has a total length of 55–65 cm (22–26 in.) and weighs 3–6 kg (6.6–13 lb.), whereas the black-tailed has a total length of 46–63 cm (18–25 in.) and weighs 2.4–4 kg (5.3–8.8 lb.).

Voice: Like all rabbits and hares, these species are normally silent but scream loudly (and frighteningly) in pain or distress, as when captured by a predator, and the young similarly scream when distressed or frightened. Females also vocalize softly to their young when assembling them, and adults utter loud *qua-qua* calls when fighting. Probably all members of the rabbit-hare family have glands under the chin and in the groin area, the secretions of which are presumably as important as vocalizations during close social interactions. Glands on either side of the anus in jackrabbits secrete a musky odor that might also be important in individual recognition (Schwartz and Schwartz 2016).

Status: The black-tailed jackrabbit is most common in western Nebraska and is now almost absent from the eastern third of the state (Benedict, Genoways, and Freeman 2000). It favors shortgrass habitat and avoids cover more than 2 feet high. The white-tailed jackrabbit has likewise disappeared from the eastern and southern parts of its once almost statewide range (Benedict, Genoways, and Freeman 2000). It is most common in mixed grass and shrub communities, as well as grassy openings in woodlands, but it avoids cultivated areas (Schwartz and Schwartz 1959).

Habitats and Ecology: Although the two jackrabbits are similar in many ecological respects, the white-tailed jackrabbit is more closely tied to native grassland vegetation than is the black-tailed, even though it seems to have a greater capacity to live in many different grassland types. Shortgrass prairie is a commonly used habitat in Wyoming, but the species also does well in tallgrass prairies of the eastern Great Plains, bunchgrass prairies of the Pacific Northwest, annual grasslands of California, and desert grasslands of the American Southwest. Both of these species consume a wide diversity of grasses, forbs, and shrubs, with shrubs being major winter survival foods, grasses favored in spring, and forbs important during summer. Like other rabbits and hares, they also consume some of their own feces, which allows them to digest the nutrients that they had earlier eaten and that were subsequently transformed by intestinal bacteria into digestible forms. Home ranges of black-tailed jackrabbits were judged to be 16–183 hectares (39–451 acres) in Colorado (Armstrong,

Fitzpatrick, and Meaney 2011), which is similar to an estimate of 89 hectares (220 acres) for the white-tailed jackrabbit (Forsyth 1999).

Breeding Biology: Both species breed primarily in spring, the white-tailed starting in early spring, and young are born as late as August. At least two litters per breeding season have been reported for the white-tailed species, and up to four black-tailed litters might be produced in Missouri, with some breeding throughout the year (Armstrong, Fitzpatrick, and Meaney 2011; Schwartz and Schwartz 2016). Litter sizes vary greatly, with one to nine young reported for the white-tailed and one to eight for the black-tailed. The largest litters occur in the middle of the breeding season, and litters average larger in northern parts of the species' range, although fewer litters per season are possible in northern regions. Gestation periods for these two species have been reported as about 40–47 days. The young are born fully furred and are weaned by four weeks of age. Black-tailed young reportedly reach full size by two months of age, and young white-tailed jackrabbits are independent by two months but do not reach full body size until about December. A few female black-tailed jackrabbits might breed in their first year, whereas white-tailed females reportedly do not breed until their second year. Perhaps these reproductive traits vary from year to year or regionally, but like several other hares, both species undergo marked population cycles of about eight to ten years for reasons that still remain elusive. Possibly, annual changes in litter size and number of litters per breeding season under different ecological conditions could have marked influences on short-term population trends.

Selected References: Plettner 1984; Clark and Stromberg 1987; Benedict, Genoways, and Freeman 2000; Armstrong, Fitzpatrick, and Meaney 2011; Schwartz and Schwartz 2016; Buskirk 2016.

EASTERN COTTONTAIL. *SYLVILAGUS FLORIDANUS.* UB, CO

DESERT COTTONTAIL. *SYLVILAGUS AUDUBONII.* SD (WE), UN

Identification: Nebraska supports two species of cottontails. The desert cottontail is pale gray and lighter in body weight but has longer ears. The eastern cottontail is the heavier, with grizzled brown pelage and shorter ears, and is widespread and common. Desert cottontails have longer ears (longer than their hind feet) and a paler (tan rather than medium-brown) pelage color. Desert cottontail: weight 755–1,250 gm (26.6–44.1 oz.); ear length 82–78 mm (2.4–3 in.). Eastern cottontail: weight 801–1,533 gm (28.2–54.1 oz.); ear 48–60 mm (1.9–2.4 in.).

Voice: Like other rabbits, cottontails are mostly rather silent except when they are in distress or pain, when they produce a loud, child-like scream. However, they also utter clucking, purring, and humming sounds, the former two when eating or content, and the last by males during courtship. During aggressive situations, cottontails might growl, snort, or hiss.

Status: The desert cottontail is largely limited to the western half of Nebraska. Home ranges in the desert cottontail have been estimated at 3–4 hectares (7.4–9.9 acres) (Clark and Stromberg 1987) and also at 0.5–6 hectares (1.2–19.8 acres), with densities of up to 16 animals per hectare (6.5 per acre) (Armstrong, Fitzpatrick, and Meaney 2011).

Habitats and Ecology: Desert cottontails are found on the short-grass plains, where they often occur with prairie dogs and use their burrows for escape hatches. They also occur in riparian woodlands, montane scrublands, juniper woodlands, and other habitats with little vegetation but with places where burrows, crevices, or other places offering escape cover are present.

Breeding Biology: Desert cottontails breed from late winter through summer in Wyoming, while eastern cottontails breed from early March through September in Missouri. Gestation in both species is 28–30 days, with one to nine (eastern) or three to four (desert) young born per litter. The young leave their nests at about two to three weeks of age and are independent by three to four weeks. There are five to eight (eastern) or two to four (desert) litters per breeding season (Clark and Stromberg 1987; Schwartz and Schwartz 2016). The average litter size is four to five in both species.

Selected References: Clark and Stromberg 1987; Armstrong, Fitzpatrick, and Meaney 2011; Schwartz and Schwartz 2016; Buskirk 2016.

Family Soricidae (Shrews)

CINEREUS (MASKED) SHREW. *SOREX CINEREUS.* UB, UN

Identification: This tiny shrew is uniformly brown above, becoming paler on the sides and underparts. Its nearly naked tail (maximum length 50 mm) is slightly bicolored and tipped with blackish. It is extremely difficult to distinguish from near relatives such as the Hayden's shrew (which is often considered a subspecies), which has a shorter (30–42 mm) brown-tipped tail and occurs mainly in northwestern Nebraska. In spite of its common name, the masked shrew's face is not obviously "masked," and its alternate English name, cinereous (meaning "ashen"), is scarcely better. Length 87–109 mm (3.4–4.3 in.); tail 35–39 mm (1.4–1.5 in.); weight 3.5–6 gm (0.1–0.21 oz.).

Voice: Although probably all shrews are highly vocal, this species' calls are weak and very high-pitched or of ultrasonic frequencies. In some species, such as the vagrant shrew, such ultrasonic vocalizations are used for echolocation (Gould, Negus, and Novicki 1964).

Status: This shrew ranges from eastern Siberia and Alaska south through Canada to New Mexico and Georgia. In Nebraska it is one of the most widespread of shrews, occurring almost everywhere except perhaps in the northwest, where it is in contact with the almost identical Hayden's shrew. Both of these species avoid dry habitats, as do most shrews. The cinereus shrew lives solitarily, with a home range of less than 0.02 hectare (0.05 acre).

Habitats and Ecology: Like all shrews, the high metabolic rate of these tiny mammals forces them to be active and forage almost constantly, and they consume any prey that they can kill, especially insects, spiders, and earthworms. Their high metabolic rate also prevents shrews from hibernating, and they remain active throughout the day, running nimbly, swimming, and even jumping over distances of as much as about 8 inches. Shrew lifespans are short, in association with their very high metabolic rates; they must consume from about half to more than their own weight daily to avoid starvation.

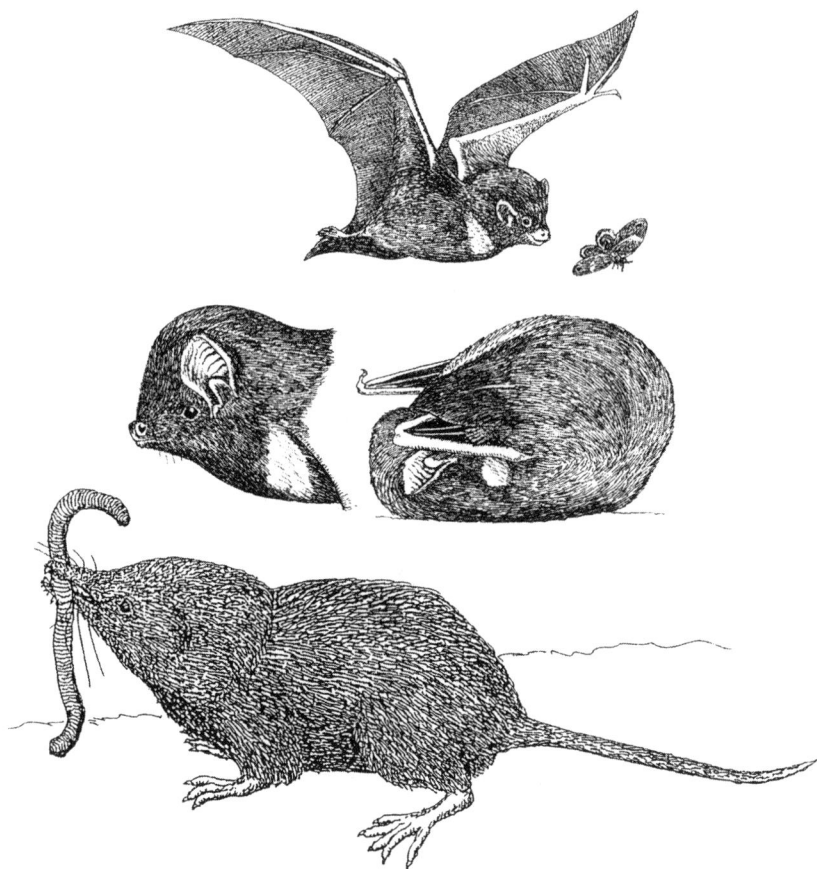

1. Eastern red bat in flight, head detail, and sleeping (above), and cinereus shrew with earthworm.

Breeding Biology: In Wyoming, females have one or two litters per year, between March and September. The male reportedly remains with his mate through pregnancy and for part of the postpartum period, but the female raises their offspring; otherwise, adult shrews tend to fight to the death when confronting one another. Some large species of shrews kill and consume smaller species. At least one genus of Nebraskan shrew (*Blarina*) produces a poisonous saliva that helps immobilize its prey.

Selected References: Clark and Stromberg 1987; Armstrong, Fitzpatrick, and Meaney 2011; Buskirk 2016.

Family Vespertilionidae (Vesper Bats)

TOWNSEND'S BIG-EARED BAT. *CORYNORHINUS TOWNSENDII.*
HL (NW), RA

Identification: This easily identified pale-brown bat has huge ears (up to 1.5 in. long) and unique fleshy glandular growths on each side of its snout. (It was once called the lump-nosed bat, and its Latin name translates as Townsend's club-shaped nose.) Total length 90–112 mm (3.5–4.4 in.); wingspan 30–34 cm (11.8–13.3 in.); weight 9–13 gm (0.3–0.4 oz.).

Voice: This species is a "whisper bat," echolocating at much lower sound amplitudes than other bats, and thus it is difficult to obtain sound recordings. For humans to hear the sounds these bats produce while echolocating, it is necessary to reduce the playback speed to one-tenth the recording speed. Their sound pulses range in duration from several to a few hundred per second. Ultrasonic frequencies above 20 kHz (20,000 cps) are used for prey detection and identification, since only extremely high frequency sounds will be reflected back from tiny objects such as moths as small as 3–10 mm (0.1–0.4 in.) in length. Lower-frequency sounds serve for general in-flight orientation, such as avoiding large objects. Vocalizations below 20 kHz are used for social interactions, such as adjusting spacing behavior in colonies, mother-offspring interactions, aggression, and warning calls.

Status: This remarkable-appearing species is rare, occurring only in northwestern Nebraska, and so far is reported only from Sheridan County. These bats might become active as early as late afternoon but usually begin feeding after dark, with most foraging occurring four or five hours after sundown. Big-eared bats are a hibernating species and often choose old mines or caves as hibernacula. A hibernaculum in Jewel Cave, Black Hills, is one of the largest known in the United States (Higgins et al. 2002). They are also fairly sedentary, not moving far from their summer home ranges to their winter quarters. Home ranges have been estimated at 200–5,900 acres (0.8–24 sq. km) (Armstrong, Fitzpatrick, and Meaney 2011).

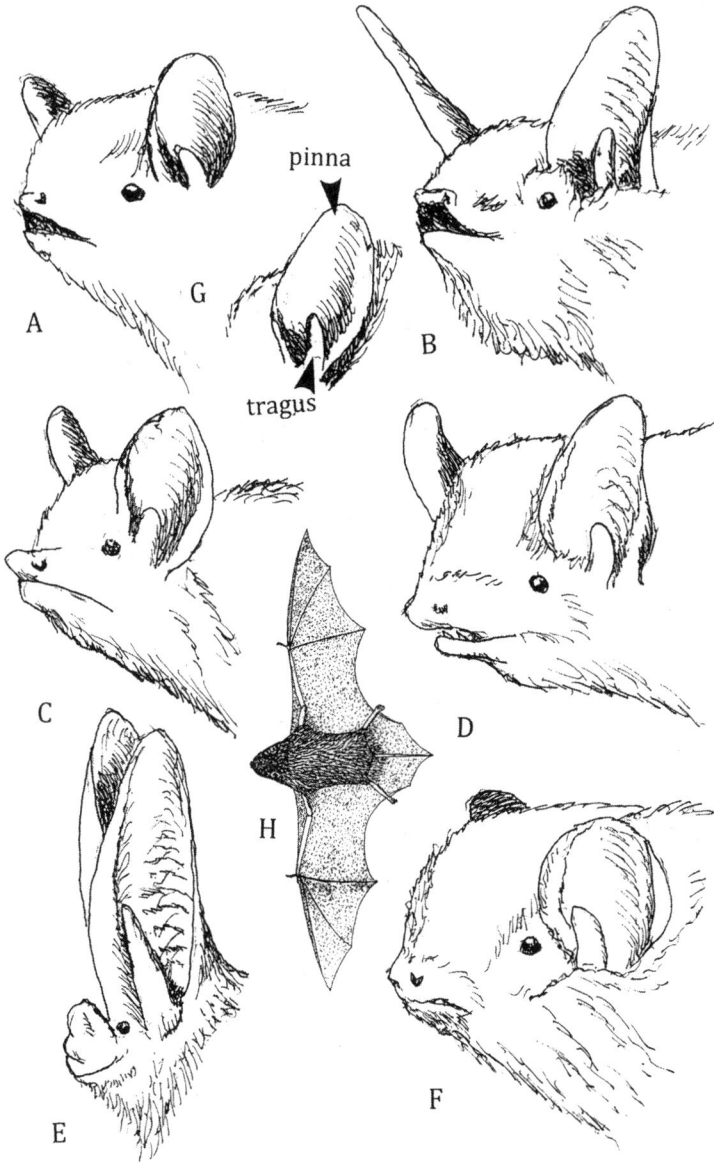

2. Heads of (A) long-legged bat, (B) long-eared bat, (C) silver-haired bat, (D) big brown bat, (E) Townsend's big-eared bat, and (F) hoary bat. Also shown are (G) the external ear of the little brown bat and (H) the flight profile of the eastern red bat. After sketches by B. Siler (A–G) (Armstrong 1975) and (H) (Johnsgard 2003).

Habitats and Ecology: This bat's huge ears not only are used to transmit sound into the bat's ear canal but also might even help to impart lift during flight. When the bat is sleeping the ears are rolled back over the head, resembling a ram's horns. Rather than specializing entirely on eating moths (about 80 percent of their diet), these bats also eat caddisflies, flies, and other insects.

Breeding Biology: The "lumps" on the side of the nose of this bat are glandular, and males scent-mark their courted females prior to copulation. Mating occurs during fall and winter while the bats are in their hibernacula, and sperm is stored until spring. Gestation lasts 56–100 days, with the single offspring being born in May or June. Initial flight occurs about three to four weeks of age, and weaning is completed by six weeks.

Selected References: Cockrum 1956; Barbour and Davis 1969; Hill and Smith 1984; Clark and Stromberg 1987; Higgins et al. 2002; Armstrong, Fitzpatrick, and Meaney 2011; Buskirk 2016.

BIG BROWN BAT. *EPTESICUS FUSCUS.* WI, CO

Identification: This large bat has uniformly yellowish-brown fur in western Nebraska and more medium-brown fur in eastern Nebraska. Among North American bats, it is second only in size to the hoary bat, and the tip of its tail is incorporated within the membrane. Total length 90–138 mm (3.5–5.4 in.); wingspan 32–40 cm (12.5–15.7 in.); weight 12–20 gm (0.4–0.7 oz.).

Voice: Thus is one of the many bats that use precise echolocation behavior for general orientation and also to locate aerial prey, which mostly consists of beetles but also includes ants, bees, and flies. Moths are infrequently taken, and it is known that some moths of the families Arctiidae (tiger and lichen moths) and Noctuidae (millers and owlet moths) can create sounds that mimic and interfere with a bat's echolocation ability (Fullard and Fenton 1979; Corcoran, Barber, and Conner 2009). Moths might also fly in loops, making noises that might startle a bat, or dive to avoid capture. When foraging, the bats utter intense search calls and fly in stereotyped flight paths. They are able to simultaneously monitor background obstacles while also tracking small, often evasive insects. One Maryland study estimated that a big brown bat maternity colony of 150 bats ate 38,000 cucumber beetles, 16,000 June beetles, 19,000 stinkbugs, and 50,000 leafhoppers in a summer! Bats have a structure at the base of each ear called a tragus, which seals off the auditory canal each time a sound signal is broadcast and quickly relaxes to allow for receiving the echo. Changes in the frequency and strength of the echo enable the bat to determine both the prey's distance and its direction of flight, and possibly even its identity.

Status: This is one of the more common bats in the state, and it breeds statewide, including the Panhandle.

Habitats and Ecology: Big brown bats are frequently found around buildings and in rock crevices, caves, and hollow trees. Buildings serve both as summer daytime roosting sites and as hibernation sites. These bats are often seen around streetlights, indicating that visual hunting might also be important. They hibernate in winter, when they might lose as much as a third of their body weight by metabolizing it to maintain a minimal survival status. They have an amazing ability to navigate, and although they are not migratory, they can return to their home roost over distances from as far away as 450 miles.

Breeding Biology: Breeding behavior in this bat begins in the fall, with mating activities occurring as late as October, although ovulation and fertilization do not occur until early spring. Their young are born in May, when colonies of up to 300 nursing females assemble. While the females are out foraging, their young are left hanging for up to an hour, although sometimes they lose their grip and fall fatally to the ground. Females rarely forage while carrying their young, which must cling tightly to their mother's teats to avoid falling. The young can fly at four to six weeks of age. Adults are known to have lived for as long as 18 years, a remarkable lifespan for an insectivorous mammal.

Selected References: Cockrum 1956; Phillips 1966; Barbour and Davis 1969; Hill and Smith 1984; Clark and Stromberg 1987; Higgins, et al. 2002; Armstrong, Fitzpatrick, and Meaney 2011; Buskirk 2016.

SILVER-HAIRED BAT. *LASIONYCTERIS NOCTIVAGANS.*
WI, UN (MIGRATORY)

Identification: This is one of the easier species of small bats to identify. Its black fur is tipped with silver, except for its black face, and its ears are edged with yellowish or have pale tips. Total length 90–104 mm (3.5–4.1 in.); wingspan 127–32 cm (5–5.2 in.); weight 7–15 gm (0.24–0.52 oz.).

Voice: Little information exists. Echolocation behavior is still unstudied, but silver-haired bat sonograms resemble those of other species known to use echolocation. Given the species' tendency to forage late at night, its echolocation skills must be very good.

Status: These bats occur statewide in a variety of habitats, are probably abundant, and are usually found in or near pine or deciduous forests close to water. This species is on the Tier 1 list of the Nebraska Natural Legacy Project's highly threatened species in Nebraska.

Habitats and Ecology: These bats forage late into the evening hours and nighttime, up to eight hours after sunset. Their flight is slow and erratic, and they forage mostly on moths but also on beetles, wasps, flies, and other winged insects. Summer roost sites are often trees, especially aspens and conifers. Mature trees or snags in early stages of decay are preferred roosting sites. Silver-haired bats are migratory in Nebraska and spend the winter as far south as northern Mexico, roosting or hibernating in diverse locations such as rock crevices, leaf clumps, woodpecker holes, birds' nests, sheds, garages, or outbuildings.

Breeding Biology: Like that of many other Nebraska bats, mating occurs in the fall, and the sperm remains viable through the winter within the female's reproductive tract. Ovulation usually occurs in April and May, and gestation lasts 50–60 days. Typically, two young are born and are able to fly by three or four weeks of age. The female's lactation period lasts about 36 days, and the precocial young are able to breed during their first summer of life. Silver-haired bats have been documented as living up to 12 years, but in one study the mean age of a population of wild individuals was only two years.

Selected References: Cockrum 1956; Barbour and Davis 1969; Hill and Smith 1984; Clark and Stromberg 1987; Higgins et al. 2002; Schmidt 2003; Geluso et al. 2004; Armstrong, Fitzpatrick, and Meaney 2011; Buskirk 2016.

EASTERN RED BAT. *LASIURUS BOREALIS.* UB, UN (MIGRATORY)

Identification: Male eastern red bats are easily identified by their unique bright-orange-red pelage; females are reddish brown, their hair tipped with white frosting. White patches are present at the shoulder and at the base of the thumb in both sexes. Females are very slightly larger than males. Total length 107–28 mm (4.2–5 in.); wingspan 28–33 cm (11–13 in.); weight 5–16 gm (0.2–0.6 oz.).

Voice: Schmidt-French, Gillam, and Fenton (2006) found that echolocation calls of adults and subadults hunting flying prey varied in their frequency components. Differences in the duration of their echolocation calls also coincided with the environment (in open versus restricted spaces). While nursing, both mothers and pups make a vibrational humming sound, the components of which differ between mothers and young. However, calls produced by pups searching for their mothers' nipples show little evidence of sonic individuality. In general, the calls of the female to the young and the calling behavior of the species, which is solitary and foliage-roosting, appear to differ from those of gregarious species that roost in more sheltered situations, since bats roosting alone in foliage might rely more on spatial memory than on acoustic cues to find their young.

Status: Red bats occur throughout Nebraska and appear to be locally common in the east but rare in the west. They are tree- or shrub-roosting bats, either as singles or in small groups of females with a litter. Sometimes tall herbaceous plants such as sunflowers are also used for roosting. This species is on the Nebraska Natural Legacy Project's Tier 1 list of threatened species in Nebraska.

Habitats and Ecology: Red bats are solitary and occur at low densities of less than 0.2 per acre (Clark and Stromberg 1987), and their home ranges are thought to be only about 2 acres in area. In winter they migrate south, where they hibernate lightly, emerging on warm days.

Breeding Biology: Mating in this species occurs during fall, often while the pair is in flight. The young are not born until about mid-June, while the female is roosting in foliage. Red bats typically have three or four young and are among the few American bats having four nipples. However, they apparently do not try to fly with their young attached, as females have been found on the ground with attached young, apparently being unable to take flight and having fallen.

Selected References: Cockrum 1956; Barbour and Davis 1969; Hill and Smith 1984; Clark and Stromberg 1987; Higgins et al. 2002; Armstrong, Fitzpatrick, and Meaney 2011; Buskirk 2016.

HOARY BAT. *LASIURUS CINEREUS*. WI, UN (MIGRATORY)

Identification: This largest bat in Nebraska is easily recognized by its size and the overall "hoary" (frosted) appearance of white-tipped but otherwise dark-brown fur. Total length 120–45 mm (4.7–5.7 in.); wingspan 380–410 mm (14.9–16.5 in.); weight 18–35 gm (0.6–1.2 oz.).

Voice: Among people with excellent hearing, hoary bats are audible when they are in flight, uttering a chattering sound. However, Corcoran and Weller (2018) documented that these bats can fly within 10 feet of sensitive microphones without producing detectable echolocation calls. Acoustic modeling indicated that hoary bats sometimes do fly without using echolocation, probably reducing echolocation output to avoid eavesdropping by conspecifics during the mating season. Corcoran and Weller's findings might help explain why tens of thousands of hoary bats are killed by wind turbines each year, if they are then unable to hear the turbines.

Status: These bats are among the most widespread of all North American bats, but they are solitary and are rarely seen except during migration. The species is a strong flier and has reached Hawaii, where it is the only native land mammal and is an endemic subspecies. The hoary bat is on the Nebraska Natural Legacy Project's Tier 1 list of threatened species in Nebraska.

Habitats and Ecology: This large bat usually roosts in trees, using tall deciduous trees in the eastern United States and favoring large ponderosa pines in the western states. Hoary bats often roost in trees near the edges of clearings, where they are protected from above by leafy cover but can easily fly into from below. Their foods are mostly insects, especially moths, but they also take beetles, flies, and wasps. Hoary bats are also known to prey on several other species of bats. The species is strongly migratory, flying seasonal distances possibly in excess of 1,000 miles. Spring migration occurs in May and June, with females arriving on the breeding grounds about a month earlier than males (Armstrong, Fitzpatrick, and Meaney 2011).

Breeding Biology: Little is known of the reproductive biology of this species, but it seems that mating probably occurs on the wintering grounds. Males evidently do not follow the females all the way to their northern breeding grounds. Twins are the usual litter number, but litter sizes range from one to four. The young are carried by the female until they are 6 or 7 days old and are able to fly by 34 days of age.

Selected References: Cockrum 1956; Findley and Jones 1964; Barbour and Davis 1969; Hill and Smith 1984; Clark and Stromberg 1987; Higgins et al. 2002; Armstrong, Fitzpatrick, and Meaney 2011; Buskirk 2016; Corcoran and Weller 2018.

LITTLE BROWN MYOTIS. *MYOTIS LUCIFUGUS.* SD (NW, EA), CO

Identification: This is one of the very large group (with about 16 U.S. species) of small-eared *Myotis* species (*Myotis* is Latin for "mouse-eared") that are probably impossible to identify without careful in-hand examination. It is very small, with glossy yellow-brown fur and yellow-buff underparts in western Nebraska and medium-brown fur in the east. Total length 90–100 mm (3.5–3.9 in.); wingspan 22–27 cm (8.7–10.1 in.); weight 4–9 gm (0.1–0.3 oz.).

Voice: This species' vocalizations are well studied. Barclay, Fenton, and Thomas (1979) determined that at least ten vocalizations are used. Echolocation calls are used for orientation, avoiding near objects and during first flights of young. Nonecholocation calls serve in agonistic encounters, for roost-space protection, in mother-infant interactions, and during mating behavior. Other studies indicate that individual identities can be detected acoustically by flying bats and that individual identity, state of lactation, and age category can all be determined from calls by nonflying individuals.

Status: This species is widespread over nearly all of North America. It often roosts in rafters and is one of the most common bats to enter houses when searching for a roosting site, especially among inexperienced youngsters. This species is on the Nebraska Natural Legacy Project's Tier 1 list of threatened species in Nebraska.

Habitats and Ecology: The little brown bat often forages over water, flying slowly and erratically, catching insects. It especially preys on moths but also eats other insects as small as mosquitoes, the latter caught at a rate of up to 600 per hour! These bats forage through the night, and the chitin in the exoskeletons of the insects they eat is partially digested with the aid of enzymes produced by symbiotic bacteria in the bats' intestines. Roosting occurs in many different kinds of sites, such as in trees and buildings and under rocks. Hibernation might begin as early as October.

Breeding Biology: Mating occurs in late fall, and both sexes are promiscuous, with males sometimes copulating with torpid females. The sperm is stored, and ovulation might occur a few days after the female's emergence from hibernation. The gestation period is 50–60 days, with a single infant being born from late May to late June. The young can fly within four to six weeks and are of adult weight by six weeks. Nursing colonies of up to as many as 800 individuals have been found during summer. These tiny bats are remarkably long-lived, with many living to 10 years of age, and there is even an almost incredible record of survival to 30 years.

Selected References: Cockrum 1956; Barbour and Davis 1969; Hill and Smith 1984; Clark and Stromberg 1987; Higgins et al. 2002; Armstrong, Fitzpatrick, and Meaney 2011; Buskirk 2016.

Family Felidae (Cats)

CANADA LYNX. *LYNX CANADENSIS.* VAGRANT

Identification: The Canada lynx differs from the bobcat in having a shorter, black-tipped tail, more obvious ear tufts, and a paler, less clearly spotted and barred pelage that is gray in winter and pale tan in summer. Length 670–820 mm (25.3–32.3 in.); tail 50–140 mm (2–5.5 in.); weight 5.1–18 kg (11–40 lb.).

Voice: The voice of a lynx is distinctly cat-like: adult males utter a deep, loud meow and females a whining purr. During the mating season, two males sometimes engage in a prolonged and almost frightening screaming contest (caterwauling) that has been described as banshee-like.

Status: This elusive cat has been reported in Nebraska only sporadically. It is rare everywhere and is mostly associated with dense coniferous forests. Over much of its range its population size is closely tied to that of the snowshoe hare, which across North America constitutes an average of 35–100 percent of the lynx's annual diet. In Colorado, preliminary data on 548 individual prey indicated that snowshoe hares comprised 35–91 percent of their annual total food, averaging 74 percent, with red squirrels of secondary importance (Armstrong, Fitzpatrick, and Meaney 2011).

Habitats and Ecology: Lynx are largely limited to coniferous forests where deep snow accumulates in winter. During summer and autumn, they might also move into riparian woods. The presence of young aspens, where the cover is not too dense for a snowshoe hare's mobility, is also possibly a factor that influences lynx habitat use. Lynx are highly mobile; some individuals of a group of several hundred lynx that were introduced into Colorado were later found as far away as Nebraska, Utah, Montana, Wyoming, Kansas, and Iowa. Home ranges in the southern Rockies have been found to vary from 39 to 506 square kilometers (15 to 195 square miles). Home ranges of males are often about twice as large as those of females (Armstrong, Fitzpatrick, and Meaney 2011).

Breeding Biology: Like other cats, male lynx are polygynous, mating with multiple females, and females are seasonally polyestrous, having several estrus cycles per season. Mating occurs during March and April, and the gestation period is about nine weeks. The kittens are born in late May and early June. Litter sizes range from one to six but average three. Juveniles disperse during their first fall, and females probably breed as yearlings (Clark and Stromberg 1987; Armstrong, Fitzpatrick, and Meaney 2011).

Selected References: Weigand 1964; Clark and Stromberg 1987; Ruggiero 1994; McKelvey 2000; Higgins et al. 2002; Armstrong, Fitzpatrick, and Meaney 2011; Buskirk 2016.

BOBCAT. *LYNX RUFUS.* UB, UN

Identification: This widely distributed but highly elusive cat occurs throughout Nebraska. It favors habitats with good hiding places, such as rock caves, brush piles, and fallen trees. It is easily distinguished from domestic cats by its bobbed tail (thus, "bobcat"), which is black-tipped. Like those of many other wild cats, this cat's ears are slightly tufted and are contrastingly white behind. (Most cat species have similar white spots behind their ears; I have long wondered if they might serve as "taillights" for kittens trying to follow their mother at night.) Males average slightly larger than females. Length 81–101 cm (32–39.7 in.); tail 100–165 mm (3.9–6.5 in.); weight 9–12 kg (19.8–26.4 lb.).

Voice: Bobcats have vocalizations much like those of domestic cats, including yowling during the breeding season, a cough-bark when threatened, and a loud scream.

Status: Bobcats almost certainly occur in every Nebraska county, although county records are lacking for many. Bobcat populations increased significantly between the 1940s and 2000, perhaps because of the increasing abundance of prey such as white-tailed deer (Landholt and Genoways 2000).

Habitats and Ecology: Bobcats are largely nocturnal, but I once saw one during the early morning hours near a Platte River crane roost, and daytime sightings are not rare. They are also solitary, forming only brief pair bonds during breeding, which usually occurs during winter and spring. Home ranges of adults are highly variable, from 0.6 to 20 square kilometers (150 to nearly 5,000 acres). Their foods are also highly diverse, but rabbits and rat-sized rodents predominate, and prey ranges in size from mice to young fawns. Fish, amphibians, reptiles, and birds are also eaten. Great horned owls have been reported to prey upon immature bobcats; coyotes and mountain lions often kill adults.

Breeding Biology: Bobcats are solitary, forming only brief pair bonds during the breeding season, which usually occurs during winter and spring. The gestation period is about 62 days, and the litter size ranges from one to eight but averages only three. Like those of other cats, the kits' eyes are closed at birth but open at about ten days, and by four weeks kits are able to eat solid foods. They are weaned by seven or eight weeks and begin to follow their mother on short trips. By about seven months juveniles start to disperse from their natal range and are fairly independent. Females become sexually mature at one or two years, but males are usually mature only in their second year.

Selected References: Epperson 1978; Clark and Stromberg 1987; Higgins et al. 2002; Hoffmann 2006; Armstrong, Fitzpatrick, and Meaney 2011; Buskirk 2016.

COUGAR (PUMA, MOUNTAIN LION). *PUMA CONCOLOR.*

DI (NW), RA

Identification: The cougar is the largest North American cat (about wolf-sized) and the only wild cat with a long tail that is tipped with black. Length 200–250 cm (78–98 in.); tail 650–800 mm (25–31 in.); weight 34–91 kg (75–200 lb.). The cougar is not closely related to the African lion; its closest relative is the New World jaguarondi (*Herpailurus yagouaroundi*). The distinctive English name "cougar" is from the Portuguese and refers to the animal's deer-like pelage color; *puma* is a Spanish word derived from the Quechuan language.

Voice: Although generally silent, male cougars are said to produce a wailing cry, and females utter a far-carrying, high-pitched yowl (caterwauling) when in estrus. Males are unable to roar like

African lions, apparently because their hyoid bone is solid and can't vibrate, and thus they cannot produce a throaty roar. Kittens produce a loud, rasping purring, and adults growl like overgrown house cats. Soft chirping sounds are used by mothers and young to locate one another, and typical cat-like meowing, hissing, and spitting sounds are also uttered.

Status: Cougars are mostly found in the Pine Ridge region of northwestern Nebraska. There are probably also a few potentially breeding animals in the Niobrara Valley and the Wildcat Hills. They generally prefer dense cover and rugged terrain where deer are abundant and are often found in shrublands and juniper woodlands. Surveys conducted between 2010 and 2019 indicated a Pine Ridge population of nearly 60 animals in 2017. The Pine Ridge region then supported a breeding population of up to 20 pairs, from which immatures often dispersed eastwardly along wooded riparian corridors. These inexperienced animals are the ones most likely to encounter humans and very rarely might seemingly threaten them, since they are ineffective predators and could be near starvation (Johnsgard 2014c ["To kill a mountain lion"]).

Habitats and Ecology: Cougars hunt by day or night but need sufficient cover for stalking prey, such as brushy, wooded vegetation or rough terrain. They also need a source of available water. Home ranges vary greatly, from as little as about 40 to more than 700 square kilometers (15 to 270 square miles) for females and from 120 to 830 square kilometers (46 to 320 square miles) for males (Armstrong, Fitzpatrick, and Meaney 2011). About 80 percent of their food intake consists of deer; it has been estimated that an adult cougar must kill a deer about every two weeks to survive the winter. Females with young have higher food requirements, and kill rates also vary with age and social structure. Cougars thus provide an important part in controlling the overpopulations of deer in more enlightened states where they are a protected species, as in California. They have sometimes been killed by Nebraskans in reputed "self-defense" but are of incalculably less danger to humans than are other humans. Only about 25 human deaths caused by cougars have been documented in about two centuries of North American history, but since about 17,000 Americans are murdered in the United States annually, any American is roughly 700,000 times more likely to be killed by another American than by a cougar.

Breeding Biology: Cougars breed at any time of the year, depending on when females come into estrus. The very large home ranges of males would indicate that males are probably polygynous. Females are polyestrous, cycling every few weeks until they are bred, and utter loud caterwauling calls to attract males. The gestation period is about 92 days, with kittens likely to be born in late summer or fall. Litter sizes range from one to six (females have six functional mammae) but average 2.6. Female reproductive rates are low, averaging about four litters per decade, so about 40 percent of the adult females produce young during any one year. Females continue to hunt for their young for as long as 22 months, or usually through the spring of their second year. This long period of juvenile dependency means that when females with dependent young are killed by "sportsmen," their entire families are likely to die too.

Hunting and Population: In Nebraska's first authorized cougar hunt of 2014, five animals were killed legally, and eleven more were killed that year by illegal hunting or by other means (Johnsgard 2014c). Ten of the sixteen killed were females, of which at least four (40 percent) might statistically have been mothers tending up to eight to ten dependent young, all of whom possibly starved when left motherless. The Pine Ridge population was estimated at 59 animals in 2018. In the state's second hunt, held in the winter of 2019, four animals were designated to be killed in each of two Pine Ridge areas, including no more than two females in each. Three males and two females were legally killed during that season. Without waiting for a determination of the effects of the 2019 season on the surviving cougar population, the Nebraska Game and Parks commissioners approved a January–February 2020 season in the Pine Ridge region, allowing for the killing of four males and four females (and potentially the deaths of as many as ten dependent kittens, if all four females killed were mothers). If the allotted number of animals had not been eliminated by Nebraska's nimrods before the end of February, a supplemental season was to open in March 2020, when cougar chasers could legally use tracking dogs to tree their prey before dispatching them. A treed cougar allows an execution-style kill that poses no danger to the shooter and requires absolutely no hunting skill. It also permits a small-caliber bullet to be used, thus avoiding much disfigurement of the animal's pelt, even if it results in a long and agonizing death for the cougar. On the brighter side, in 2017 some 25,000 Nebraskans decided they would pay up to forty dollars beyond the usual car-licensing fees to have a cougar-conservation motif displayed on their plates. The extra income of more than $100,000 was directed by state legislation to found conservation education of the general public, albeit apparently not of the commissioners.

Selected References: Nowak 1974; Clark and Stromberg 1987; Anderson and Lindzey 2005; Higgins et al. 2002; Hoffmann 2002; Beckoff and Lowe 2007; Armstrong, Fitzpatrick, and Meaney 2011; Johnsgard 2014c; Buskirk 2016.

Family Canidae (Dogs)

COYOTE. *CANIS LATRANS.* UB, UN

Identification: Coyotes are very dog-like but are uniformly grizzled gray overall, except for a black-tipped tail and rusty to yellowish legs. The ears are prominent, and the tail is bushy and is held almost between the legs when the animal is running. Coyotes typically weigh 20–35 pounds; coydogs (coyote-dog hybrids) might be much larger. Length 110–22 cm (44–48 in.); tail 330–400 mm (13–20 in.); weight 10–20 kg (22–44 lb.).

Voice: The communal howling calls of coyotes are familiar to farmers, ranchers, and lovers of the outdoors. The calls are prolonged vocalizations, usually uttered during evening or early morning hours. A call typically consists of a few sharp barks, followed by a prolonged mournful howl, and ending with several short sharp yips. Barking calls are used in threat situations. Other vocalizations and social posturing are essentially like those of domestic dogs.

The coyote's Latin name translates as "barking dog." Coyotes occasionally hybridize with large dogs, producing so-called coy-dogs. Its English name is from the native Aztec (Nahuatl) language, *coyotl*.

Status: In spite of constant persecution from humans, coyotes have managed to survive across most of their historic range and probably still are present in every Nebraska county.

Habitats and Ecology: Coyotes are highly adaptable, occurring in North American habitats ranging from deserts to woodlands and even extending into tundra and tropical forests. They are mobile daytime predators, with long-term ranges of up to 400 miles. They can trot at speeds of 25–30 miles per hour, run at speeds up to 40 miles per hour, and leap as far as 14 feet. Although most ranchers hate them, coyotes are highly beneficial to them overall (Hyde 1986). In a major study of 8,265 coyote stomach contents, rabbits comprised 33 percent of the food intake by volume, rodents 17 percent, carrion 13 percent, and livestock 14 percent (Johnsgard 2005). The livestock that they eat are mostly scavenged or involve weakened animals. Coyotes form monogamous pair bonds, and females come into estrus for only a few days once per year, between February and March, following a prolonged courtship period. There is a gestation period of 63 days and an average litter size of six pups. The young are born blind and helpless but are weaned after being nursed for six or seven weeks, while the male brings food to the pair's den. The pair bond lasts through the period of reproduction, and family ties thereafter are the basis for autumn and winter social hunting parties, as in wolf packs. Lifespans in the wild rarely exceed ten years; given America's gun-addicted culture and general ignorance of predator values, most Nebraska coyotes are probably very lucky if they reach five years of age.

Hunting and Population: Unlike most furbearers, coyotes in Nebraska can be shot or trapped at any time, without a state permit. Good quality finished pelt values were $55 to $85 in 2019. Based on trapping records from 1930 to 2000, there was a probable long-term increase in coyote populations during that period (Landholt and Genoways 2000), but in recent years sarcoptic mange has had serious effects on coyote survival and pelt quality. No tallies of coyotes shot in Nebraska are available; in some years as many as 12,000 coyotes have been trapped in Nebraska. In 2003 a coyote hunter shot a gray wolf near Spaulding, Nebraska, the first record of wolves in the state since 1913. DNA evidence indicated that it came from the Great Lakes population.

Selected References: Fichter 1950; Ryden 1977; Pringle 1977; Camenzind 1978; Hyde 1986; Clark and Stromberg 1987; Laydet 1988; Higgins et al. 2002; Doby 2006; Armstrong, Fitzpatrick, and Meaney 2011; Buskirk 2016.

GRAY FOX. *UROCYON CINEREOARGENTEUS.* DI (EA), OC

Identification: This fox closely resembles the substantially smaller swift fox but has a distinctive black stripe extending along the top of its bushy tail. A cinnamon-orange tint usually extends along the bottom of the tail and underparts and up over the sides of the shoulders and neck to the back of the ears, but sometimes these areas are whitish. Gray foxes somewhat resemble coyotes, which are substantially larger and lack the gray fox's just-mentioned strong cinnamon-orange tints. Length 80–112 cm (31.5–44.3 in.); tail 275–343 mm (10.8–13.5 in.); weight 3–7 kg (6.6–15.4 lb.).

Voice: Gray foxes utter loud barks or yips during the breeding season and produce a variety of other chuckles, growls, and squeals. Probably their vocabulary is much like that of the red fox, but no detailed comparisons seem to be available.

Status: Gray foxes are rare and have a limited Nebraska range. They mostly occur in eastern Nebraska, but records locally extend west in the Platte Valley to the Wyoming border, and there are a few records from Sioux to Cherry County (Benedict, Genoways, and Freeman 2000). They are often found along stream courses lined with deciduous riparian woods, as well as in brushy woodlands among areas of hilly or irregular topography. In some areas they have also been found close to human habitation. They den in rocky outcrops, brush piles, burrows, and hollow trees. Unlike other foxes, they are able to climb trees with ease by grasping with their forefeet and long front claws and using their hind legs for propulsion. Tree dens up to 7.6 meters (25 feet) above the ground have been found (Clark and Stromberg 1987). In the western states, gray foxes have large home ranges, estimated to range from about 30 to 350 hectares (74 to 864 acres), and in the East, their home ranges are often even larger, sometimes exceeding 500 hectares (1,235 acres) (Armstrong, Fitzpatrick, and Meaney 2011).

Habitats and Ecology: Gray foxes are too rare in Nebraska to report on their ecology here, but in general they are much like other foxes and eat a wide variety of plants and animals. Their prey includes many species of mammals up to the size of muskrats and jackrabbits, as well as (presumably small) domestic dogs and cats, some prey as small as mice, and even insects. Birds, nestling birds, and bird eggs of species such as crows, ducks, and songbirds are sometimes eaten. All sorts of vegetable matter are also consumed. Their predators include eagles, cougars, bobcats, and coyotes. In some areas they have been affected by rabies (Armstrong, Fitzpatrick, and Meaney 2011).

Breeding Biology: Gray foxes are monogamous and breed from January to April over much of their range. The gestation period is about 59 days, with one to seven pups being born, the average litter size being about four. The young remain in their den for the first month but begin to disperse that fall. Sexual maturity among females occurs at ten months, but many females probably do not breed during their first year (Armstrong, Fitzpatrick, and Meaney 2011).

Selected References: Menzel 1968; Clark and Stromberg 1987; Higgins et al. 2002; Moehrenschlager et al. 2004; Armstrong, Fitzpatrick, and Meaney 2011; Buskirk 2016.

RED FOX. *VULPES VULPES.* WI (EA), UN

Identification: This rust-colored canid is easily recognized by its long ears, pointed snout, and large bushy tail, which is tipped with white. Its pelage is usually rufous brown overall, but variations often occur. Melanistic (black) pelage occurs commonly in western populations.

A so-called silver variant is basically melanistic, but white tips on the long blackish guard hairs give the overall pelage a silvery cast. There is also a "cross fox" variant, which is yellowish dorsally but has a cross-like pattern of darker fur along the dorsal midline and down the shoulders. The two latter pelage types are most common in northern populations. Length 83–101 cm (32.5–40 in.); tail 291–461 mm (11.4–18.1 in.); weight 3–7 kg (6.6–15.4 lb.).

Voice: The most common red fox vocalization is a series of rapid, high-pitched, yip-like barks, generally thought to be individual identification signals, as it is known that foxes can recognize other individuals by barking. Another vocalization is a loud "screamy howl," sounding like a human baby in distress, uttered by estrus vixens to attract males. The remaining vocalizations are much softer and are used in close-up communication. One is "gekkering," a rapid chattering series of notes with occasional inserted howls that is used during aggressive encounters by adults and during play-fighting by juveniles. Parents chortle to their young when bringing food to them.

Status: Red foxes occur statewide, mostly in open country, such as scattered woodlands, wooded riparian corridors in grasslands, farmlands, pastures, and brushlands. Occasionally they are found inside towns, where they gain protection from coyotes and sometimes even steal food put out for pet dogs. Grasshoppers, other insects, and crayfish comprise about a quarter of the normal summer diet, and during winter mice, rabbits, squirrels, and other small mammals are important foods. They also eat a wide variety of plant materials, including grass, berries, acorns, and fruits. Caching of excess food is typical. Their hearing is highly sensitive to low-frequency sounds, such as the noises produced by subterranean mammals or those beneath snow cover, which they can pounce upon with a surprising degree of accuracy.

Habitats and Ecology: Home ranges of red foxes are highly variable and have been estimated to be as little as 57 hectares (116 acres) in urban areas to 6,000 hectares (14,800 acres) in less food-rich habitats. Daily movements of an individual might exceed 10 kilometers (6 miles). Territories of a mated pair are defended during the breeding season and are scent-marked with urine and feces (Armstrong, Fitzpatrick, and Meaney 2011).

Breeding Biology: Breeding occurs from December to March; in Colorado it primarily occurs in January and February. Estrus lasts one to six days, and gestation requires about 52 days. The average litter size is 5.5 pups but ranges from 1 to 7. Both members of the pair provision the pups, which remain in the den for about a month. They are fully grown by six months of age, and by late September the juveniles begin to disperse. Females can breed at about ten months of age, although about 10 percent of the females might not breed as yearlings (Clark and Stromberg 1987; Armstrong, Fitzpatrick, and Meaney 2011).

Trapping and Population: Based on trapping data, red fox populations increased significantly between the 1940s and 2000 (Landholt and Genoways 2000). Their recent incursions into town and cities suggest a continuing increase in populations since then.

Selected References: Jochum 1980; Schmidt 1981; Powell and Case 1982; Clark and Stromberg 1987; Gese, Stotts, and Grothe 1996; Higgins et al. 2002; Armstrong, Fitzpatrick, and Meaney 2011; Buskirk 2016.

SWIFT FOX. *VULPES VELOX.* LO (WE) (CC)

Identification: This tiny (Yorkshire terrier–sized) fox closely resembles the gray fox, which is unlikely to occur in the swift fox's shortgrass range. Both species have black-tipped tails, but the top of the tail is not black in swift foxes. They are also generally paler than gray foxes and are only slightly cinnamon tinted along the flanks and legs. Length 735–880 mm (28.9–34.6 in.); tail 240–350 mm (9.4–13.8 in.); male weight 2.2–2.9 kg (4.8–6.4 lb.); female weight 1.8–2.3 kg (4–5.1 lb.).

Voice: Swift foxes utter rapid sequences of 4–25 barks, mainly during the breeding season. Their barks are thought to be contact calls for maintaining social units and are associated with mating and territoriality. These calls have been found to be acoustically

3. Swift fox, adult.

unique to individuals (Darden, Dabelsteen, and Pedersen 2003). No doubt other dog-like vocalizations also are uttered, but they apparently remain undescribed.

Status: The swift fox occurs in the Kimball County grasslands, Oglala National Grassland, Panhandle grasslands, the Pine Ridge, the upper Niobrara River Valley, and the Wildcat Hills (Panella 2010). Clark and Stromberg (1987) noted that swift foxes have been widely subjected to predator poisoning, overtrapping, and habitat losses and are often hit by cars. Coyotes commonly kill them and at times have been judged to be the species' most significant source of mortality. Annual survival rates in Wyoming have been estimated at only 40–69 percent (Olson and Lindzey 2002), so few wild animals are likely to survive beyond their third year.

Population: The swift is considered to be generally stable or declining across its overall range and is on the Nebraska Natural Legacy Project's Tier 1 list of threatened species in Nebraska. No hunting or trapping is allowed. It was a candidate for federal listing as endangered or threatened until 2002, when that status was unfortunately terminated. The very similar kit fox (*Vulpes macrotis*) of the American Southwest has often been considered a subspecies of the swift fox, but recent genetic data support its species-level recognition.

Habitats and Ecology: It has been reported that in southeastern Wyoming, the foxes, which are associated with shortgrass prairies, have long been locally abundant, with densities of a pair per 5–8 square kilometers (3–5 square miles) typical in wetter shortgrass prairies. Based on studies in five states, arthropods (mainly insects) are their primary food sources by number of items eaten, although by volume of consumption, mammals such as jackrabbits and cottontails are probably most important (Johnsgard 2005).

Breeding Biology: Swift foxes mate monogamously, although in some areas males have been found paired with two females sharing the same den. They breed from early March through July, and the gestation period is 51–53 days. The litter size ranges from one to eight pups (eight mammae are present) but typically is four to five. The pups' eyes are open by two weeks of age, and they begin to leave their den by four to five weeks. Females start to breed during their first year of life; captive females have bred when up to six years of age, and one male reportedly bred at ten years of age (Armstrong, Fitzpatrick, and Meaney 2011).

Selected References: Weigand 1965; Hines, Case, and Lock 1981; Clark and Stromberg 1987; Hines and Case 1991; Covell 1992; Andelt 1995b, 1997; Higgins et al. 2002; Schauster, Gese, and Kitchen 2002; Dark-Smiley and Keinath 2003; Armstrong, Fitzpatrick, and Meaney 2011; Buskirk 2016.

Family Mustelidae (Weasels)

AMERICAN BADGER. *TAXIDEA TAXUS.* UB, UN

Identification: The badger is another of the Great Plains grassland mammals that is instantly recognizable, having a unique black-patterned face with a white line up the midline of the head and broader buffy stripes extending from the mouth diagonally back to the ears. Badgers have long grizzled grayish-brown dorsal fur, and a distinctive low-slung body profile. Adults weigh 8–25 pounds; males are larger than females. Badgers have powerful forelegs, with

4. Badger, adult.

partially webbed front toes and long claws, which provide for highly efficient digging. Length 600–730 mm (23.6–28.7 in.); tail 105–35 mm (4.1–5.3 in.); weight 6.4–11.5 kg (14–25 lb.).

Voice: A badger utters various hissing, growling, snarling, and grunting noises when threatened, but these are generally very silent animals.

Status: Badgers have a widespread occurrence in Nebraska but are most common on the open plains, where rodent populations are high. They are sometimes shot for "recreation," and at times bounties have been paid for them, but probably most are killed in traps set for other mammals. However, they are effective rodent eaters and killers of venomous snakes; they are relatively immune to rattlesnake venom. Their digging activities help in soil development, and the holes they produce are used for escape routes or as dens by many other animals. Their digging activities also break up the distribution of the most dominant prairie plants, making room for other less common ones to survive and increasing plant diversity.

Habitats and Ecology: Badgers range across many habitats and ecosystems but favor open areas of meadows, prairies, steppe grasslands, or other places where subterranean dens for breeding and semihibernation during winter can be dug. They are extremely adept at excavating ground squirrel and prairie dog tunnels; at times, two badgers will collaborate, with one digging and the other waiting at another tunnel entrance to catch any escaping animals. They sometimes also cooperate with coyotes in this activity. Both sexes are highly mobile as adults. Males have enormous home ranges of from about 600 to 4,000 acres (0.9 to 6.25 square miles). Home ranges of males often overlap those of several females; anal scent glands are present that might help in social communication and coordinating sexual activities.

Breeding Biology: Badgers breed in the summer or fall, but the pair bond lasts only until the female is fertilized. After conception she takes on all the responsibilities for reproduction, including protecting the offspring. The young are not born until March or April owing to delayed implantation of the embryos (see mink profile). Two to four babies typically comprise a litter; they are fur covered but are helpless and blind at birth. Young badgers are weaned when they are about two-thirds grown, and year-old females can become sexually mature. Badgers can potentially live for up to 14 years, but there is a high mortality among the young, probably mainly as a result of starvation. Judging from state trapping data, badger populations did not change significantly between the 1940s and 2000 (Landholt and Genoways 2000).

Selected References: Clark and Stromberg 1987; Neal 1996; Goodrich and Buskirk 1998; Higgins et al. 2002; Armstrong, Fitzpatrick, and Meaney 2011; Buskirk 2016.

AMERICAN MINK. *NEOVISON VISON.* UB, UN

Identification: This is the largest of Nebraska's weasels and has the most uniformly dark-brown pelage, except for some white on the chin and chest. Its tail is long, up to half the animal's total length. Length 460–700 mm (18.1–27.6 in.); tail 150–230 mm (5.9–9 in.); weight 0.9–1.6 kg (1.98–3.5 lb.).

Voice: Mink produce a wide variety of vocalizations that include aggressive hissing, screaming, chuckling, squeaking, barking, and purring. Screaming is used as a defensive threat, and squeaking is used by both sexes when in pain or fearful. Chuckling is uttered by both sexes during the breeding season and is associated with sexual stimulation (Gilbert 1969). No analysis of the mink's other calls and their functions is apparently available.

Status: Mink have a statewide distribution in Nebraska but are always found close to streams, rivers, lakes, or ponds. Their home ranges tend to be linear along river or stream courses and in one study averaged 1,630 meters (5,350 feet) (Clark and Stromberg 1987). Together with the long-tailed weasel, mink populations decreased significantly between the 1940s and 2000. These long-term declines in mink and other weasel populations have probably been the result of environmental contaminants affecting prey populations (Landholt and Genoways 2000).

Habitats and Ecology: Mink are active throughout the year, becoming more diurnal during winter months and increasing their use of fish as prey. However, they are opportunistic predators, often feeding on rodents, and are evidently able to hear the ultrasonic sounds often made by their prey. Compared with river otters, they are more likely to take terrestrial prey such as mammals and birds and feed less on fish. In a Colorado study, their density was estimated as 1.7 mink per square mile (1 per 376 acres) (Armstrong, Fitzpatrick, and Meaney 2011).

Breeding Biology: Breeding occurs from late February to April. Male are polygynous, and females might also mate with multiple males, so that a single litter might have more than one father. The gestation period is highly variable, from 40 to 74 days owing to variably delayed implantation, in which the ovum is fertilized but does not immediately begin to divide and form an embryo, thus delaying birth of the young until the most favorable birthing season arrives, usually spring. Litter sizes range from one to eight but average five, and young are born in late April or May. They reach sexual maturity at about ten months and might remain reproductively active for about seven years (Armstrong, Fitzpatrick, and Meaney 2011).

Selected References: Hall 1951; Clark and Stromberg 1987; Higgins et al. 2002; Armstrong, Fitzpatrick, and Meaney 2011; Buskirk 2016.

NORTHERN RIVER OTTER. *LONTRA CANADENSIS.*

LO (REINTRODUCED)

Identification: River otters are well named; they are mostly found along rivers, are substantially larger than mink, and have a stout, gradually tapering tail. They swim very rapidly, often while totally submerged, and are dark brown overall, except for lighter underparts. Length 915–1,346 mm (36–53 in.); tail 352–510 mm (13.8–20 in.); weight 5–13.7 kg (11–30 lb.).

Voice: Like other members of the weasel family, river otters are highly vocal. Their utterances are said to include buzzes, whistles, chirps, twitters, growls, and a staccato chuckle, as well as a hair-raising scream that is audible up to 1.5 miles away. Because river otters are highly social animals, it is not surprising that a wide diversity of vocalizations is present, although they have seemingly not yet been fully functionally analyzed.

Status: Otters are now well distributed in all of Nebraska's major rivers except for the Republican and Missouri as a result of extensive reintroduction efforts involving several hundred animals (Bischof 2003c; Panella 2010). Water quality and quantity are important features of the habitat that influence otter numbers, as well as an abundance of fish and crustaceans and the presence of ice-free stretches of river during winter. Crayfish and slow-moving fish such as suckers are important foods, but amphibians, birds, and insects are also consumed (Clark and Stromberg 1987; Armstrong, Fitzpatrick, and Meaney 2011).

Habitats and Ecology: Home ranges of otters vary greatly and have been reported in Colorado to be from 2 to 78 kilometers (1.2 to 48 miles) in length, with a mean of 32 kilometers (20 miles). However, in Alberta male home ranges were sometimes found to be greater than 200 kilometers (124 miles) long, whereas female home ranges averaged about 70 kilometers (49 miles). Otters also tend to form social groups that include bachelor males, maternal families composed of females and their young, and mixed-sex congregations of up to 30 animals (Armstrong, Fitzpatrick, and Meaney 2011).

Breeding Biology: Otters reach sexual maturity at two years of age, but some males might not breed successfully until they are at least five years old. Mating occurs shortly after the birth of a female's litter, during March and April in Colorado. A female's estrus period might last more than 40 days, during which she releases scent tracks that can be followed by males. Apparently ovulation is induced by copulation, and, because of delayed implantation (see mink account), the gestation period might last 290–375 days. The average litter size is about three pups, ranging from one to six. Youngsters begin to leave their den at two months and are weaned by about three months. They remain with their mother for about seven months, and siblings might stay together for more than a year (Armstrong, Fitzpatrick, and Meaney 2011).

Selected References: Clark and Stromberg 1987; Ruggiero 1994; Higgins et al. 2002; Kruuk 2006; Wengeler, Kelt, and Johnson 2010; Armstrong, Fitzpatrick, and Meaney 2011; Buskirk 2016.

LONG-TAILED WEASEL. *MUSTELA FRENATA.* UB, UN

Identification: This is the largest of the typical (*Mustela*) weasels. It is much larger than the least weasel, and its much-longer tail is tipped with black. During winter it is entirely white, except for its black-tipped tail. Length 300–450 mm (13.7–17.7 in.); tail 110–75 mm (4.3–6.9 in.); weight 130–316 g (0.2–0.7 lb.).

Voice: Svenden (1976) described three basic weasel vocalizations: the trill, the screech, and the squeal. The trill is produced under calm conditions, as when the animal is at play or hunting. The screech occurs when an individual is disturbed or as a defensive signal. A squeal is uttered when the animal is in distress.

Status: Judging from the number and distribution of records, this is probably the most common and widespread of Nebraska's weasels, although it is no longer present in eastern parts of Nebraska (Benedict, Genoways, and Freeman 2000). All three of Nebraska's weasels subsist primarily on small mammals, including species up to the size of rabbits, and long-tailed weasels take birds at least as large as grouse. They also eat waterfowl and upland game-bird eggs. They have been reported to consume 20–40 percent of their body weight daily, indicating a very high metabolic rate for a mammal of comparable size. Long-tailed weasels are generally solitary except when breeding but probably become more social when food sources are abundant and more solitary when they are rare. Together with muskrat, mink, eastern spotted skunk, and striped skunk, long-tailed weasel populations decreased significantly between the 1940s and 2000. Declines in the long-tailed weasel, other mustelids, and the two skunks have probably been the result of environmental contaminants such as pesticides that affected prey populations (Landholt and Genoways 2000).

Habitats and Ecology: During summer these weasels are mostly diurnal and spend much of the time hunting and searching for mates, primarily during early mornings and late afternoons. Home ranges of males are larger than those of females, often of about 12–16 hectares (30–40 acres) in favorable habitats, but in areas of fragmented agricultural land ranges might be as large as 51 hectares (126 acres) for females and 180 hectares (445 acres) for males. In Colorado a weasel density of 0.8 per square kilometer (2.1 per square mile) was estimated, but in the eastern United States some much higher density estimates have been suggested (Clark and Stromberg 1987; Armstrong, Fitzpatrick, and Meaney 2011). In Nebraska, the species seems to be most common in grasslands, especially the Sandhills prairies (Russell Benedict, personal communication).

Breeding Biology: Breeding occurs during summer, mostly in July and August, the female mating with the male whose home range includes hers. Males might form transitory pair bonds with several females but play no part in parental care. The gestation period lasts 220–37 days as a result of greatly delayed implantation of the fertilized ovum, so young are mostly born in April and May. A single litter is produced per year, which averages about seven young. By six weeks of age the young are weaned. Females reach sexual maturity at 3–4 months of age, whereas males don't begin to reproduce until about 15 months of age (Armstrong, Fitzpatrick, and Meaney 2011).

Selected References: McGrew 1958; Svenden 1976; Clark and Stromberg 1987; Farrar 1999; Higgins et al. 2002; Armstrong, Fitzpatrick, and Meaney 2011; Buskirk 2016.

BLACK-FOOTED FERRET. *MUSTELA NIGRIPES.* EXTIRPATED

Identification: This rare weasel has a distinctive facial pattern of a black mask on a mostly white face, a black-tipped tail, and black legs on an otherwise yellowish-tan body. The species is essentially limited in distribution to large prairie dog colonies and has long been extirpated from Nebraska. Length 480–567 mm (18.9–22.3 in.); tail 114–27 mm (4.5–5 in.); weight 530–1,300 gm (1.2–2.9 lb.).

Voice: Clark et al. (1986) listed the vocalizations that they detected in this species: the bark, chattering bark, bluff-hiss, growl, and an *ungh.* These utterances were contextually classified as variously associated with threat, defense, greeting, and mating, plus some additional calls that are uttered by the young. A few other vocalizations have been mentioned in the literature, but no comprehensive analysis of the species' vocalizations and their situational functions seems to be yet available.

Status: Extirpated. This is perhaps America's rarest wild mammal and was among the first species selected for listing as federally endangered in the 1973 Endangered Species Act. At that time this ferret was generally believed to already be extinct, but in 1981 a small Wyoming population was discovered near Meeteetse, in the Big Horn Basin. That population was immediately placed under the strictest protection but was nearly lost in 1985 as the result of a canine distemper outbreak. However, 17 still-surviving animals were live-trapped and placed into captivity. This small group became the reproductive nucleus of the entire population that is alive today, having produced a genetic line of several hundred highly inbred animals. Reintroduction efforts have proven to be extremely difficult, and it is impossible to judge their long-term prospects for success. As of 2019 some 300 ferrets were in captivity. However, plague has been a serious problem for both reintroduced ferrets and prairie dogs; in South Dakota the entire ferret population in Wind Cave National Park died out in 2019.

Habitats and Ecology: Black-footed ferrets are almost wholly dependent on prairie dogs for their food intake; about 90 percent of their food intake is from that source. To sustain itself, an adult ferret must eat one prairie dog every 2–6 days, or about 100 per year. Assuming typical prairie dog densities, each adult ferret needs 19–38 hectares (47–94 acres) of prairie dog habitat to survive, if no other foods are consumed (Armstrong, Fitzpatrick, and Meaney 2011). In a reintroduced South Dakota ferret population, the prairie dog colonies averaged 19 hectares (47 acres) in area and were 1.5 kilometers (0.9 mile) apart, while in a Wyoming study they averaged 322 hectares (795 acres) and were 0.6 kilometer (0.4 mile) apart (Clark and Stromberg 1987).

Breeding Biology: Like other weasels, male ferrets are probably polygynous, and females probably mate with males sharing their overlapping home ranges. They mate during spring (mid-March to early April), and the gestation period is about 42–45 days. Most young are born in May. Litter sizes average 3.5 young but range from 1 to 5, and there is a single litter per breeding season. The young emerge from the burrow when they are about three-fourths grown and begin to disperse by September or October. They become sexually mature in their first year (Clark and Stromberg 1987; Armstrong, Fitzpatrick, and Meaney 2011).

Selected References: Fichter and Jones 1953; Homolka 1967; Lock 1973, 1978; Clark 1986; Clark and Stromberg 1987; Clark 1989; Seal et al. 1989; Miller, Forrest, and Reading 1996; Armstrong, Fitzpatrick, and Meaney 2011; Buskirk 2016.

Family Mephitidae (Skunks)

STRIPED SKUNK. *MEPHITIS MEPHITIS.* UB, CO

Identification: Striped skunks are easily (and thankfully) identified by their unique black-and-white-striped pelage pattern. A narrow white stripe extends from the nose to the crown, and two black-bordered white stripes extend back from the nape or forehead along the lower back to the tail. The rest of the pelage is black. Length 520–770 mm (20.4–30.3 in.); tail 170–400 mm (6.7–15.7 in.); weight 1.8–4.5 kg (3.9–9.9 lb.).

Voice: Skunks produce a wide variety of calls, described as including growling, purring, hissing, whining, twittering, churring, barking, squealing, and screaming. They also spit, sniff, and click their teeth, and they stamp their feet loudly when they are confronted with a threat and about to retaliate by spraying musk. No complete inventory of their vocalizations and sound production is available.

Status: Skunks occur throughout Nebraska and in most terrestrial habitats. They are very common in grasslands but avoid dense forests. They have a keen sense of smell but rather poor eyesight, which might account for the frequency with which they are killed by moving vehicles. Although they can run at speeds of up to about 10 miles per hour, skunks will rarely retreat from danger, relying on their foul-smelling mercaptan-based spray for protection. They can send this powerful fluid out for several feet ahead of them with good accuracy and powerful effect, but they usually provide fair warning with their foot-stamping and conspicuous tail-erection signaling. Along with the eastern spotted skunk, the state's striped skunk population decreased significantly between the 1940s and 2000, probably as a result of environmental contaminants such as pesticides affecting prey populations (Landholt and Genoways 2000).

Habitats and Ecology: Striped skunks favor mixed woodlands, brushlands, and open fields with wooded ravines and rocky outcrops. They are abundant in cultivated fields and near farmsteads. They are opportunistic foragers, with vegetable materials compris-

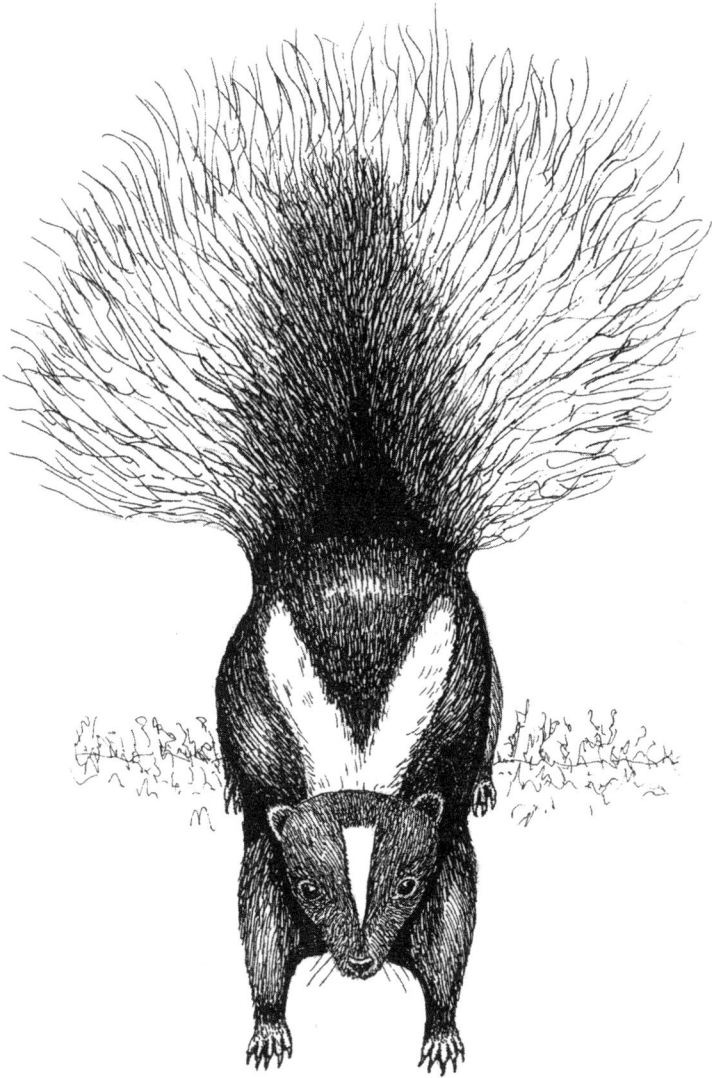

5. Striped skunk, adult threat posture.

ing up to 20 percent of their diet, and are active during both day and night. During the summer months, insects are important foods, while during fall and winter, carrion, small mammals, other verte-brates, and plant materials are of increased importance. Like bears, skunks move into dens during winter but do not hibernate. Males' home ranges are larger than those of females, and in various studies those of both sexes have varied from 0.4 to 12 square kilometers (0.15–4.6 square miles) (Armstrong, Fitzpatrick, and Meaney 2011).

Breeding Biology: Breeding occurs in February or March, with ovulation occurring about 48 hours after copulation. The gestation period ranges from 59 to 77 days, with births occurring in May or early June. Litter sizes range from two to ten, averaging six; females have 12 mammae. The young are able to breed in the spring following their birth, at about ten months of age. In spite of their seemingly strong defense mechanism, most skunks do not live to reach a year of age, with badgers, coyotes, foxes, and large owls all preying on them.

Selected References: Verts 1967; Clark and Stromberg 1987; Higgins et al. 2002; Armstrong, Fitzpatrick, and Meaney 2011; Buskirk 2016.

EASTERN SPOTTED SKUNK. *SPILOGALE PUTORIUS.* UB, OC

Identification: The spotted skunk has a complex pattern of white spots and stripes on its head and body. Besides a white forehead spot, a white stripe begins below each ear and extends back to the shoulders. Two other white stripes begin at the base of the forelegs and flanks and extend upward toward the middle of the back. Variably smaller spots are present near the tail, which is mostly black but white-tipped. Length 426–567 mm (16.7–21.9 in.); tail 140–235 mm (5.5–9.2 in.); weight 425–661 g (0.9–1.5 lb.).

Voice: Growling, purring, and hissing sounds are used in threat or defense by both of Nebraska's skunk species. They also use foot pattering as a warning signal when threatened. Unlike striped skunks, spotted skunks perform a unique handstand posture before emitting their defensively effective musky spray. The spray is chemically much like that of the striped skunk, and affected clothes or skin can reputedly be deodorized by tomato juice, although personal experience with the striped skunk and tomato juice leads me to judge this treatment as wishful thinking.

Status: Spotted skunks are entirely nocturnal and are also more arboreal and woods-dwelling than are striped skunks, although they also tend to occupy agricultural lands. Their population is in long-term decline (Landholt and Genoways 2000), and spotted skunks no longer occur in eastern Nebraska (Benedict, Genoways, and Freeman 2000). This species is on the Nebraska Natural Legacy Project's Tier 1 list of threatened species in Nebraska (Schneider et al. 2018).

Habitats and Ecology: Spotted skunks consume a very wide variety of plant and animal foods, much like those of the striped skunk. Spotted skunks are usually found in relatively arid habitats, such as shrublands, where they can den in rock crevices, under shrubs, or in human-made structures or woodpiles.

Breeding Biology: The eastern spotted skunk breeds in the spring and has a somewhat delayed implantation of the fertilized egg, delaying birth by about 14 days. The gestation period lasts 50–65 days, and the young are born in May or June. Their litter size ranges from three to six young, averaging about four. The young become sexually mature and able to breed by about four or five months of age (Clark and Stromberg 1987; Armstrong, Fitzpatrick, and Meaney 2011).

Selected References: Clark and Stromberg 1987; Higgins et al. 2002; Gompper and Hackett 2005; Armstrong, Fitzpatrick, and Meaney 2011; Buskirk 2016.

Family Procyonidae (Raccoons)

NORTHERN RACCOON. *PROCYON LOTOR.* UB, CO

Identification: Another easily recognized mammal species, raccoons have a white-bordered black-masked face and a tail with four to seven brown and black rings. They also have white-bordered black patches behind their ears but are otherwise rather uniformly brown, grayish, or reddish brown, varying somewhat with the season and the regional population. Adults sometimes weigh up to 23 kilograms (50 pounds). Northern populations are the largest, and unlike southern ones they accumulate and store body fat for winter survival. Length 60–105 cm (23.6–41 in.); tail 200–400 mm (7.8–15.7 in.); weight 3.6–9 kg (8–19.8 lb.).

Voice: Highly vocal, raccoons produce a very wide variety of sounds, including whimpering, purring, screaming, growling, hissing, snarling, and chuckling. Adults utter a shrill, tremulous whistle during autumn for long-distance communication.

Status: Raccoons are widespread, intelligent, and highly adaptable, and they are able to survive almost as well in cities as in the countryside. Their foods are equally adaptable; depending on their habitats, raccoons vary opportunistically from being carnivores to vegetarians. They consume such diverse items as insects, fish, crustaceans, bird eggs, mollusks, earthworms, grain, fruit, and discarded kitchen wastes. Their front feet are relatively hand-like and are well adapted for holding and manipulating objects. Raccoon populations increased significantly between the 1940s and 2000, perhaps because of the increasing abundance of desirable foods, such as corn (Landholt and Genoways 2000).

Habitats and Ecology: Raccoons in the northern Great Plains are probably most abundant in riverine woodlands, such as floodplain forests having mature oaks, elms, and sycamores, and they use squirrel nests, tree hollows, stumps, or rotten logs for dens. They are most common where water is nearby. Adult raccoons often maintain several dens within their home range, which in males might exceed 4,800 acres (7.5 square miles) and in females up to 2,400 acres (3.75 square miles). Mostly nocturnal, raccoons can move at speeds of up to 24 kilometers per hour (15 miles per hour), and long-term movements of more than 281 kilometers (175 miles) have been reported. Raccoons do not establish territories but instead form family groups of up to about six animals.

Breeding Biology: Males become sexually active in January and February, moving from den to den in search of receptive females. Gestation requires 63–65 days, with most litters of three or four young being born in late April or early May; rarely, births occur as late as October. Newborns are blind at birth, and their eyes open at 18–29 days of age. The youngsters are weaned by eight to ten weeks and soon begin to follow their mother on foraging searches. They become sexually adult as yearlings. Average longevity in the wild is usually only a few years, although lifespans of up to 12 years have been reported (Clark and Stromberg 1987).

Selected References: Goldman 1950; Clark and Stromberg 1987; Higgins et al. 2002; Zeveloff 2002; Armstrong, Fitzpatrick, and Meaney 2011; Buskirk 2016.

Family Cervidae (Deer)

ELK. *CERVUS CANADENSIS.* LO (REINTRODUCED, WE, NO), OC

Identification: Elk are easily distinguished from other North American deer by their large size and a creamy-white rump area surrounding a relatively small tail of the same color. The antlers of males seasonally have a pair of single large, vertically oriented tines, with up to five smaller ones projecting forward and upward (figure 6). Length 108–234 cm (3.5–7.6 ft.); tail 8–14 cm (3.1–5.5 in.); weight 118–497 kg (260–1,095 lb.).

Voice: The elk's most distinctive vocalization is the "bugling" of rutting males during September and October; it begins with a low grunt and shifts to a high-pitched whistling scream that might last for several seconds. Bugling is most frequent in early morning and late afternoon. It might serve as a cow attractant, a threat to other bulls, or a herd-spacing mechanism (Armstrong, Fitzpatrick, and Meaney 2011). Females (cows) utter barking sounds when alarmed and sometimes also bugle. Cows also vocalize softly to their calves; the calves likewise call to their mothers and squeal when alarmed.

Status: As a result of range expansion from South Dakota and reintroductions, elk are distributed locally but widely in Nebraska on various preserves, such as Fort Niobrara National Wildlife Refuge; on a few wildlife management areas, such as Wapiti WMA in Lincoln County; and in the Pine Ridge (Grier 1985; Cover 2000). Except for the population in Lincoln County and adjacent Hayes and Frontier Counties, most elk occur in the Panhandle region. Some elk, probably from the Rosebud Indian Reservation in South Dakota, also migrate south annually to locally winter along the middle and lower stretches of the Niobrara River, east to Knox County.

Habitats and Ecology: In the Great Plains, grasslands and sagebrush-

6. Elk, adult male.

dominated shrublands are primary habitat for elk. Grasses and forbs (broad-leaved herbaceous plants) are primary winter foods, grasses are the most important food source in spring, and forbs become increasingly important in summer. Browsing on willows, aspen, and other woody materials might be significant during winter, when grasses become unavailable. Summer home ranges are fairly small, of about 8–250 square kilometers (3–96 square miles), but seasonal migrations in the Greater Yellowstone region of Wyoming sometimes extend over distances of more than 96 kilometers (60 miles).

Breeding Biology: Elk breed in the fall after being stimulated by decreasing day lengths. Rutting bulls attempt to gather as many cows as possible under their control by threats, sparring contests, and intense fights that might become deadly at times. Dominant males might thereby gather up to 30 cows for breeding along with their calves. Females undergo several estrus cycles until they are bred; single calves are born the following spring. Calves often become sexually mature within a year, but males that are at least three years old are likely to do the majority of breeding. Although captive elk might survive for up to 20 years, mortality in the wild is much higher and is strongly influenced by hunting (Clark and Stromberg 1987; Armstrong, Fitzpatrick, and Meaney 2011).

Hunting and Population: Elk have been legally hunted in Nebraska since 1986, with a total of over 2,000 animals having been killed as of 2017. The state's elk population was estimated at about 1,400 head in 2007 and is still increasing, with the prospect of it eventually reaching a goal of 2,500–3,500 animals.

Selected References: Houston 1968, 1982; Van Wormer 1969; Cole 1969; Boyce and Hayden-Wing 1979; Clark and Stromberg 1987; Royce 1989; Stillings 1999; Higgins et al. 2002; Toweill and Thomas 2002; Hoffmann 2005; Armstrong, Fitzpatrick, and Meaney 2011; Buskirk 2016.

WHITE-TAILED DEER. *ODOCOILEUS VIRGINIANUS*. UB, CO

Identification: Compared with mule deer, white-tailed deer have longer and wider tails, which are brown above and white below and are raised in alarm, exposing their white undersides (tail-flagging). Mule deer have notably larger and slightly longer ears (about as long as the distance from the base of their ears to their nostrils) and a rather narrow white tail, which is tipped with black. Adult male white-tailed deer have paired antlers with four or more tines that rise vertically from curved and forward-pointing main tines, whereas mule deer have antlers with both main and secondary tines that tilt upward and fork in a Y-like pattern (figure 7). The mule deer is further differentiated from the white-tail by its stiff-legged bounding style of running, unlike the white-tail's smooth gallop. Length 134–215 cm (4.9–7 ft.); tail 15–36 cm (5.9–14 in.); weight 40–215 kg (88–473 lb.).

Voice: Males utter grunts in low-level antagonistic situations, grunt-snorts at stronger intensities, and grunt-snort-wheezes during high-level dominance interactions. Rutting males utter grunts and perform a "flehmen sniff" (a lip-curling and inhaling behavior common to many ungulates) during rutting (Atkeson, Marchinton, and Miller 1988). Mothers and fawns utter several maternal-neonatal sounds, and females utter contact calls when separated from a group. Females utter "breeding bellows" to attract males when in estrus, and males in search of females utter "bawling" calls. Both sexes perform foot-stamping when alert for possible danger.

Status: White-tailed deer occur statewide, occupying many habitats, but especially dense deciduous riparian corridors in eastern Nebraska. They are attracted to agricultural lands with access to corn, wheat, fruits, and other crops and also favor wetland areas with dense cover. They avoid dense coniferous forests, open prairie, and very dry habitats and are least common in southwestern Nebraska.

Habitats and Ecology: White-tailed deer have a highly diverse diet that tends to be higher in grasses and forbs than is typical of mule deer. Unlike mule deer in some western states, white-tailed deer do not undertake seasonal migrations but often have permanent home ranges covering up to several square miles. Male home ranges average larger than those of females (Armstrong, Fitzpatrick, and Meaney 2011). White-tailed deer in Nebraska average slightly heavier than mule deer and have gradually displaced them over large areas of contact in central and western Nebraska, especially where extensive tall woody cover is present.

Breeding Biology: Male white-tailed deer do not assemble harems, but during the fall rutting season they mark parts of their home range with signposts. These include scrapes made on the earth by pawing an area, urinating on it, breaking nearby branches, and leaving scent marks from the metatarsal glands on adjacent shrubs or other vegetation. They also polish areas on the trunks of nearby trees with their antlers by rubbing and scraping them, thus removing the bark. Males find and follow estrus females by smell; dominant males are most likely to succeed in mating. The estrus period in females lasts about 24 hours and is repeated about a month later if mating does not occur. The gestation period is 201 days; fawns are born the following spring. Yearling females produce single fawns, but among older females two young are usually born, and triplets are not rare. The fawns are highly precocial and can run soon after birth. They begin to eat solid foods by about two to three weeks after birth and are fully weaned by their fifth month. Males probably are sexually mature as yearlings but do not reach full size until they are three or four years old, and only by then are they able to compete effectively for mating opportunities (Clark and Stromberg 1987; Armstrong, Fitzpatrick, and Meaney 2011).

Hunting and Population: By 1900 deer had nearly been extirpated from Nebraska due to unlimited hunting. Hunting was prohibited from the early 1900s until 1945. During the first legal deer hunt of 1945, 275 mule deer and 2 white-tailed deer were killed. In 2018 about 26,500 male white-tails were killed during the regular rifle season. By then about 77,000 white-tailed deer were being killed annually as a result of special seasons and the legal killing of females. No estimate of the state's current white-tailed deer population could be located, but it must number in the hundreds of thousands. Such large numbers sometimes cause problems, because the deer invade suburban gardens and cornfields, and a few thousand animals are killed annually by collisions with vehicles, sometimes causing major accidents. Diseases in Nebraska's deer population have also been a periodic worry. Deaths from chronic wasting disease were first discovered in 2002 and had become a serious problem a decade later, but apparently the disease has more recently declined in frequency.

Selected References: Taylor 1956; Jones 1964; Menzel 1975; Wallmo 1981; Clark and Stromberg 1987; Putnam 1988; Atkeson, Marchinton, and Miller 1988; Bouc 1998; Higgins et al. 2002; Armstrong, Fitzpatrick, and Meaney 2011; Buskirk 2016.

MULE DEER. *ODOCOILEUS HEMIONUS.* WI (WE, CEN)

Identification: See the white-tailed deer account for tips on how to distinguish these two species, which occasionally hybridize in their zone of overlap. One means of distinguishing hybrids is to examine the metatarsal gland on the outside of the hind legs. On mule deer these glands are situated high on the lower leg (tarsus), are 4–6 inches long, and are surrounded by brown fur. The glands of whitetails are at or below the midpoint of the tarsus, are usually less than 1 inch long, and are surrounded by white hairs. Those of hybrids measure between 2 and 4 inches long and are sometimes encircled with white hair. Length 116–200 cm (3.8–5.6 ft.); tail 10–23 mm (3.2–7.5 in.); weight 50–200 kg (110–440 lb.).

Voice: Vocalizations in the mule deer are essentially like those of the white-tailed deer, but mule deer are reportedly less vocal generally than are white-tails.

Status: Mule deer are the more common deer in westernmost Nebraska, favoring drier and brushier habitats than the woodland- and forest-inhabiting white-tailed deer of eastern Nebraska. However, the eastern limit of primary mule deer range occurs west of a line extending from eastern Cherry County south to Furnas County. Some white-tailed deer occur west to the Wyoming border, but no mule deer occur in the eastern quarter of the state.

Habitats and Ecology: Mule deer are more arid adapted than white-tailed deer and are most abundant in shrublands with broken topography. Winter diets in the Rocky Mountains consist mostly of browse (74 percent), with forbs comprising 15 percent of that amount. Browse (the eating of twigs and tree buds) comprises half of the total in spring, with grasses and forbs adding 25 percent each. During summer, browse again makes up about half of the diet and forbs 46 percent, while in fall browse consumption increases to 60 percent and forbs decline to 30 percent. Unlike white-tailed deer, mule deer seem to be able to survive in the absence of free water, except in very arid habitats. Their annual home ranges are highly variable, from 3.9 to 22 square kilometers (1.5 to 8.5 square miles), and in some regions they make seasonal migrations of up to 160 kilometers (100 miles) (Sawyer, Lindzey, and McWhirter 2005; Armstrong, Fitzpatrick, and Meaney 2011).

Breeding Biology: Breeding occurs late in the fall and early winter, when the polygynous males detect and seek out estrus females. They also aggressively interact with other competing males, as described for the white-tailed deer. Gestation likewise lasts about 200 days, and two fawns are typically born to mature females. Captive does might live for up to 22 years and bucks to 16 years. However, lifetimes in the wild are much shorter, with average annual mortalities of 28–43 percent common in stable herds and with fawns accounting for about half of the mortality (Armstrong, Fitzpatrick, and Meaney 2011).

Hunting and Population: Nebraska sport hunters kill about 8,500 male mule deer annually, compared with about 25,000 male white-tailed deer, suggesting an approximate 3:1 statewide abundance ratio. Most hunters probably reside in eastern Nebraska and often prefer to hunt local white-tails, which are somewhat larger and are often corn-fed. No estimate of the state's current mute deer population could be located, but it is probably under 50,000.

Selected References: Taylor 1956; Jones 1964; Menzel 1975; Wallmo 1981; Clark and Stromberg 1987; Putnam 1988; Bouc 1998; Higgins et al. 2002; Armstrong, Fitzpatrick, and Meaney 2011; Buskirk 2016.

Family Antilocapridae (Pronghorns)

PRONGHORN. *ANTILOCAPRA AMERICANA.* WI (WE)

Identification: Both sexes of pronghorns are mostly sandy brown in color, with a white rump, underparts, and lower legs and white patches on the neck and cheeks. Adult males have a broad black

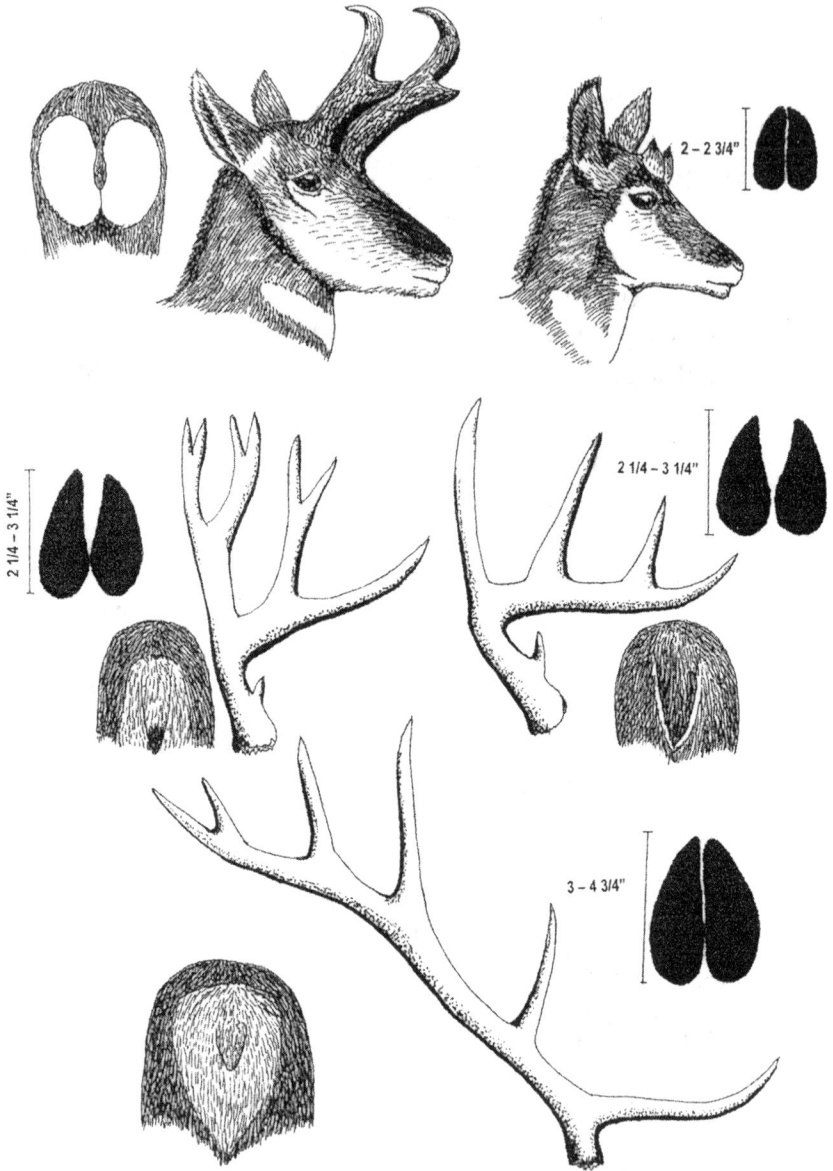

7. Pronghorn male (left) and female heads, rump, and hoofprint (top), and rumps, hoofprints, and antlers of male mule deer (middle left), white-tailed deer (middle right), and elk (bottom).

streak down the muzzle, black cheek patches, and a short black mane. Adult males also have short black horns (not antlers) that are slightly forked and curve backward toward their tips (figure 7). The outer fibrous sheaths of the males' horns are shed annually. The smaller females lack black facial markings and have short, nub-like horns, as do juvenile males. Length 124–47 cm (4.1–4.8 ft.); tail 97–178 mm (3.1–5.8 in.); weight 40–70 kg (88–154 lb.).

Voice: Pronghorn vocalizations include the male's advertisement or territorial snort-like roar, which is uttered when a male enters another's territory. A similar short chuckle is used by males during antagonistic encounters. There is also an alarm snort-wheeze that is uttered by both sexes when threatened, and males emit a high-pitched whine during courtship. Fawns bleat when seeking attention.

Status: Like elk, bighorn sheep, and bison, pronghorns were virtually extirpated from Nebraska by the 1920s, but through later reintroductions into the Sandhills and elsewhere they have become well established in western parts of the state (Grier 1998b). The state population numbered several thousand by 2020. The highest populations are in Sioux and northern Dawes Counties (the Pierre Hills and Box Butte tableland), but they extend south through the Panhandle and east commonly to Lincoln County and northeast to Custer County and the central Sandhills. In good range the pronghorn density is about one animal per square mile, but it might locally reach 3.7 per square mile in the Pierre Hills. They are present in smaller numbers in the Sandhills, where the sandy substrate probably restricts rapid running.

Habitats and Ecology: In Colorado, shrubs comprise more than 90 percent of fall and winter foods on the pronghorn's sagebrush-dominated range, whereas in spring and summer forbs account for 60–84 percent of their diet. In a statewide study of year-round foods in Colorado, forbs and shrubs each accounted for 43 percent, with cactus adding 11 percent and grasses 3 percent. Although fairly drought tolerant, pronghorns prefer to drink water daily during warmer months, so surface water sources influence their local distribution. Overall, home ranges vary greatly, from 65 to 2,300 hectares (160 to 5,700 acres), within which defended areas (territories) range from 25 to 400 hectares (64 to 990 acres) on good ranges (Armstrong, Fitzpatrick, and Meaney 2011). Pronghorns are highly social; herds in Wyoming historically ranged in size up to a thousand or more individuals. Their densities there now vary from 0.6 to 3.3 pronghorns per 100 hectares (1.5 to 8.3 per square mile) (Clark and Stromberg 1987).

Breeding Biology: Pronghorn males are polygynous, and females become sexually mature at 16 months. They breed in the fall, from mid-September to mid-October, when males attempt to gather harems or try to mate with any females entering their territories. In Nebraska the peak period of conception is October 10–20. The gestation period lasts 230–40 days, with maximum numbers of fawns being born from June 10–15. Usually females give birth to twins, which soon after birth are able to run swiftly. Within a few months the young can reach the maximum speed of adults, up to more than 60 miles per hour (Clark and Stromberg 1987; Armstrong, Fitzpatrick, and Meaney 2011).

Hunting and Population: Pronghorns have been legally hunted since 1958. In recent years, nearly 1,000 animals of both sexes have been legally killed annually. No estimate of the state's current pronghorn population could be located, but it might approach 10,000.

Selected References: Grier 1998b; Menzel and Suetsugu 1966; Sundstrom, Hepworth, and Diem 1973; Kitchen 1974; Suetsugu 1975; Clark and Stromberg 1987; Turbak 1995; Byers 1997; Higgins et al. 2002; O'Gara and Yoakum 2004; Hoffmann and Taylor 2007; Armstrong, Fitzpatrick, and Meaney 2011; Buskirk 2016.

Family Bovidae (Bison, Sheep, and Goats)

AMERICAN BISON. *BISON BISON.* EXTIRPATED

(ONLY CONFINED HERDS EXIST IN NEBRASKA)

Identification: The bison is the largest North American mammal and is unique in having black tapering horns that curve upward and outward. Males have massive shoulders, a shaggy beard, and dense woolly forehead and shoulder pelage that extends back over

8. Bison, adult male head.

the forelegs. Females are smaller and more cow-like in appearance. Length 198–380 cm (6.7–12.5 ft.); tail 43–81 cm (14.1–26.7 in.); weight 410–910 kg (900–2,000 lb.).

Voice: The deep roaring bellow of a rutting male bison is distinctive. It lasts up to about six seconds or more and has a base-dominated frequency range of 1,500–7,500 Hz. Cows also roar, but less loudly than males. Their usual vocalizations are soft, guttural grunts. Calves produce higher-pitched grunts, as well as bleating (Gunderson and Mahan 1980).

Status: Bison are now well established in managed herds in places such as the Nature Conservancy's Niobrara Valley Preserve. Large privately owned captive herds of bison are also present in Nebraska, such as in the several ranches owned by Ted Turner.

Habitats and Ecology: Bison were historically mainly associated with shortgrass and mixed-grass prairies, meadows, and, to a limited extent, shrub-grass and desert grasslands, plus taller grasses eastwardly. In Yellowstone National Park, bison favor sedge- and grass-dominated meadows interspersed with pine forests. In the fall and winter, sedges make up 37–50 percent of their diet, with grasses comprising somewhat smaller percentages. They also eat some shrubs and browse, mostly in summer (Clark and Stromberg 1987).

Hunting and Population: Nearly all North American bison are on private lands, state and national parks, ranches, and preserves. By 2018, free-ranging but managed bison populations in North America numbered over 300,000, and there were about 500,000 bison in privately owned herds. No sport hunting is allowed in state or federally owned herds, but Native American and private herd owners are allowed to sponsor hunts. As of 2018, Ted Turner had ranch holdings of more than 500,000 acres in Nebraska, all of which are directed toward bison rearing and which might easily support about 20,000 bison.

Breeding Biology: Bison breed over a long period, mostly from July to October. Then mature males often engage in pushing or fighting contests for access to females, utter loud bellowing, and closely follow (tend) females approaching or in estrus. In Yellow-

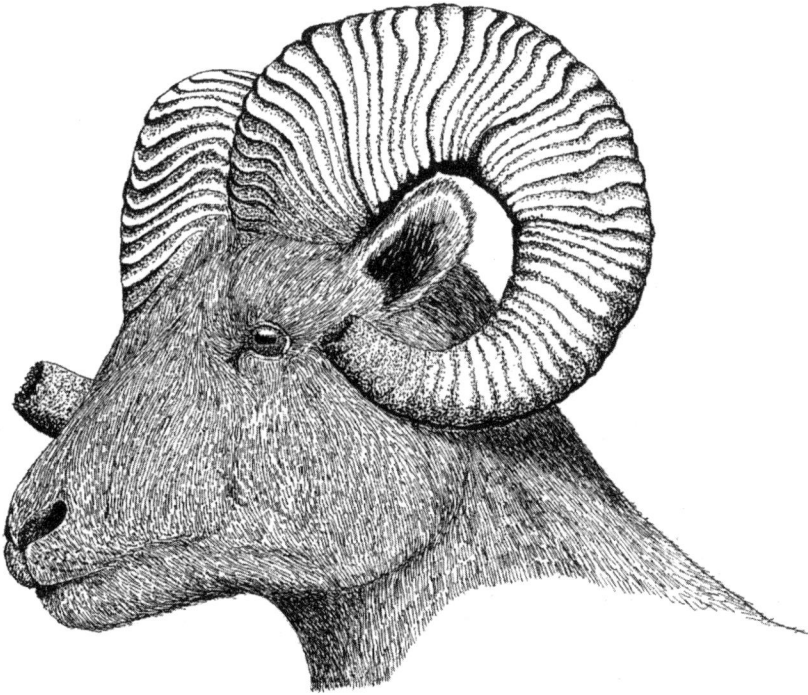

9. Bighorn sheep, adult male head.

stone National Park, about half of the females first become pregnant at 3.5 years of age, and males over 8 years old are the most sexually active. Gestation lasts 270–85 days, and single (rarely twin) calves are born from April to June. In nonhunted situations, bison might live to 41 years of age (Clark and Stromberg 1987).

Selected References: McHugh 1958; Mahan 1978; Farrar 1974b; Clark and Stromberg 1987; Wolff 1998; Danz 1997; Irby and Knight 1998; Higgins et al. 2002; Rinella 2009; Armstrong, Fitzpatrick, and Meaney 2011; Buskirk 2016.

BIGHORN (MOUNTAIN) SHEEP. *OVIS CANADENSIS.*

RA (REINTRODUCED, WE)

Identification: Bighorn sheep are the only North American sheep having massive recurved horns. In adult males these can form almost a complete circle, but in females and juvenile males the horns are shorter, thinner, and only slightly recurved. Both sexes have a medium-brown pelage, except for a large white rump contrasting with a short brown tail. Length 149–95 cm (4.9–6.4 ft.); tail 80–127 mm (2.6–4.5 in.); weight 75–168 kg (165–370 lb.).

Voice: Adult sheep snort or make cough-like noises when alarmed but are otherwise quite silent. Adults bleat when searching for other group members, such as when females are searching for lambs. The crashing sounds made by males when they head-butt is surprisingly loud and can be heard from several hundred yards away.

Status: Bighorn sheep are usually associated with high mountains, but a plains-dwelling race (*Ovis canadensis audubonii*) survived until the early twentieth century from the western Dakotas south to northwestern Nebraska (Jones, Armstrong, and Choate 1985). Bighorn were historically present in the steppe-badlands of northwestern Nebraska but had been extirpated by the 1920s. Bighorns of the Rocky Mountain race were successfully reintroduced into the Wildcat Hills and Pine Ridge in 1981 (Grier 1998a).

Habitats and Ecology: Bighorns are quite mobile and have been found to move up to 5 kilometers (3 miles) per day in Yellowstone National Park. In winter they might descend in elevation to as much as several hundred feet lower and up to 50 kilometers (30 miles) in distance to areas where snow is less deep and their home ranges are smaller. During spring and summer the sexes separate, with older males forming small bachelor bands, and females, their lambs, and immature males gathering in larger assemblages (Armstrong, Fitzpatrick, and Meaney 2011).

Breeding Biology: Male bighorn sheep are polygynous, seeking multiple matings, and they compete strongly for mating rights. Estrus females tend to seek out and mate with the largest rams. Males at least seven years old obtain most of the breeding, which occurs during November and December in Colorado. Gestation lasts about 170–80 days, with lambs being born during May or June. Usually only a single lamb is born per pregnancy, but they are highly precocial and are weaned by five to six months. Many diseases affect bighorn populations, and mortality is high during the first year of life. However, lifespans of up to 17 years have been reported (Armstrong, Fitzpatrick, and Meaney 2011).

Hunting and Population: Sport hunting of bighorns was first allowed on a lottery basis in 1998, nearly 20 years after reintroduction efforts began. One or two lottery- or auction-based permits have since been issued annually, and a total of 25 animals had been legally killed as of 2018. There was a Panhandle population numbering about 350 animals in 2018.

Selected References: Cowan 1940; Geist 1971; Fairbanks 1985; Clark and Stromberg 1987; Grier 1998a; Higgins et al. 2002; White 2001a; Armstrong, Fitzpatrick, and Meaney 2011; Buskirk 2016.

Family Sciuridae (Squirrels, Marmots, and Prairie Dogs)
BLACK-TAILED PRAIRIE DOG. *CYNOMYS LUDOVICIANUS*,
DI (WE), UN

Identification: Prairie dogs are the most gregarious of Nebraska's rodents and are always found within colonies that are marked by scattered mounds indicating burrow entrances. They resemble large gophers but have black-tipped tails. Length 312–410 mm (12.3–16.1 in.); tail 72–90 mm (2.8–3.5 in.); weight 680–1,500 gm (24–52.9 oz.).

Voice: The adult prairie dog's most common call is a sharp, whistled *wee-ooo* call, which is uttered during a "jump-jip" display. It reputedly serves as a territorial call and as an "all-clear" signal. It certainly has other meanings, as it is highly contagious between colony and family members and probably functions in large part in maintaining group cohesion or aiding social familiarization (Waring 1970). The black-tailed prairie dog has an estimated eight vocalizations plus tooth-chattering, whereas the closely related but considerably less gregarious white-tailed prairie dog (*Cynomys leucurus*) has only six described vocalizations (Waring 1970).

Status: The black-tailed prairie dog has suffered greatly at the hands of humans; its total North American population probably has been reduced by 99 percent during historic times, from one that once might have exceeded 3 billion animals (Johnsgard 2005). In Nebraska, prairie dogs are still regularly found in a few Panhandle

10. Black-tailed prairie dog, jump-yip call.

counties, such as Morrill County and the Pine Ridge grasslands. Isolated protected populations also exist in a few locations, such as Fort Niobrara National Wildlife Refuge and Scotts Bluff National Monument.

Habitats and Ecology: Estimates of home ranges and densities of black-tailed prairie dogs are complicated by the highly social nature of the species, with "towns" divided into "wards," and wards in turn divided into individual family units called "coteries." Densities vary greatly by season, but averages ranging from 10 to 50 animals per hectare (4 to 20 per acre) have been reported. In one study a ward was estimated to occupy 3 hectares (7.4 acres), within which eight coteries occupied 0.21 hectare (0.5 acre) each (Armstrong, Fitzpatrick, and Meaney 2011). Prairie dogs are associated with many commensal species that variously benefit from their presence. Black-footed ferrets cannot survive in the absence of large populations of prairie dogs, and other associates such as burrowing owls, prairie rattlesnakes, grasshopper mice, and thirteen-lined ground squirrels also benefit. Nearly 150 species have been found to be facultative associates of prairie dogs (Clark et al. 1982; Johnsgard 2005).

Breeding Biology: Prairie dog species breed during early spring (March and April), and the gestation period lasts 28–32 days. Litter sizes vary greatly, from two to ten pups, averaging about five. Young appear aboveground during early June at about 40 days of age. By 90 days they are fully grown, although they might not attain their adult weight until they are about 120 days of age. Prairie dogs might hibernate from 30 to 120 days in regions with very cold winters, although in Nebraska they exhibit little or no true dormancy. Sylvatic plague (caused by a bacterium that is closely related to the one causing deadly bubonic plague in humans but that is not similarly air-transmitted) has had significant effects on prairie dog and other mammalian populations across the western states. It is transmitted by flea bites, so direct contact with wild prairie dogs and related rodents such as ground squirrels, as well as their burrows, should be strictly avoided.

Hunting and Population: A state survey of Nebraska colonies in 2000 estimated that about 130 square miles of active prairie dog colonies still then existed in the state. This area is slightly less than that of Omaha (140 square miles), or about 0.16 percent of the state's land area, nearly all of which was probably once occupied by prairie dogs. Nebraska's prairie dogs are currently perhaps most common on the Oglala National Grassland, Sioux and Dawes Counties, where they can be hunted without limits. During the late 1990s, about 300,000–350,000 prairie dogs were being killed annually in Nebraska by recreational shooters (Johnsgard 2005). In eastern Nebraska, a few small prairie dog colonies were still surviving near Kearney and Hastings as of 2018. In that year a population of several hundred animals on a college-owned prairie research area near Grafton, Nebraska, the easternmost remaining colony in the United States, was destroyed when the property was sold and converted to cropland.

11. Southern flying squirrel, adult.

Selected References: Farrar 1974a; Clark and Stromberg 1987; Foster 1990; Hoogland 1995, 2006; Bischof 2003b; Johnsgard 2005; Slobodchikoff, Perla, and Verdolin 2009; Armstrong, Fitzpatrick, and Meaney 2011; Buskirk 2016.

SOUTHERN FLYING SQUIRREL. *GLAUCOMYS VOLANS.*
LO (SE), OC

Identification: This is a very easy species to identify, but it is nocturnal and very shy, so the chances of seeing it in the wild are virtually nil. At rest its enormous ebony-black eyes and its sharply divided brown upperparts and white underparts, separated by a loose skin fold, are unique, as is its flattened, bushy tail. Total length 290–315 mm (11.4–12.4 in.); tail 129–42 mm (5.1–5.6 in.); weight 105–70 gm (3.7–6 oz.).

Voice: In addition to some audible vocalizations, including *chucks* and chattering notes, flying squirrels also emit ultrasonic calls ranging up to 51 kHz (Murrant et al. 2013). These consist of frequency-modulated (FM) calls of several types, although their possible functions are still not well understood. There is no evidence that these calls are used for navigation by echolocation while gliding, but there is the possibility that they serve for cryptic communication above the hearing range of owls, which are probably the flying squirrels' major predators.

Status: In Nebraska, flying squirrels occur only as isolated populations in southeastern Missouri Valley forests of Richardson, Nemaha, and Otoe Counties and as of 1993 had not been reported north of Nebraska City (Benedict, Genoways, and Freeman 2000; Bischof 2003a). Almost nothing is known of regional population densities. Home ranges are generally believed to average about 0.8–1.2 hectares (2–3 acres) but might be as large as 15–20 hectares (37–50 acres). Density estimates have ranged from less than 1 to more than 10 per hectare (less than 2.5 to 25-plus per acre). This species is on the Nebraska Natural Legacy Project's Tier 1 list of threatened species in Nebraska (Schneider et al. 2018).

Habitats and Ecology: The gliding abilities of flying squirrels are remarkable; they can readily glide over distances in excess of 20 meters (65 feet), and there are reports of flights as long as 90 meters (300 feet), no doubt in areas of very tall trees (Wilson and Ruff 1999). One of their major sources of food are fungi such as mushrooms (Dubay 2000). In the Pacific Northwest, they are believed to play an important role in the dispersal of spores of fungi that symbiotically help forest trees absorb nutrients. They also eat seeds, nuts, fruits, insects, and other invertebrates, as well as birds' eggs and some carrion.

Breeding Biology: Flying squirrels do not hibernate and are able to be active at temperatures as low as -20° C (-4° F). They breed during spring, in Wyoming from late March to May, and perhaps produce a second litter during summer. The gestation period is 37–42 days, and the litter size varies from two to six. Newborns weigh only 5–6 grams, but by 32 days their eyes are open, and they begin exploring by 40 days. They are weaned by 60 days of age but remain with their mother for several additional months before gaining independence. Typical longevity is three to four years (Wilson and Ruff 1999).

Selected References: Woods 1980; Clark and Stromberg 1987; Goldingay and Scheibe 2000; Bischof 2003a; Armstrong, Fitzpatrick, and Meaney 2011; Murrant et al. 2013; Buskirk 2016.

THIRTEEN-LINED GROUND SQUIRREL.
ICTIDOMYS TRIDECEMLINEATUS. UB, CO

Identification: This is the familiar "gopher" that is often seen at the edges of country roads, where its striped and spotted body is easily recognized. Its upper parts consist of a series of about 13 alternating dark brownish and buff stripes, the dark stripes marked with regularly spaced buff spots. The legs are short, the ears are small, and the tail is fairly long but not bushy. Adults weigh as little as 3 to a maximum of 10 ounces, the heaviest weights occurring just prior to fall hibernation. Length 220–87 mm (8.7–11.3 in.); tail 82–103 mm (3.3–4 in.); weight 83–185 gm (2.9–6.3 oz.).

Voice: When frightened, this ground squirrel utters a loud bird-like trill or tremulous whistle, which is also used as a warning call. Another common Great Plains grassland species, the Richardson's ground squirrel (*Urocitellus richardsonii*), is known to produce ultrasonic alarm calls.

Status: This is the most common ground squirrel in Nebraska. Population densities in the northern plains range from about 0.4 to 4 per hectare (1 to 10 per acre), but higher numbers occur late in the breeding season as young emerge. In one Great Plains study, the home ranges of males averaged 4.4–4.8 hectares (11–12 acres) and were largest during the breeding season, while those of females averaged 1.2–1.6 hectares (3–4 acres) and were largest during pregnancy and lactation.

Habitats and Ecology: Much more associated with shortgrass and mixed-grass habitats than tallgrass prairie, this species is common statewide in grazed pastures, golf courses, and roadsides where the

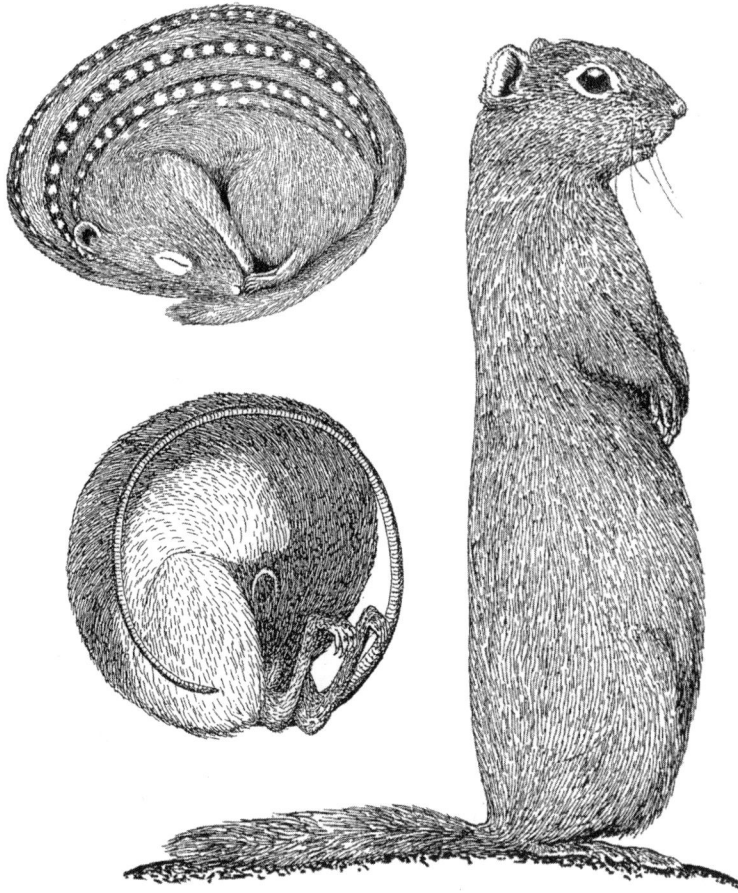

12. Hibernating adults of thirteen-lined ground squirrel (upper left) and meadow jumping mouse (lower left), and adult Franklin's ground squirrel, alert posture (right).

soil is well drained and can be easily excavated. Burrows might be up to 20 feet long and vary in depth from a few inches to as much as 4 feet. There are also short escape burrows scattered through the animal's home range. Except during spring, piles of soil do not reveal the entrances of burrows. Aboveground, there are grassy runways that are used to travel from one burrow to another or for foraging. This species is among the least social of the ground squirrels, but adults have a greeting ceremony during which they touch noses, and they make scent markings by rubbing their faces over objects in their environment, depositing odors from their oral glands.

Breeding Biology: Males emerge from their hibernation during late March or April, ready to mate with females when they emerge a week or more later. After mating, there is a gestation period of 27–28 days, and litters of 5–13 young are born during May or early June. A single litter per year is produced. Hibernation begins in late September or early October, although in some arid regions adults might become dormant during midsummer to avoid excessive heat stress (estivation).

Selected References: Schildman 1955; Clark and Stromberg 1987; Armstrong, Fitzpatrick, and Meaney 2011; Buskirk 2016.

LEAST CHIPMUNK. *TAMIAS MINIMUS.* SD (NW PAN), UN

Identification: This small chipmunk is easily recognized by its long tail (almost as long as its head–body length) and the prominent striping on its head and back. There are five dark and four light stripes on the back and sides, the middle dark stripe extends to the tip of the tail, and the outermost pair of dark and white body stripes are very distinct. It has a white belly, and it holds its tail vertically when it runs. Total length 190–212 mm (7.5–8.3 in.); tail 74–91 mm (2.9–3.6 in.); weight 29–55 gm (1–1.9 oz.).

Voice: Little has been written on the vocalizations of this species, apart from its frequent chipping calls. It closely resembles and co-occurs with up to about ten other similar western chipmunks. Its calls are known to advertise territory, indicate alarm, and attract mates. Vocalizations are uttered in synchrony with body posturing and rhythmic movements of its tail. Apart from the *chip*, which is uttered almost constantly, there are also *chipper* and *chuck* calls and a bird-like trill. Youngsters squeal, as do adults occasionally (Wilson and Ruff 1999).

Status: This is a local species in the Pine Ridge and in adjacent badlands of Sioux and Dawes Counties, east to about Chadron. It is most common in sagebrush and on dry, rocky slopes. This species is on the Nebraska Natural Legacy Project's Tier 2 list of threatened species in Nebraska (Schneider et al. 2018).

Habitats and Ecology: All chipmunks are seed eaters, and the least chipmunk especially favors conifer seeds but secondarily consumes grass and forb seeds. It also eats flowers, leaf buds, leaves, mushrooms and other fungi, and even insects, birds' eggs, and carrion. It is solitary and territorial but is able to coexist with larger chipmunk species. Least chipmunks are diurnally active, only infrequently climb trees or shrubs, and spend much of their lives in underground burrows and rock crevices. They dig winter hibernacula, provisioned with a good supply of seeds to eat when they periodically arouse. Hibernation extends from about September until April, greatly reducing their metabolic rate and perhaps allowing them to live longer lives than some other nonhibernating mammals of similar size (Wilson and Ruff 1999).

Breeding Biology: Least chipmunks have a single litter per season, with breeding extending in Wyoming from mid-March through mid-May. Females enter estrus almost immediately after emergence from hibernation, and mating soon follows. Gestation lasts four weeks, and the litter size is four to six. Within two weeks the eyes of the young have opened, and they are well furred. By 30 days they are weaned and emerging from their natal burrow. Lifespans are relatively long for these tiny animals; some individuals are known to have survived for as long as six or seven years (Wilson and Ruff 1999).

Selected References: Skryja 1970, 1974; Clark and Stromberg 1987; Armstrong, Fitzpatrick, and Meaney 2011; Buskirk 2016.

Family Castoridae (Beavers)

BEAVER. *CASTOR CANADENSIS.* UB, CO

Identification: America's largest native rodent is easily recognized by its size, aquatic behavior, and flattened tail. The muskrat is much smaller and has a rounded tail. Length 850–1,000 mm (33–40 in.); tail 258–325 mm (10–13 in.); weight 11–35 kg (24–77 lb.).

Voice: Although often assumed to lack vocalizations, beavers do have a substantial vocabulary and also produce gnawing and chewing sounds (Novakowski 1969). Adults hiss when confronted with other animals and defending their territory. Young beavers are particularly vocal and utter a soft repetitive whine when soliciting food or when in uncomfortable social situations.

Status: Beavers are common in rivers, ponds, and other aquatic habitats across Nebraska as a result of reintroductions following earlier extirpation. Their densities are hard to judge, but in Colorado they have been found to have colony territories of about 0.4–8 hectares (1–20 acres) and nearest-neighbor distances of 0.7–1.5 kilometers (0.3–0.9 mile). Densities of beavers have also been estimated as ranging from 0.3 to 1.5 families per kilometer (0.5 to 2.4 families per mile) of stream. Distances between colonies vary greatly and in part are related to the abundance of willows along the river or stream. Aspens and cottonwoods are also important winter foods, and alders, birches, and conifers are also used to a limited extent (Armstrong, Fitzpatrick, and Meaney 2011). Beaver populations increased significantly between the 1940s and 2000 perhaps because of the increase in woodland along Nebraska rivers and streams (Landholt and Genoways 2000).

Habitats and Ecology: Beavers are master water engineers; they build sometimes massive dams that often exceed 100 feet in width and establish canal routes to influence water flows and provide channels for transporting foods such as tree branches. Their dams might be as much as several hundred feet long and up to about 8 feet high and sometimes last for decades as a result of constant maintenance and repair. Most dam construction is done during spring and fall, and during fall a colony might fell a mature aspen every other night while assembling a winter food supply. They also construct wood-and-earthen lodges and dig bank burrows for safety, food storage, and reproduction. Both sexes have castor glands, which secrete scents that are spread on mud mounds and used to mark the boundaries of their territory.

Breeding Biology: Beavers are monogamous. A family typically consists of an adult pair plus their yearlings and juvenile offspring, totaling from about four to eight animals. Pair bonds last an average of 2.5 years, with the death of a mate the most common reason for mate replacement (Armstrong, Fitzpatrick, and Meaney 2011). Mating behavior occurs very early, from January to February, and gestation lasts 104–11 days. In Colorado, litter sizes vary with altitude; females living at elevations above 5,000 feet have smaller litters than those living at lower elevations (average 4.4 versus 2.7). The young (kits) are born fully furred and are weaned by about two months. They become sexually mature by their second winter and soon begin to disperse, the males more likely to disperse than the females. Maximum longevity is about 20 years, but average lifespans are 8–10 years (Clark and Stromberg 1987; Armstrong, Fitzpatrick, and Meaney 2011).

Trapping and Population: During the years 2000–2008 the estimated state "harvest" of trapped beavers ranged from about 12,000 to 16,000 animals, so the total state population might have then exceeded 50,000 animals.

Selected References: Wood 1966; Novakowski 1969; Clark and Stromberg 1987; Armstrong, Fitzpatrick, and Meaney 2011; Müller-Schwarze 2011; Buskirk 2016.

Family Heteromyidae (Pocket Mice and Kangaroo Rats)

ORD'S KANGAROO RAT. *DIPODOMYS ORDII*. WI (WE, SH), CO

Identification: Nebraska's only kangaroo rat, this remarkable saltatory (hopping) mammal has highly developed hind limbs that allow for long horizontal and high vertical leaps. Its forelimbs are correspondingly small and mostly used for holding and manipulating foods. Its tail is long (longer than its head–body length), doubly striped with black and white, and noticeably tufted. Its dark eyes are very large and marked with whitish "eyebrow" markings. The upperparts are otherwise rather uniformly tan. Length 249–80 mm (9.8–11 in.); tail 138–63 mm (5.4–6.4 in.); weight 45–100 gm (1.6–3.6 oz.).

Voice: Kangaroo rats vocalize rarely, but foot-drumming and tooth-chattering are common during aggressive encounters. However, there are a few accounts of squeaks or squeals uttered by animals

13. Bushy-tailed woodrat (above) and Ord's kangaroo rat, adults.

under duress, and purring growls, clucks, and grunts are produced by captive animals (Eisenberg 1963). The somewhat large ears and specialized middle ear structures of the species in this mostly desert-dwelling rodent group reflect the importance of hearing in these nocturnal mammals. Their large outer ears (pinnae) help localize sound sources, and their highly enlarged tympanic bullae (bony chambers that enclose the middle ears and probably help dampen extraneous noises) allow for the detection of extremely weak and very low frequency sounds, such as those that might be made by approaching predators. One study of hearing in a kangaroo rat (Heffner 2005) found that the animals could respond to frequencies ranging from 50 Hz to 65 kHz and could locate the sources of very brief sounds.

Status: Kangaroo rats are very common in the Nebraska Sandhills and occur elsewhere where the soils are moderately sandy. They are highly mobile, and home ranges of from 2.3 to 3.3 hectares (5.7 to 8.1 acres) and a density of about 32 per hectare (14 animals per acre) have been reported (Clark and Stromberg 1987). Lower density estimates in Texas have ranged from 0.04 to 15.6 animals per hectare (0.01 to 6.3 per acre). In Wyoming, Maxell and Brown (l968) found the highest densities to occur in sand dunes, with yucca grasslands and sage grasslands supporting progressively lower densities.

Habitats and Ecology: Because of their efficient seed-gathering abilities, kangaroo rats are among the rodent species that are important seed dispersers. One study in Utah suggested that 68 percent of foods taken there was seeds, 25 percent was green vegetation, and 7 percent was arthropods, whereas in the Pawnee National Grassland (Colorado), 85–95 percent of the winter diet consisted of seeds. Kangaroo rats can survive in the absence of free water, urinating very little, so these visually endearing "roo-rats" were favorite pets of students at Cedar Point Biological Station until, as a precaution against disease transmission, they were no longer allowed to capture and keep them.

Breeding Biology: The breeding season is extended, with activity in some regions occurring during all months except December. In the northern plains, there are only two litters per year, with one born during late winter or early spring and another during late summer. In Nebraska, pregnant females have been trapped from May through September. Gestation lasts about 30 days, and litter sizes range from one to six. The eyes of the young are open at about two weeks, and youngsters are independent of their mother by two months of age. However, youngsters might remain with their mother for several months, when males become aggressive. The animals are active year-round and breed as yearlings (Clark and Stromberg 1987; Armstrong, Fitzpatrick, and Meaney 2011).

Selected References: Eisenberg 1963; Garner 1974; Lund and Farney 1975; Farrar 1987; Clark and Stromberg 1987; Armstrong, Fitzpatrick, and Meaney 2011; Buskirk 2016.

OLIVE-BACKED POCKET MOUSE. *PEROGNATHUS FASCIATUS.*
SD (PAN), CO

PLAINS POCKET MOUSE. *PEROGNATHUS FLAVESCENS.*
WI (EA, CEN), CO

Identification: Of Nebraska's four species of pocket mice, the olive-backed occurs over the Panhandle's high plains and extends eastward on the Crookston tableland of Cherry County. The abundant plains pocket mouse has a broad distribution in eastern and central Nebraska, west to the northwestern Panhandle. All pocket mice are similar in size and appearance, and all have fur-lined cheek pouches. These pouches allow the mice to carry seeds back to their burrows without getting the seeds wet and subject to mold during storage. The pouches can even be everted for emptying out their contents, and the mice can clean the pouches with their forepaws. All pocket mice also have much larger hind legs than forelegs, large heads, large eyes, and long, bicolored tails. The olive-backed pocket mouse has a dark olive-brown, almost greenish, upper body pelage tint that is separated from the white underparts by a yellowish lateral line, sometimes lined with black, along its flanks. Three other species of pocket mice occur in Nebraska that are quite similar in appearance. The more widespread plains pocket mouse is very similar in size and appearance to the olive-backed but has a whiter belly, a paler buffy lateral line between the belly, and brown upperpart pelage that is not so olive tinted as the olive-backed. One of the other two Nebraska species, the hispid pocket mouse, is substantially larger (minimum length 200 mm [7.9 in.]) and has noticeably coarse dorsal pelage that is harsh to the touch,

14. Hispid pocket mouse, adult.

whereas the smaller silky pocket mouse (maximum length 117 mm [5.5 in.]) has notably soft, silky pelage that is tinted with reddish to yellowish buff. Olive-backed pocket mouse: length 127–37 mm (5–5.4 in.); tail 57–68 mm (2.2–2.7 in.); weight 8–14 gm (0.3–0.5 oz.). Plains pocket mouse: length 114–30 mm (4.5–5.1 in.); tail 52–65 mm (2–2.6 in.); weight 7–12 gm (0.3–0.5 oz.).

Voice: No specific information is available on the vocalizations of Nebraska's species of pocket mice. In a related species (*Perognathus inornatus*), the vocalizations include growls, squeals, and low grunts; tooth-chattering and foot-drumming serve for nonvocal communication.

Status: The olive-backed pocket mouse is fairly widespread in western Nebraska, but its status and biology are little known. It is especially adapted to shrub-steppe and to shortgrass prairies having sandy to gravelly soils, but in some areas it extends into the ponderosa pine zone. Sites with loamy sand to clay soils, low vegetation, and a good deal of bare ground are preferred, allowing for bipedal hopping when escaping danger. The plains pocket mouse favors Sandhills prairies, sand sage grasslands, and shortgrass prairies. The olive-backed pocket mouse is on the Nebraska Natural Legacy Project's Tier 2 list of threatened species in Nebraska (Schneider et al. 2018).

Habitats and Ecology: All pocket mice are heavily dependent on seeds for their diets; they eat a wide variety of grass and forb seeds, and it is likely that they also consume some insect materials, based on studies of captives. In one Utah study of the plains pocket mouse, seeds were found to make up 81 percent of the stomach contents, while in a New Mexico study, 91 percent of the diet consisted of seeds and 9 percent was of arthropods (Armstrong, Fitzpatrick, and Meaney 2011). It is known that in several species, such as the plains pocket mouse, access to drinking water is not needed, as water is generated from the carbohydrates that they consume (Clark and Stromberg 1987). They evidently do not hibernate but undergo periods of torpor, periodically waking to eat some of their stored seeds. These mice dig long burrow systems, often under a yucca or shrub, with side tunnels for food storage. Summer nests are about 300–375 millimeters (12–15 inches) below the ground, but winter or summer dormancy (estivation) burrows might be as much as 2 meters (6 feet) deep. Densities of pocket mice on the plains of Montana and North Dakota have ranged from 0.6 to 4 animals per hectare (0.2 to 1.6 per acre) (Armstrong, Fitzpatrick, and Meaney 2011).

Breeding Biology: Not studied in Nebraska, breeding in pocket mice probably occurs from May to July. In Colorado, one or two litters are probably produced during spring and early summer, with litters typically consisting of four to six young (six teats are present in females). Gestation lasts 30 days, but almost nothing is known of the development of the young.

Selected References: Eisenberg 1963; Pefaur and Hoffman 1974; Clark and Stromberg 1987; Beebe 2004; Armstrong, Fitzpatrick, and Meaney 2011; Buskirk 2016.

Family Dipodidae (Jumping Mice)
MEADOW JUMPING MOUSE. *ZAPUS HUDSONICUS.*
WI (EA, CEN), CO

Identification: The meadow jumping mouse is widespread in Nebraska and differs from all other native mice in having an extremely long tail (comprising over half the animal's total length).

Length 180–220 mm (7.1–8.7 in.); tail 115–36 mm (4.5–5.3 in.); weight 12–22 gm (0.4–0.8 oz.).

Voice: Meadow jumping mice utter bird-like chirps, chatter their teeth, and squeak when fighting or disturbed. They also vibrate their tails against the substrate, producing a drumming sound (Schwartz and Schwartz 2016).

Status: The meadow jumping mouse is a grassland species that is most common in taller grasses near water and in herbaceous understory vegetation of wooded areas.

Habitats and Ecology: The jumping mice are well named; they reportedly can make long jumps of up to about 12 feet in distance when frightened. It has even been claimed that they can alter course somewhat while in the air, using the long tail as a rudder. They often perform a series of short jumps and then hide motionless in tall grass. Home ranges for female meadow jumping mice average about 0.6 hectare (1.5 acres) and those of males about one per hectare (2.5 acres), while densities range from 7.4 to 48 per hectare (2.3 to 19.5 per acre) (Clark and Stromberg 1987). These mice eat a variety of seeds, insects, and fungi but do not store food. They are also deep hibernators (figure 12), being dormant for nearly half of the year and spending their summer activity in nocturnal foraging and fat accumulation in preparation for their next hibernation. They must accumulate about 6 grams of fat to survive hibernation, sometimes doubling their posthibernation spring body weight; heavier animals have the best chance of winter survival (Armstrong, Fitzpatrick, and Meaney 2011).

Breeding Biology: Jumping mice begin to breed shortly after emerging from hibernation, probably during May, and might have two or three litters during their summer breeding period. Gestation lasts 17–20 days, and litter sizes usually number four to six. The young are weaned by four weeks and reach adult size by three months of age. They begin breeding the following spring and have average longevities of about three years (Clark and Stromberg 1987).

Selected References: Quimby 1951; Clark 1971; Clark and Stromberg 1987; Armstrong, Fitzpatrick, and Meaney 2011; Buskirk 2016.

Family Cricetidae (New World Mice)

PRAIRIE VOLE. *MICROTUS OCHROGASTER.* UB, CO

MEADOW VOLE. *MICROTUS PENNSYLVANICUS.* UB, CO

Identification: Typical *Microtus* voles are stocky grassland- and meadow-dwelling mice with short legs, large heads, and relatively small eyes. The term "vole" is derived from a Scandinavian name for a meadow mouse, and *Microtus* translates as "small-eared," but these mice simply have most of their moderately large, rounded ears hidden by long fur. Meadow and prairie voles appear very similar, being uniformly grayish brown to reddish brown, with variably grayish underparts. They are most easily distinguished by their pelage; the prairie vole has salt-and-pepper (grizzled) brown upperparts, paler sides, and buffy to ochraceous underparts, whereas the meadow vole has chestnut-brown upperparts, pale-brown sides, and dusky-gray to silvery-white underparts. Meadow vole: length 140–90 mm (5.5–7.5 in.); tail 32–52 mm (1.25–2 in.), averaging 42 mm in western Nebraska; weight 30–75 gm (1.2–2.9 oz.). Prairie vole: length 110–88 mm (3.9–6.6 in.); tail 33–45 mm (1.3–1.8 in.), averaging 37 mm in central Nebraska; weight 30–70 gm (1.1–2.5 oz.).

Voice: Both of these species are known to emit ultrasonic vocalizations as infants, and at least the prairie vole also produces ultrasonic vocalizations as adults (Lepri, Theodorides, and Wysocki 1988). The functions of these calls are still poorly understood, but in the case of infants, the calls appear to be related to begging behavior associated with sibling competition for access to their mother's nipples (females have four pairs of mammae) or are uttered when the infant is cold and in distress.

Status: The prairie vole is a common mammal over nearly all of Nebraska and overlaps with the meadow vole in grasslands over northeastern parts of the state. Prairie voles occur in drier areas where shrubs are not codominants, but they favor relict mixed-grass and tallgrass prairies. They also extend into riparian habitats of willows and cottonwoods and shrub-woodland habitats. Meadow voles are found in areas close to water, such as riparian meadows, orchards, and other habitats supporting thick grass. They favor riparian areas with cattails and other aquatic vegetation, and the animals can swim very well. Both species make tunnel trails through the grass that provide visual camouflage and ease of movement.

Habitats and Ecology: Home ranges of the prairie vole average about 0.1–0.25 hectare (0.04–0.6 acre). Unlike the meadow vole and nearly all other *Microtus*, the prairie vole is rather social, and adults form strong monogamous pair bonds, a behavior pattern controlled by the female's secretion of hormones that control male behavior (Russell Benedict, personal communication). Both parents tend to their young and help maintain trails and burrows. Social groups form by the addition of offspring and nonrelated animals; communal nests of 20 or more animals have been found in some areas (Clark and Stromberg 1987; Armstrong, Fitzpatrick, and Meaney 2011).

Breeding Biology: Both of these species are polyestrous, with breeding occurring repeatedly throughout the year. Gestation lasts about 21 days, and litter sizes are typically three or four but range up to seven. Young are weaned by three weeks of age and can breed by only 30 days of age. Longevity in both species is very short, usually averaging well under a year (Armstrong, Fitzpatrick, and Meaney 2011).

Selected References: Meserve 1971; Clark 1973; Clark and Stromberg 1987; Lepri, Theodorides, and Wysocki 1988; Armstrong, Fitzpatrick, and Meaney 2011; Buskirk 2016.

BUSHY-TAILED WOODRAT. *NEOTOMA CINEREA.* SD (PAN), CO

EASTERN WOODRAT. *NEOTOMA FLORIDANA.* LO (NO, SW, SC, SE), CO

Identification: The bushy-tailed woodrat, with its light-brown to blackish body pelage, white underparts, and well-furred, bicolored tail, more closely resembles an overgrown white-footed mouse rather than a Norway rat (*Rattus norvegicus*). They have huge eyes and large ears, and the tail is about three-fourths the length of the head and body. The eastern woodrat is very similar, but its tail is not thickly furred and is almost unicolored. Bushy-tailed woodrat: length 350–470 mm (13.8–18.5 in.); tail 135–223 mm (5.3–8.7 in.); weight 240–90 gm (8.5–10.3 oz.). Eastern woodrat: length 350–400 mm (13.8–15.7 in.); tail 135–80 mm (5.3–7.1 in.); weight 300–400 gm (10.6–14.1 oz.).

Voice: In addition to aggressive foot-thumping and tooth-chattering, several vocalizations are used by bushy-tailed woodrats. These include a buzzing sound uttered by the male prior to copulation and sometimes also by females in sexual situations. Animals in distress or during fights utter a loud scream, and they emit a short squeal in less extreme situations (Escherich 1981).

Status: The bushy-tailed woodrat favors rocky habitats, from semi-desert scrubs and arid grasslands to forests. They are also attracted to abandoned buildings and occupied cabins, where their tendency to chew on isolated wires can be troublesome. They are entirely vegetarian but consume a very wide variety of leaves, twigs, berries, bark, and other plant materials. They leave foods out to dry and then store them for winter consumption. Eastern woodrats favor wooded ravines or river valleys with heavy undercover and construct large stick houses that might be used for several generations. These might be placed under rocks, under downed timber, or under roots. The bushy-tailed woodrat is fairly widespread in the western Panhandle, but the eastern woodrat occurs as an endemic Nebraska population (*Neotoma floridana baileyi*) along the Niobrara River and its immediate tributaries. It is well separated geographically from the species' larger population, which extends north to the Republican River and is a probable Pleistocene relict. The endemic northern subspecies is on the Nebraska Natural Legacy Project's Tier 2 list of the state's threatened taxa (Schneider et al. 2018).

Habitats and Ecology: Bushy-tailed woodrats, or "packrats," are famous for finding and carrying off small items to a storehouse of both edible and inedible objects around its nest. Such middens might remain active for decades. While camping in rocky scablands of Washington State as a graduate student, I often lost coins that I happened to leave in my tent and was in constant fear that a woodrat might find my hidden car keys and leave me stranded, as its huge nest was located in an inaccessible rock crevice. Densities in Wyoming have been estimated at 0.8–4.6 animals per hectare (0.3–1.9 per acre) (Belitsky 1981; Clark and Stromberg 1987).

Breeding Biology: These woodrats breed over spring and summer months and possibly produce two litters per season. Both sexes scent-mark and strongly defend individual territories, although males will allow females into their territories. Gestation lasts 27–35 days, and litter sizes vary from two to five young, usually three or four. No pair-bonding occurs, and only females tend the offspring. The young are independent by two months and begin to breed at two years. Their average longevity in the wild is about three to four years.

Selected References: Robertson 1968; Escherich 1981; Clausen 1983; Clark and Stromberg 1987; Farrar 2007; Armstrong, Fitzpatrick, and Meaney 2011; Buskirk 2016.

MUSKRAT. *ONDATRA ZIBETHICUS.* UB, CO

Identification: Muskrats are easily distinguished from much larger beavers, the only other large swimming mammal in Nebraska. Muskrats are about the size of a prairie dog (12–13 inches long exclusive of tail), and the tail is long and rat-like but laterally flattened, whereas the beaver's body is 24 inches long, exclusive of its flattened tail. Muskrats are never found far from water, and they usually build associated dome-like "houses" of cattails in shallow water. Muskrats also sometimes dig shoreline burrows. Length 420–620 mm (16–24 in.); tail 200–290 mm (7.9–11.8 in.); weight 700–1,500 gm (24–53 oz.).

Voice: Muskrats utter low squeals, loud squeals, snarls, and moans, and they chatter their teeth when cornered or fighting. Both sexes utter high-pitched *n-n-n-n* sounds during the breeding season, and young utter a squeaky cry (Schwartz and Schwartz 2016).

Status: Muskrats probably occur in every Nebraska county, wherever there is standing water and emergent vegetation; the state's population possibly numbers in the hundreds of thousands. In good wetland habitats, populations might exceed 55 animals per hectare (22 per acre).

Habitats and Ecology: Like beavers, muskrats are herbivorous but eat mostly aquatic vegetation, especially cattails, as well as crayfish, fish, and mollusks. They are active throughout the year and are largely nocturnal but often can be seen during late afternoon hours, and sometimes they are active even near midday. Home ranges of females and young are small and center on their lodge or burrow.

Breeding Biology: Both sexes are territorial and strongly defend breeding territories, which are scent-marked. Muskrats are promiscuous or loosely monogamous; the male will take over care of the young if the female dies. Females are polyestrous, with repeated cycles of about 30 days, so two or perhaps three breeding cycles might be completed within a single breeding season. Gestation lasts 25–30 days, six or seven young are the usual litter, and the young are independent by about two months of age. They become sexually mature during their second spring. Typical lifespans are three to four years (Clark and Stromberg 1987).

Trapping and Population: Nebraska's muskrat population possibly numbers in the hundreds of thousands. During the years 2000–2008 the estimated state total of trapped muskrats ranged from about 14,000 to 30,000 animals, but recent numbers have probably declined, as the value of muskrat pelts has diminished. Muskrat populations apparently decreased between the 1940s and 2000, but short-term environmental changes such as water-level fluctuations and disease incidence make muskrat population data hard to interpret (Landholt and Genoways 2000).

Selected References: Sather 1959; Errington 1963; Clark and Stromberg 1987; Armstrong, Fitzpatrick, and Meaney 2011; Buskirk 2016.

NORTHERN GRASSHOPPER MOUSE.

ONYCHOMYS LEUCOGASTER. UB, CO

Identification: Grasshopper mice generally resemble deer mice but have much shorter tails (less than half the length of the head and body) and are noticeably stockier in build. They are often medium-gray dorsally and have large blackish eyes, furred ears that often have a patch of white at their bases, and very long whiskers. Length 120–55 mm (4.7–6.1 in.); tail 29–50 mm (1.1–2 in.); weight 35–45 gm (1.2–1.6 oz.).

Voice: The grasshopper mouse is the most vocal of Nebraska mice. Adults "sing" in an upright posture and with an open mouth (figure 15); the vocalization is an easily heard wail of about 12 kHz that lasts about a second and identifies the individual, its sex, and its location and probably imparts other information. These are highly mobile mice, so long-distance sound communication is probably highly adaptive.

Status: This unusual mouse occurs statewide but has been reported from the easternmost counties only in recent decades (Benedict, Genoways, and Freeman 2000). It is most common in overgrazed pastures, shortgrass prairies, and grassy sandhills. Although mostly nocturnal, grasshopper mice are carnivorous and sometimes actively hunt grasshoppers (a favorite prey), other insects, and small vertebrates. They have a powerful bite for their size, as my young son Scott once discovered when, seeing one hiding in Sandhills grasses, he impulsively grabbed it and stuffed it in a pocket. Scott soon yelped in pain, quickly pulled the mouse from his pocket, and dropped it back into the grass. Grasshopper mice readily kill other mice and even have been known to kill mammals up to the size of hispid cotton rats (*Sigmodon hispidus*), which average three to four times the grasshopper mouse's own weight (Armstrong, Fitzpatrick, and Meaney 2011).

Habitats and Ecology: Grasshopper mice have large home ranges that average about 2.3 hectares (5.7 acres) but range up to 3 hectares (7.4 acres) and include scent-marked and defended territories. In one Colorado study, six male home ranges averaged 1.7 hectares (4.2 acres) and overlapped with the home ranges of several females. Like many carnivores, they are monogamous, with both parents tending their young. They do not hibernate, but in winter they shift to a diet in which seeds predominate (Armstrong, Fitzpatrick, and Meaney 2011).

Breeding Biology: Grasshopper mice breed in spring; young are born from March to September in Colorado, where three or four litters per season might be produced. Gestation lasts from 27 to 47 days, being longer for lactating females than for nonlactating ones. Typically, three or four young are born, but litters of up to six young sometimes occur. By a month of age the young are weaned, and they reach sexual maturity at three to four months. In captivity, grasshopper mice have lived for more than four years, but most wild animals do not survive much longer than two years (Clark and Stromberg 1987; Armstrong, Fitzpatrick, and Meaney 2011).

Selected References: Ruffer 1965; Hurt 1969; Clark and Stromberg 1987; Armstrong, Fitzpatrick, and Meaney 2011; Buskirk 2016.

WHITE-FOOTED DEER MOUSE. *PEROMYSCUS LEUCOPUS.* UB, CO

AMERICAN DEER MOUSE. *PEROMYSCUS MANICULATUS.* UB, CO

Identification: White-footed mice and deer mice are so similar in appearance that an in-hand inspection is needed. Both are variable geographically, as well as by season and by age, in their measurements and upperpart colors, which vary from gray in immatures to rich buff to brownish-black hues in adults. However, invariably they have sharply contrasting white underparts. This bicolored pattern includes a long, sparsely haired tail. Their ears are very large, and their eyes are similarly large and protruding. At least in Nebraska's populations, the white-footed mouse can be tentatively separated from the deer mouse by the white-footed mouse's slightly longer tail; in Nebraska, it is usually longer than 70 millimeters (2.75 inches) in the white-footed mouse, while the deer mouse's tail is usually less than 65 millimeters (2.5 inches). The hind toes (including claws) of white-footed mice usually measure 21 millimeters or more, and those of deer mice are usually 20 millimeters or less. Deer mice vary geographically from blackish brown dorsally in the east to grayish buff in western regions. The white-footed mouse is generally more grayish overall, its shorter tail is less strongly bicolored than that of the deer mouse, and it is also rare in the western Sandhills and southwestern Nebraska (Benedict, Genoways, and Freeman 2000).

15. White-footed deer mouse, adult dozing (above) and adult nose-to-nose encounter (upper middle), and northern grasshopper mouse, male howling (lower middle) and adult crouching (bottom).

White-footed mouse (north-central Nebraska): length 161–86 mm, average 173 mm; tail 68–81 mm, average 74.7 mm; weight 27–36 gm, average 30.5 g. Deer mouse (north-central Nebraska): length 130–64 mm, average 148 mm; tail 50–71 mm, average 59.6 mm; weight 19–25 gm, average 21 gm (Jones et al. 1983).

Voice: These mice are not notably vocal (at least to the human ear), but young deer mice of two different subspecies were found to produce distress calls at frequencies of 3,600–26,500 Hz (Hart and King 1966). It is also known that male deer mice use ultrasonic calls (35,000 Hz) both prior to and after copulation (Pomerantz and Clemens 1981). White-footed mice rapidly drum their forefeet when alarmed.

Status: The deer mouse is widespread across Nebraska and is probably the most common North American rodent. The white-footed mouse is also locally abundant in the state, at least in wooded habitats.

Habitats and Ecology: In Nebraska, white-footed mice are mostly found in wooded areas, such as riparian and upland forests, shelterbelts, and fencerows, whereas the deer mouse is more adaptable and occurs in open meadows, prairies, badlands, croplands, shelterbelts, hedgerows, and coniferous woodlands. Where white-footed mice are present in woodlands, deer mice are rare or absent, but they are found in woodlands that lack white-footed mice. Both species live in burrows, the deer mouse's often under rocks, among debris, or under fallen logs, where they make vegetation-lined nests that are for sleeping, protection from cold, and rearing their young. They are mostly nocturnal. The home ranges of white-footed mice vary greatly but rarely are larger than an acre, and females usually have smaller ranges than males. Deer mouse home ranges in the northern Great Plains vary widely, from 0.04 to 4 hectares (0.1 to 10 acres). Both species eat a variety of berries, seeds, fruits, nuts, and other vegetation, as well as insects, other invertebrates, and occasionally small vertebrates.

Breeding Biology: Both species are active year-round, but the period of reproduction is mostly during spring and fall. The gestation period of both species is 22–23 days, and litter sizes in both range up to nine but average about four. The babies' eyes open at about two weeks, and the youngsters are weaned at about four weeks. Both sexes are mature by about seven to eight weeks of age, so by fall several generations might be present in the population.

Selected References: Dice 1941; King 1968; Pomerantz and Clemens 1981; Jones et al. 1983; Clark and Stromberg 1987; Armstrong, Fitzpatrick, and Meaney 2011; Buskirk 2016.

WESTERN HARVEST MOUSE. *REITHRODONTOMYS MEGALOTIS.* UB, CO

PLAINS HARVEST MOUSE. *REITHRODONTOMYS MONTANUS.* UB, CO

Identification: Harvest mice closely resemble several other grassland mouse species but are very small and long-tailed and, uniquely, build bird-like nests. The more widespread western harvest mouse has a very long, bicolored tail (as long or longer than the head–body length) and a dark-brown back that is grizzled with black hairs. The plains harvest mouse is limited to the eastern fourth of the state, has a shorter tail (less than the head–body length), and lacks a black grizzled dorsal appearance. The dorsal dark stripe on the tail is about one-half the diameter of the tail on the western harvest mouse; on the plains species it is about one-fourth the tail's diameter. Western harvest mouse: length 122–55 mm (4.8–6.1 in.); tail 56–73 mm (2.2–2.9 in.); weight 11–17 gm (0.4–0.6 oz.). Plains harvest mouse: length 110–40 mm (4.3–5.5 in.); tail 40–63 mm (1.9–2.5 in.); weight 10–12 gm (0.35–0.42 oz.).

Voice: Little specific information on these species' vocalizations is available, but the western harvest mouse "sings" at a very high pitch almost inaudible to a human's ear (Schwartz and Schwartz 2016). Studies of several other harvest mice indicate that their vocalizing includes both audible and ultrasonic components and probably represents announcement calls (Miller 2010).

Status: Both species of harvest mice are nearly ubiquitous in Nebraska and are relatively active during daylight hours, as well as nocturnally. The western harvest mouse is common in wetter, denser, and taller grassy environments of western Nebraska. The plains harvest mouse occurs statewide in drier and more open habitats, such as where there is more than 50 percent bare soil, and less often occurs in shrubby areas than does the western species (Higgins et al. 2002). Densities of harvest mice vary greatly, from about 2.5 to 60 per hectare (1 to 24 per acre) and reportedly even up to 123 per hectare (50 per acre) (Clark and Stromberg 1987; Armstrong, Fitzpatrick, and Meaney 2011).

Habitats and Ecology: Home ranges of these tiny mice are rather small, from 0.2 to 0.56 hectare (0.08 to 0.20 acre). Harvest mice are active nocturnally all year, eating a wide variety of seeds, insects, and various vegetable matter, which they gather (harvest) from growing plants and carry back down to the ground in their mouths. Harvest mice are remarkable in that they construct nests that are spherical, baseball-sized structures comprised of woven grasses and other fibrous vegetation and lined inside with soft plant materials. These might be placed on the ground or elevated in vegetation and have a small entrance hole at the bottom. The mice huddle there socially in cold weather and in some regions might even hibernate or at least become torpid for short periods (Armstrong, Fitzpatrick, and Meaney 2011).

Breeding Biology: Harvest mice breed from spring to fall, with many litters produced per season. In the wild these probably total about three or four, and gestation takes 24 days. From three to nine young are born, but typically about five. By five weeks they are fully grown, and they can breed by two months of age (Clark and Stromberg 1987).

Selected References: Lerass 1938; Robertson 1965; Clark and Stromberg 1987; Miller 2010; Armstrong, Fitzpatrick, and Meaney 2011; Buskirk 2016.

Family Erethizontidae (Porcupines)

PORCUPINE. *ERETHIZON DORSATUM.* WI (WE), UN

Identification: Easily identified by its array of sharp, barbed quills, which are modified hairs and are mostly obscured by the animal's long body fur. The quills vary in length, up to several inches long, and become longer as the animal matures. They are raised during defensive threat, exposing their white bases, which results in a white skunk-like stripe appearing on the animal's otherwise dark back and perhaps helping to deter further aggression. During an attack, the porcupine lashes its tail from side to side, deeply embedding the quills into any nearby animal or object. There might be as many as 30,000 quills present on a single porcupine, not all of which are barbed. They are hollow, and the body heat of the victim causes the embedded quills to expand and become harder to extract. Cutting off the end of an embedded quill releases the air pressure inside and makes it somewhat easier to remove. Length 79–103 cm (31–40 in.); tail 145–300 mm (5.7–11.8 in.); weight 3.5–18 kg (7.7–39 lb.).

Voice: Normally silent, porcupines perform a prolonged tooth-chattering prior to a defensive attack and also emit a pungent odor as a secondary warning. Their generic name appropriately translates as "one who rises in anger." Various observers have subjectively described many vocalizations, including squeaking, moaning, grunting, hooting, sobbing, wailing, shrieking, and howling. A detailed inventory and functional analysis of their obviously very complex vocabulary apparently remain to be done.

Status: Porcupines are most common in northwestern Nebraska, but recent records locally extend to eastern and even to southeastern counties (Benedict, Genoways, and Freeman 2000). They are rarely found far from trees, which they use as a refuge, as a den if there is a cavity, and as a source of food, eating the tree's inner bark, leaves, needles, buds, and twigs, as well as other vegetable materials. Maples, aspens, basswoods, ashes, and apples are among their favorite food sources. Cellulose-digesting bacteria in an intestinal caecum (a blind pouch similar structurally to the human appendix but functional in digestion) allow porcupines to exploit this normally indigestible energy source (Whittaker 1980).

Habitats and Ecology: Porcupines are nocturnal but at times can be seen in daytime resting or sleeping in trees. They move slowly and are unlikely to retreat or flee from any source of possible danger. Among mammalian predators, only the fisher (*Martes pennanti*) is efficient at flipping a porcupine on its back so that its vulnerable belly is exposed to a lethal attack. In the wild, porcupines might live up to ten years, and captive individuals have lived up to 20 years. They are reported to make intelligent, albeit potentially dangerous, pets.

Breeding Biology: The answer to the most common question asked about porcupine reproduction is, "Very carefully." The actual answer is that the female relaxes and lowers her quills before raising her tail to the male. Prior to that there is an extended courtship, and before mating the male squirts a spray of urine over the female. The breeding season occurs during autumn, and the gestation period lasts 29–31 weeks. Typically, only a single, relatively precocial offspring is born whose soft spines at birth quickly harden and in less than an hour are functional weapons. Nursing lasts two or three weeks, and the youngster remains with its mother for several months (Clark and Stromberg 1987; Whittaker 1980).

Selected References: Gunderson 1983; Clark and Stromberg 1987; Roze 1989; Farrar 1996; Armstrong, Fitzpatrick, and Meaney 2011; Buskirk 2016.

CHAPTER 3

Birds

THE RECENT HISTORY OF NEBRASKA'S BIRDS

Lying near the center of the Great Plains of North America, Nebraska exhibits a variety of geographic and ecologic influences on its bird fauna. The avian ecology and natural history of Nebraska have been discussed previously (Johnsgard 2001), as have the biology and birds of the Platte Valley (Brown and Johnsgard 2013; Johnsgard 2008a, 2008b), the Niobrara Valley (Johnsgard 2007b), the Nebraska Sandhills (Johnsgard 1995) and Nebraska's wetlands (Johnsgard 2012a, 2012b).

Of Nebraska's more than 200 breeding bird species (Johnsgard 1979b; Mollhoff 2016), the largest single component is arboreal, or adapted to living in trees, woodlands, and forests, while limnic (aquatic and shoreline-adapted) species make up the second largest component. Species primarily associated with grasslands comprise a still smaller breeding component, and xeric-adapted forms associated with semidesert scrub are the least numerous (Johnsgard 1987b).

Nebraska's arboreal species of birds, the passerine or perching birds, comprise about 45 percent of the state's total bird species. They are mostly eastern or northern in their geographic breeding affinities, while a small percentage are western or southern in origin. Of the aquatic-adapted forms, which make up about 32 percent of the state's total avifauna, a considerable proportion are either northern or widespread (pandemic) in breeding distributional affinities, and many of these are only migrants in the state. Bird species especially associated with natural grasslands, which once made up the largest historic vegetational component in the state, now comprise less than 10 percent of the state's total avifauna, and the remaining species are mostly rather general in their ecological requirements.

DECLINING AND EXTIRPATED SPECIES

Several species of Nebraska birds have become extirpated as an apparent result of human activities since settlement times, and many relatively specialized aquatic species have undergone considerable retraction of ranges as wetlands have disappeared and natural vegetation has given way to agriculture, urbanization, and other disturbances. Other, more generalist species, such as various "blackbirds" (including grackles, starlings, and cowbirds) and crows, have benefited from these same changes. The increasing development of riparian woodlands along the Platte and other river systems crossing the plains has also facilitated east–west distributional changes in forest-adapted birds, and reservoirs such as Lake McConaughy have attracted many new or rare species (especially gulls and ducks) to the state in recent years (Johnsgard 2018a).

Changes in the abundance and occurrence of Nebraska's birds have been great during the past century. In 1900 the Eskimo curlew was still a common migrant, and the passenger pigeon was seen in small migrating flocks. The Carolina parakeet was by then probably extirpated from Nebraska and within a few decades of extinction. The bald eagle, osprey, trumpeter swan, greater sandhill crane, pileated woodpecker, and common raven were probably also extirpated as breeding birds by then. Several warblers, including the blue-winged, prairie, and northern waterthrush, also disappeared around the turn of the century as breeders, as well as the yellow-bellied sapsucker and Baird's sparrow. The Chihuahuan raven was probably gone by 1950.

Among game birds, the wild turkey and ruffed grouse had probably also disappeared from Nebraska by 1900; no breeding records exist for either. One of the last Eskimo curlews to be collected in North America was shot near Hastings in April 1915; market hunters had once killed thousands annually in Nebraska (Johnsgard 1980b).

The greater sage-grouse has also disappeared from the Panhandle and also was never proven to be a state breeder. The lesser prairie-chicken disappeared from southwestern Nebraska in the 1920s or 1930s, but breeding in the state was never documented (Johnsgard 2002a). The greater prairie-chicken population peaked in Nebraska during the early 1900s and then began a gradual decline as natural grasslands increasingly gave way to croplands, but it still survives in relict tallgrass prairies and the eastern Sandhills (Johnsgard and Wood 1968).

INTRODUCED AND INVADING SPECIES

The ring-necked pheasant was successfully introduced from China into Nebraska in 1914 and reached its highest populations during the 1940s, but it has been in slow decline for many decades. Similar gray partridge (*Perdix perdix*) introduction efforts were only marginally successful in northern Nebraska, where a few small populations still persist as outliers of a South Dakota population. Several other alien game-bird introduction attempts, such as the chukar (*Alectoris chukar*) and crested tinamou (*Eudromia elegans*), were failures. Several races of the native but once-extirpated wild turkey were successfully reintroduced by state-funded efforts started in 1959. This species is now established in most counties, is widely hunted, and is nearly a pest in some suburban areas. Also aided by state-funded restoration efforts that began in the 1960s, Canada geese have become common to abundant breeders across Nebraska. By the early 2000s they were probably breeding in at least 72 counties (Mollhoff 2016). They have also become semiurbanized and by far the most common bird species seen during the most recent (2018) Christmas Bird Counts in both Lincoln and Omaha, outnumbering even starlings!

In some less fortunate introduction programs, the European house sparrow was already well established in Nebraska by 1900, having been introduced into the United States in the early 1850s. It might have arrived in Nebraska by the 1870s, anecdotally as a result of releases by farmers who vainly hoped it might help control the locust plagues occurring then. The European starling did not appear in eastern Nebraska until the 1930s, about 60 years after it was released in New York City and had gradually expanded westwardly.

The self-introduced Eurasian collared-dove appeared in eastern Nebraska in 1997 after a very rapid expansion from an invasion originating in Florida during the 1980s. Its population has since exploded in both towns and agricultural areas across Nebraska. The rock pigeon was probably brought into North America with the earliest colonists as a semidomesticated species, but records of its first Nebraska appearance are lacking.

The house finch, native to the western states, reached eastern Nebraska from westward expansions during the 1980s, only four decades after some captive birds had been illegally released from a pet store in New York. From there the birds soon adapted to their new landscape and quickly populated the eastern states. However, an indigenous western population of house finches had already been present in western Nebraska since at least the early 1900s, and its range has slowly progressed eastwardly since then. The two populations eventually met somewhere along the Platte Valley during the late 1900s or early 2000s. House finches have since gradually become established in most Nebraska counties, especially around humans (Mollhoff 2016).

Other recent self-introduced arrivals that are now breeding in Nebraska include the cattle egret, present since 1965, and the great-tailed grackle, reported since the mid-1970s. Both species are associated with wetlands, and both are still expanding their Nebraska ranges, as are some other wetland birds such as the white-faced ibis and glossy ibis.

The trumpeter swan, extirpated from the Great Plains in the late 1800s, likewise resumed breeding in the northern Nebraska Sandhills in 1968 as a result of southward expansion from reintroductions in South Dakota and has become a regular breeding and overwintering species in central and eastern Nebraska. The bald eagle began nesting successfully in Nebraska during 1991 after more than a century of absence. It is now breeding in many counties after recovering from the disastrous reproductive effects of DDT. Greater sandhill cranes have also been breeding locally since 1999, perhaps at a rate of one or two pairs per year, although the birds are so secretive when nesting that their numbers could be somewhat higher.

The growth of trees along our major waterways has improved foraging habitat and opportunities for east–west dispersal among many woodland-dependent birds. The chuck-will's-widow apparently had invaded extreme southeastern Nebraska from Kansas by 1965 and has recently entered the lower Platte Valley. The wood duck, northern cardinal, and red-billed woodpecker are among the many species that have gradually moved westward across the entire state along the Platte Valley in recent decades, and breeding American woodcocks have also moved west along the Platte to at least as far as Kearney.

Eastern bluebirds and tree swallows have recently become more common across Nebraska owing to birdhouse erection and monitoring programs. As a result of planned-release programs (hacking) of hand-reared birds at various potential nesting locations, a few peregrine falcons have been nesting locally in Lincoln and Omaha since 1992 (Johnsgard 2010b). When provided with suitable nesting platforms, the osprey first nested successfully in western Nebraska in 2008, probably as a result of restoration efforts in South Dakota that may have aided in their pioneering into Nebraska. Independently of human help, Mississippi kites have moved north from Kansas, and since 1994 a few have bred locally in western Nebraska.

OVERALL POPULATION TRENDS

Since 1972 several Nebraska bird species have been identified as threatened or endangered either statewide or nationally, such as the whooping crane (state and federally endangered), the interior race of the least tern (state and federally endangered), and the piping plover (state and federally threatened). As a result, several new refuges and nature sanctuaries were established during the 1970s, including two important migratory stopover sites for sandhill and whooping cranes along the Platte River, which have also benefited geese and ducks. Minimum flow rates for the Platte have thankfully been established, and many wetlands in the Rainwater Basin have come under better protection, also helping water and shoreline birds. Although general shoreline development and human recreational usage have reduced habitat for barren shoreline–dependent species, the construction of flood-control dams and resulting reservoirs has increased surface waters and shoreline acreage in the state. These developments have also strongly affected wintering and migration patterns of fish-eating and other water-dependent species, such as waterfowl, eagles, grebes, and loons. Milder recent winters have also allowed many migratory species that once rarely, if ever, wintered here to winter farther north, or at least to migrate south later during fall (Johnsgard 2015a). During the 2018–19 Nebraska Christmas Bird Counts, 26 species of waterfowl were observed; during the 1950s, only the mallard was regularly seen.

Although many Neotropical woodland-adapted and insect-eating migrants have declined in recent decades, many city-adapted and crop-dependent species such as the rock pigeon, European starling, American crow, brown-headed cowbird, common grackle, and red-winged blackbird have vastly increased and have sometimes become serious pests. The major bird additions to Nebraska's avifauna by introductions or invasions from abroad (European starling, house sparrow, rock pigeon, Eurasian collared-dove, cattle egret, ring-necked pheasant, gray partridge) or by major range expansions from other states (cattle egret, white-faced ibis, Mississippi kite, great-tailed grackle), when balanced against three extinctions (passenger pigeon, Carolina parakeet, Eskimo curlew) and the extirpations of eight or nine of Nebraska's breeding birds, leave us both aesthetically and numerically poorer than we were a century ago. I have provided a review of the major changes in Nebraska's bird populations during the twentieth century (Johnsgard 2001a).

Even Nebraska's state bird, the western meadowlark, is currently in decline, as are nearly 30 other prairie-dependent birds. Another iconic grassland bird, the burrowing owl, might well disappear from

Nebraska during the next few decades as prairie dogs become ever rarer. In general, insect-dependent species of birds have most consistently declined in Nebraska, whereas grain- and seed-eating species have more often increased, as have fish-eaters and scavengers. Among the nearly 200 species profiles in this book, North American Breeding Bird Surveys suggest that nearly two-thirds of them have undergone national population declines since the late 1960s.

In spite of all these losses, countless skeins of Arctic-bound snow geese still can etch a March Nebraska sky from dawn to dusk, greater prairie-chickens still annually greet spring sunrises with their ancestral rituals on relict tallgrass prairies, and the spine-tingling cries of sandhill cranes coming to roost on the Platte still bring with them distant echoes of thundering bison and trumpeting mammoths and even of times before recorded time. We can still totally lose ourselves in their grace and beauty, imagining that we have discovered some other Eden and hopefully resolving to act in such a way that these birds might survive to cast their marvelous spells just as strongly as they did a century or two ago.

THE GEOGRAPHY OF NEBRASKA'S BIRD FAUNA

The westernmost part of Nebraska, its geographic "Panhandle," is largely a semiarid ridge-and-canyon region interspersed with High Plains topography and shortgrass steppe vegetation. The major birding attractions in the Panhandle include the biologically diverse and scenic Pine Ridge region and the Wildcat Hills, located southeast of Scottsbluff. The pine-covered hills and escarpments of the Pine Ridge and Wildcat Hills remind one of the Black Hills, and about 3.5 percent of the region's land has been covered by wooded habitats, especially ponderosa pine, although recent forest fires have substantially reduced that coverage.

Several pine-adapted bird species that are common in the Black Hills breed only or primarily in the northwestern corner of Nebraska, such as the pinyon jay, plumbeous vireo, dark-eyed junco, western tanager, yellow-rumped (myrtle) warbler, Townsend's solitaire, Swainson's thrush, red-breasted nuthatch, and red crossbill.

The Clark's nutcracker, Cassin's finch, and Lewis's woodpecker are all apparently limited to Sioux County as breeding species. Some of these same species, as well as the Cassin's kingbird, cordilleran flycatcher, violet-green swallow, white-throated swift, and pygmy nuthatch, also breed in the pine forests of the Scottsbluff area and the Wildcat Hills region.

The Panhandle region also has over 5.6 million acres of mostly arid grasslands that support a few local shortgrass or High Plains bird species, such as the McCown's and chestnut-collared longspurs and the mountain plover. There is also a relatively small area of big sagebrush (*Artemisia tridentata*) in Sioux County (mainly in the Oglala National Grassland), the probable breeding habitat of a few sage-dependent or arid shrub–related species such as the Brewer's sparrow, the sage thrasher, and the now-extirpated greater sage-grouse. Western Nebraska is also a land rich in the fossil remains of early mammals and birds, including an eight-million-year-old fossil arm bone (humerus) that closely resembles that of a modern sandhill crane but cannot be specifically attributed to it. Sandhill cranes in the tens of thousands still pass through far western Nebraska each spring and fall, but their major migratory pathway lies well to the east, in the central Platte Valley.

Typical open-country and arid-land Panhandle birds include the mountain plover, prairie falcon, ferruginous hawk, golden eagle, merlin, Say's phoebe, rock wren, and McCown's longspur. The Cassin's kingbird, western wood-pewee, pinyon jay, mountain bluebird, yellow-rumped warbler, western tanager, and pygmy nuthatch are more closely associated with ponderosa pines or other coniferous trees, and the plumbeous vireo is associated with shrubs and saplings.

The area around Lake McConaughy, Nebraska's largest reservoir near Ogallala, has one of the very few bird lists exceeding 360 reported species for any location north of Texas (Brown, Dinsmore, and Brown 2012). It is especially notable for its rare gulls and diving ducks, as well as up to 40,000 western grebes that stage there during their fall migration. Nearby is Cedar Point Biological Station, a University of Nebraska facility of nearly 400 acres that has been a center of ecological research and summer classes since the late 1970s, especially ornithology (Brown et al. 1996; Scharf et al. 2008).

About 75 miles northwest of Lake McConaughy is Crescent Lake National Wildlife Refuge, a wilderness refuge in the western Sandhills that has the second-largest local bird list for the state, with 273 species. To the north of Crescent Lake, in northern Garden County and southern Sheridan County, are hundreds of highly saline Sandhills marshes that abound during May with waterfowl, shorebirds, and marshland birds and support local breeders such as black-necked stilts, American avocets, cinnamon teal, long-billed curlews, and Wilson's phalaropes.

The west-central portion of Nebraska includes two of the very best bird-finding localities in the state, namely, Valentine and Fort Niobrara National Wildlife Refuges. These two locations have bird lists of nearly 250 species each. Additionally, the region includes those parts of the Niobrara and Platte Valleys that lie in the middle of the transition zone between the Rocky Mountain coniferous forest and eastern deciduous forest biogeographic regions. This transition zone includes areas of hybridization between several species or nascent species pairs of birds that are now in secondary contact, as mentioned earlier. There are also a few species of cool-adapted plants that have been isolated in the shady canyons of the Niobrara Valley since Pleistocene times.

This important east–west transition zone is very wide in the Platte Valley but is compressed to a distance of less than 100 miles in the Niobrara Valley. Much of this zone is now included within the boundaries of the Fort Niobrara National Wildlife Refuge, a 30-square-mile preserve 4 miles east of Valentine, and the Nature Conservancy's Niobrara Valley Preserve, a nearly 90-square-mile preserve in Brown County north of Johnstown that borders several miles of the Niobrara River (Kaul, Kantak, and Churchill 1988).

The lower Niobrara Valley also supports breeding populations of several eastern wooded habitat species that are otherwise mostly limited to Nebraska's Missouri Valley, including the wood thrush, American redstart, ovenbird, and scarlet tanager. Along the upper

Niobrara Valley, several western or northern bird species likewise extend eastwardly, including the common poorwill, red-breasted nuthatch, chestnut-collared longspur, red crossbill, western wood-pewee, spotted towhee, black-headed grosbeak, lazuli bunting, and Bullock's oriole. I have provided a survey of the Niobrara River Valley's natural history, with checklists of major vertebrate and invertebrate groups (see Johnsgard 2007b).

The Nebraska Sandhills region represents the largest natural ecosystem in the state, covering nearly 19,000 square miles, or almost a quarter of the state, and located north of the Platte River and extending locally into southernmost South Dakota. It is also the largest remaining grassland ecosystem in the United States that is still virtually intact both faunistically and floristically. It is a land with far fewer people than cattle where the roads are even fewer and where tourist facilities and accommodations are almost nonexistent. Those roads that do exist are little traveled and mostly consist of only slightly improved sandy trails leading to ranches.

However, the Sandhills region is filled with breathtaking vistas, spectacular bird populations in its hundreds of marshes, and a pioneer spirit that requires everyone to stop and help anyone who happens to get into trouble while on the road. It is a land designed for naturalists who would like to study virtually unaltered grassland ecosystems and who are prepared to deal with nature on its own terms. Many Nebraska waterbirds and shorebirds nest primarily or only here in the state, such as the American wigeon, canvasback, redhead, ruddy duck, long-billed curlew, willet, Wilson's snipe, Forster's tern, black tern, marsh wren, and swamp sparrow.

The central Platte Valley and nearby western Rainwater Basin (a region rich in playa wetlands located mostly between the Platte and Republican Rivers) provide some of the best spring birding opportunities in all of North America. For most of March about ten million waterfowl and more than half a million sandhill cranes pour into the region, remaining until late March in the case of the waterfowl and until about the second week of April in the case of sandhill cranes. As the last sandhill cranes are leaving, whooping cranes begin to arrive, as do the earlier shorebirds, continuing the amazing spring spectacle until about the end of April. Brown and Johnsgard (2013) documented 373 bird species from this region of nearly 10,000 square miles.

Birding in the central Platte Valley during March is a chancy affair in terms of weather; late winter snowstorms might blanket the entire area in a foot of snow, which when melting leaves country roads slippery at best, and thus driving requires a good deal of care. This is especially true in the Rainwater Basin, an area of clay-rich soils that prevent water from percolating down and thus generate temporary wetlands (locally called "lagoons") at the peak of spring waterfowl populations. This is only true during those years when abundant winter snowfalls or spring rains allow the shallow basins to fill. During drier years only the deepest basins, or those that are kept full by pumping, can accommodate the hordes of ducks and geese passing through. Then the stresses caused by bad weather and overcrowding can set off outbreaks of fowl cholera and might kill thousands of birds in a short time. Some of these birds are consumed by wintering bald eagles, hundreds of which concentrate along ice-free areas of the Platte from late fall until early spring.

Gravel roads on the south side of the Platte River are better for crane-watching during March and early April than are roads on the north side of Interstate 80. Perhaps the best is the Platte River Road, which parallels the Platte River from Doniphan (3 miles south of Grand Island) west and, with several southward jogs and a few confusing road name changes, extends nearly to Elm Island Road and Rowe Sanctuary. Cranes usually can be found foraging along this road during early morning and late afternoon hours, but the most rewarding way to watch cranes is from riverside blinds near roosting locations. Crane-viewing blind reservations are available from the Crane Trust's Visitor Center, immediately south of I-80 Exit 306 west of Grand Island, and from the Audubon Society's Rowe Sanctuary, 4 miles southwest of the I-80 Exit 285, on Elm Island Road (see chapter 6 for detailed information). Johnsgard (2008b) provides a regional natural history of the central Platte

Valley, while Brown and Johnsgard (2013) documents the birds of the central Platte Valley. The cranes of the Platte Valley have also been extensively described (Johnsgard 2003a, 2009, 2011b, 2015b).

The western Rainwater Basin region, which extends east from Gosper to Clay County, is just as attractive for birding as is the Platte Valley during early spring. Then, snow meltwaters accumulate temporarily in the clay-rich lowlands, and an estimated seven to nine million ducks and at least five million geese pass through. These flocks have been estimated to include up to 90 percent of the midcontinental greater white-fronted goose population, 50 percent of the midcontinental mallard population, and perhaps 30 percent of the continent's rapidly declining northern pintail population. Millions of snow geese and hundreds of thousands of Ross's, Canada, and cackling geese traverse the region each spring, the total state numbers of waterfowl possibly exceeding ten million birds during peak migration in early to mid-March. Jorgensen (2012) documented the birds of the entire Rainwater Basin, which totaled 259 species.

The shallow Rainwater Basin wetlands are also of great importance to migrant shorebirds. Jorgensen (2004) analyzed shorebird migration patterns in the eastern Rainwater Basin, which extends east from Clay County east to Saunders and Lancaster Counties. He determined that, in descending order, the most numerous spring shorebirds (out of 38 total observed species) seen there over a several-year period were the white-rumped sandpiper, Wilson's phalarope, semipalmated sandpiper, long-billed dowitcher, lesser yellowlegs, least sandpiper, and Baird's sandpiper during spring. During fall the descending order of species abundance was pectoral sandpiper, long-billed dowitcher, lesser yellowlegs, least sandpiper, and stilt sandpiper. The region is of hemispheric migratory importance to the very localized buff-breasted sandpiper, which stages in various mixed-grass sites around the eastern Rainwater Basin, its only known major staging area between the species' wintering and breeding grounds.

The easternmost region of Nebraska is a land that once was ruthlessly scraped over by glaciers, sprinkled with wind-blown loess, and finally occupied and mantled by tallgrass prairies and deciduous forests having eastern biogeographic affinities. It is bounded to the east by the Missouri River, which has sadly been deepened, narrowed, and straightened almost beyond recognition, and has been greatly degraded in terms of riverine wildlife habitat, as well as becoming much more flood prone. The immediate Missouri Valley corridor is nevertheless still a major migratory pathway for Arctic-breeding waterfowl such as snow geese, which alone now number over a million birds using this narrow migratory route, as do myriads of forest-adapted Neotropical migrants, especially warblers and vireos.

Some stretches of the Missouri River provide a good idea of how the river may have once appeared and what its riparian forest might have resembled. Remnant stands of mature deciduous forest still exist at Indian Cave State Park in Richardson County, Fontenelle Forest Nature Center, and Neale Woods in the Omaha-Bellevue area, DeSoto National Wildlife Refuge east of Blair (and partly in Iowa), and Ponca State Park in Dixon County. These are among the best places that can be visited in early and mid-May to see woodland songbirds as they journey north to breeding grounds in the upper Midwest and southern Canada. No checklist exists specifically for Indian Cave State Park, but an official county checklist totals 156 species (https://noubirds.org/Birds/CountyChecklists.aspx). A recent bird list for DeSoto National Wildlife Refuge includes 240 species, a list for Fontenelle Forest has 246 species, and one for Ponca State Park and its surrounding counties totals nearly 300 species.

The broader Missouri Valley and its tributaries are the most heavily populated part of the state and thus have the fewest areas of native tallgrass prairie vegetation persisting. However, on some relict prairies, such as in Gage, Johnson and Pawnee Counties, greater prairie-chickens still gather at sunrise every spring on traditional courtship display grounds they have used for decades, if not centuries. A few tallgrass-dependent prairie species such as the

Henslow's sparrow are highly localized, but other more generalized grassland dependents, such as western meadowlarks, dickcissels, and grasshopper sparrows, still commonly announce their territories from fence posts and shrubs across eastern Nebraska. Audubon's Spring Creek Prairie, 18 miles south of Lincoln, is a mostly virgin prairie of about 850 acres, one of the state's largest remaining relict tallgrass prairies and one of a few places in Nebraska where there is a good chance of seeing the Henslow's sparrow and others of its 240 documented bird species (Johnsgard 2018b).

Many peripheral species with eastern or southeastern affinities also occur and locally breed in easternmost Nebraska, including occasional red-shouldered and broad-winged hawks. Chuck-will's-widows nest in the wooded habitats of the southeastern corner of the state, and pileated woodpeckers also nest there, making it an area of special interest to birders. Other southeastern Nebraska breeders that are primarily associated with and in some cases limited to the forested Missouri Valley include the American woodcock, barred owl, whip-poor-will, ruby-throated hummingbird, Acadian flycatcher, yellow-throated vireo, tufted titmouse, Louisiana water-thrush, prothonotary warbler, northern parula, Kentucky warbler, and summer tanager (Mollhoff 2016). The cerulean and yellow-throated warblers have also been known to breed there but are very rare.

SELECTED SPECIES PROFILES

The following narrative profiles of 173 species include nearly half of the approximately 365 regularly occurring bird species in Nebraska (Brogie 2017; Johnsgard 2018). They were chosen to represent nearly all the most common and widespread species, most of the rare and threatened species, and some of the notable ("charismatic") species that tourists with a high interest in birds might like to see and learn about.

Unlike the mammal and herpetile profiles, identification criteria, measurements, and vocalizations are not included, because many plumage variations related to sex, age, season, and geography make brief distinguishing descriptions impossible, and most bird vocalizations are too complex to allow for brief descriptions. Also, because most birds in the state are migratory, and breeding areas and abundance often vary widely, no coding is provided at the start of each species profile. The suggested viewing sites mentioned in the species profiles are very brief; for more detailed information on Nebraska's birding locations and opportunities, see Johnsgard (2005) or the Nebraska Birding Trails website (http:// nebraskabirdingtrails.com/).

At the end of each profile is a citation that refers to the relevant species account in *The Birds of North America* monograph series, produced by the American Ornithologists' Union and the Academy of Natural Sciences from 1992 to 2002. Accounts published through 2002 were published in hard-copy form only, and most major research libraries also have this series in hard copy. Many of them are also still available from used book dealers. An online version of these accounts was begun in 2004 (*Birds of North America Online*, https//birdsna.org.), and revisions and content changes continue to be made to it. Digital versions of both the revised accounts and the still-unrevised accounts are available at minimal expense through Cornell University's Laboratory of Ornithology (https://www.birds.cornell.edu/). These species accounts are individually cited as BNA in the "Selected References," followed by a number indicating their original publication sequence, plus (parenthetically) the names of their authors and the year of publication. These citations are not repeated in the references section.

For persons seeking the most up-to-date bird occurrence information, NEBirds is a list-serve for reporting recent Nebraska bird sightings and sharing current news on birds. It is accessible by free subscription by going to the following website: https://groups.yahoo .com/group/NEBirds/. Click on the link labeled "Join Group" and provide the requested information. You will have to become a Yahoo member before joining the group. When you fill out the Yahoo membership information, note the boxes relating to receiving information from Yahoo if you want to avoid spam from Yahoo advertisers.

The international eBird website (ebird.org) has also become a favorite way of inserting personal records into a worldwide permanent database and allows one to learn quickly where and when rare birds have been reported. Another very useful source of distributional information on the distribution, abundance, and recent sightings of rare Nebraska birds is the *Birds of Nebraska—Online* website (https://birds.outdoornebraska.gov/). Up-to-date information on the sightings of unusual birds in Nebraska and elsewhere is also available on the Internet via the general birding website www .birdingonthe.net/.

CLASS AVES (BIRDS)

Family Anatidae (Ducks, Geese, and Swans)

SNOW GOOSE. *ANSER CAERULESCENS*

ROSS'S GOOSE. *ANSER ROSSII*

Status: The snow goose is an uncommon to rare migrant throughout the state, especially eastwardly. It has become increasingly common nationally in recent years. The dark-plumaged "blue" morph (formerly called a "phase") and the less common intermediate (heterozygotic) phenotypes collectively comprise about a fourth of the total snow goose population in eastern Nebraska but are progressively less frequent westwardly. Migrants are abundant throughout the Great Plains states during spring and fall, east to the Missouri River Valley. A probable midcontinental population of at least 7 million birds existed by the early 2000s, and estimates of about 10–12 million continental snow geese have been advanced more recently. The smaller but closely related Ross's goose is now a regular and increasingly abundant migrant throughout the Great Plains states and by 2017 made up at least 3–5 percent of Nebraska snow goose flocks (Johnsgard 2018). During the five Nebraska Christmas Bird Counts ending in 2017–18, a total of nearly 200,000 snow geese and nearly 7,800 Ross's geese were observed, or about 3.75 percent of the total were Ross's geese. These tiny geese are easily overlooked among the vast flocks of snow geese with which they usually associate. Apparent hybrids between the species have been seen often, and the identity of the very rare "blue" morph of the Ross's goose (figure 16) might simply be the result of Ross's geese mating with blue-morph snow geese or the effects of occasional mutations that affect plumage pigmentation (Johnsgard 2014, 2016a).

Habitats and Ecology: Snow geese are generally associated with large marsh and wetland habitats. Feeding in agricultural fields is done less frequently by snow geese than by Canada geese; rootstalks and tubers of marshland plants were the traditional primary food source.

Suggested Viewing Opportunities: More common during fall than spring; wetlands and agricultural fields are good places to find snow geese. Almost any flock of snow geese is also likely to have some Ross's geese, but they might be hard to find because of their small size.

Selected References: BNA 514 (T. D. Mobray, F. Cooke, and B. Ganter 2000); Johnsgard 1975b, 2016a; Poulin et al. 2010.

CACKLING GOOSE. *BRANTA HUTCHINSII*

Status: A common migrant through the state, probably more abundant than is generally realized owing to its strong similarity to small races of the Canada goose.

Habitats and Ecology: Cackling geese are most likely seen in wetlands of central and eastern Nebraska during migration in March and April and October and November. Because the cackling goose was fairly recently (2004) recognized as specifically distinct from the small races of the Canada goose, reliable records for it are still quite limited. During the five Nebraska Christmas Bird Counts ending in 2017–18, a total of about 330,000 Canada geese and about 16,000 cackling geese were observed, or about 4.6 percent of the total were cackling geese.

Suggested Viewing Opportunities: This tundra breeder is most often seen in central Nebraska alone or in company with migrating Canada geese, and it is difficult to distinguish visually from smaller

16. Ross's goose (bottom left), probable blue-morph Ross's goose (upper left), blue-morph Ross's × snow goose hybrid (upper right), and blue-morph lesser snow goose (bottom right). Drawn to scale from specimens and photographs.

races of that species. Typical cackling geese phenotypes are very common in the Rainwater Basin during migration, and some wintering probably occurs in southern Nebraska, but most cackling geese probably winter from Kansas to Texas and west to eastern Colorado and eastern New Mexico.

Selected References: BNA (supplemental) (T. B. Mobray, C. R. Ely, J. S. Sedinger, and R. E. Trost 2002); Johnsgard 1975b, 2016a.

CANADA GOOSE. *BRANTA CANADENSIS*

Status: The Canada goose is a common, virtually pandemic resident throughout the state, whereas the cackling goose (above) is an Arctic-breeding spring and fall migrant, especially in the plains region. Canada geese of the larger races have been reared and released widely by conservation agencies and private breeders since the 1960s, and breeding now occurs throughout the state. Overwintering by larger Canada geese is now normal in the Platte Valley and around some cities such as Lincoln and Omaha. North American Breeding Bird Surveys between 1966 and 2015 indicate that the species underwent a significant population increase (9.7 percent annually) during that period, one of the largest estimated rates of annual increase reported for waterfowl (Sauer et al. 2017).

Habitats and Ecology: The extremely adaptable Canada goose sometimes nests within the city limits of large cities in parks and on private lakes, but it also occurs on prairie marshes, farm ponds, and lakes. Muskrat houses provide safe and favored nest sites in many areas.

Breeding Biology: Canada geese have strong, permanent pair bonds, and most begin to breed when two or three years old. Males establish fairly large territories in marshes, usually including the same area and often the same nest site as in previous years. Pair bonds are maintained by mutual displays, especially a mutually performed triumph ceremony, which usually follows aggressive encounters with other geese. Unless nest sites are limited or predator pressures are present, the nests tend to be well scattered. Primarily the female constructs the nest, with the male standing guard and helping to some extent. Copulation occurs on the water, primarily during the egg-laying period, and incubation does not begin until the clutch of about five eggs is complete. Males remain close to the nest and take the major responsibility for guarding it but do not help incubate. Incubation lasts 25–30 days, and both sexes tend the young. During the fledging period of about 70 days, both parents undergo a several-week flightless period while molting their flight feathers. Thereafter the family might leave the area, with family bonds persisting through the first year of the young birds' lives.

Suggested Viewing Opportunities: Canada geese occupy nearly all wetlands in the state, especially protected wetlands, and probably breed in every county.

Selected References: BNA 682, rev. ed. (T. B. Mobray, C. R. Ely, J. S. Snedinger, and R. E. Trost 2002); Johnsgard 1975b, 2016a.

TRUMPETER SWAN. *CYGNUS BUCCINATOR*

Status: A local resident in the Sandhills of Nebraska, a small but still-increasing population that resulted from expansion of introductions at LaCreek National Wildlife Refuge, South Dakota, during the early 1960s. Some limited winter movements are made to ice-free waters of reservoirs and spring-fed rivers. During the five Nebraska Christmas Bird Counts ending in 2017–18, observers recorded an average of 159 trumpeter swans. Trumpeter swans have undergone major population increases in major decades owing to introduction programs, and as of 2010 the North American population was estimated at more than 46,000 birds, compared with the 69 believed to exist in 1932 (Johnsgard 2016d). This species is on the Nebraska Natural Legacy Project's Tier 2 list of threatened species in Nebraska (Schneider et al. 2018).

Habitats and Ecology: Breeding by this species is mostly limited to fairly large ponds having considerable aquatic vegetation and relative seclusion from disturbance by humans. Breeding occurs on large shallow marshes or lakes having abundant submerged vegetation, emergent plants, and stable water levels. Wetlands used during the breeding season average about 180 acres, with about 75 percent open water and having slight to (infrequently) medium salinity levels. There were 15 confirmed Nebraska nestings during the 2006–14 atlasing period (Mollhoff 2016), and about 700 birds were probably resident in the state by 2019.

Breeding Biology: Trumpeter swans pair for life, and each pair returns to its nesting area in spring as soon as the weather allows. Territories are established that average more than 30 acres, sometimes more than 100 acres, and are vigorously defended; the adults even exclude their own offspring of previous years. The male performs such territorial defense, but the female participates in mutual triumph ceremonies after territorial disputes and also helps defend the nest site. Both sexes construct the rather bulky nest of marsh vegetation, which might require a week or more. The eggs are laid at two-day intervals, and no incubation is performed until the clutch of about five eggs is complete. Thereafter the female performs all the incubation (a few records of males incubating are considered abnormal), while the male defends the nest. Incubation lasts about 33 days. Most of the cygnets hatch within a few hours of each other and are led from the nest within 24 hours of hatching. The nest might later be used for resting or brooding, but often the brood is led some distance from the nest for rearing on quiet and secluded ponds. The fledging period is approximately 100 days in Montana and Wyoming, which occupies the entire summer and makes it impossible for birds to renest after nest failure. Fledging periods in Canada and Alaska are progressively shorter.

Suggested Viewing Opportunities: The largest local populations are found during winter on reservoirs and on spring-fed and ice-free rivers such as the North Loup and Middle Loup in the Sandhills. Several populations are often present at migratory or overwintering locations such as Omaha's Carter Lake, DeSoto National Wildlife Refuge, Calamus Reservoir, and the Snake River. Breeding pairs have occupied a marsh just west of Hyannis along U.S. Highway 2 for decades, and for several years a pair nested on a South Loup River backwater at the southern edge of Ravenna, one of the state's easternmost breeding sites. Swan pairs or families can often be seen

17. Wood duck, pair.

on marshes visible along Nebraska Hwy 83 in Valentine National Wildlife Refuge.

Selected References: BNA 105, rev. ed. (C. D. Mitchell and M. W. Eichholz 2004); Banko 1960; Johnsgard 1975b, 2016d.

WOOD DUCK. *AIX SPONSA*

Status: A local uncommon summer resident across Nebraska, these birds are found among woodlands having fairly large trees offering nesting holes and frequently those having acorns or similar nutlike foods in abundance. Possible to confirmed breeding records (291) were reported for all but five counties during fieldwork for *The Second Nebraska Breeding Bird Atlas* (Mollhoff 2016).

Habitats and Ecology: Even outside the breeding season these birds are usually associated with flooded woodlands rather than open marshes. North American Breeding Bird Surveys between 1966 and 2015 indicate that the species underwent a population increase (1.63 percent annually) during that period (Sauer et al. 2017).

Breeding Biology: Pair bonds are established each year after a prolonged period of courtship displays (Johnsgard 1965). No definite territorial behavior exists, but males assist females in seeking out suitable nest sites, which might take several days. Competition for nest sites is frequent, and thus collective "dump nests" having eggs of several females are locally prevalent. The usual clutch size is of eight to ten eggs. The female does the incubation, and males normally desert their mates before hatching. Incubation lasts 28–32 days. The female alone raises the brood, which fledges at about 60 days of age. Renesting after loss of the first clutch is fairly frequent.

Suggested Viewing Opportunities: Wooded streams in eastern Nebraska are the best place for seeing wood ducks, especially where nesting boxes are present, but on migration the birds occur fairly far from trees. During the breeding season they are most often seen on wooded habitat "patches" of 2–10 acres. Based on breeding records alone, wood ducks are the state's second most common breeding duck (Mollhoff 2016), but this figure is misleading owing to the ease of counting active nest boxes.

Selected References: BNA 169, rev. ed. (G. R. Hepp and F. C. Bellrose 2013); Shurleff and Savage 1996; Johnsgard 1965, 1975b, 2017a.

BLUE-WINGED TEAL. *SPATULA (ANAS) DISCORS*

Status: A common summer resident throughout the state. A total of 231 possible to confirmed breeding records was reported for all but a few counties during fieldwork for *The Second Nebraska Breeding Bird Atlas* (Mollhoff 2016). It is common on the grasslands and foothills, especially in the prairie pothole country. North American Breeding Bird Surveys between 1966 and 2015 indicate that the species underwent a slight population increase (0.49 percent annually) during that period (Sauer et al. 2017).

Habitats and Ecology: This species favors relatively small, shallow marshes over larger and deeper ones, especially those that are surrounded by grass or sedge meadows. Migration in spring occurs fairly late, as does pair formation; nonetheless, renesting efforts are fairly common following nest failure. The species is both a late spring and early fall migrant and sometimes winters as far south as Central America and northern South America.

Breeding Biology: Pair bonds are formed fairly late in blue-winged teal, mainly during the northward migration, but some courtship might occur on the nesting grounds. Pairs are relatively tolerant of other pairs and often center their home ranges on very small ponds or even roadside ditches. The female chooses the nest site and builds the nest, while the male waits nearby. The usual clutch size is eight to ten eggs. After incubation begins the pair bond is dissolved, and males often fly elsewhere to complete their summer molt. Incubation lasts 23–24 days. Females take their broods to water within hours after hatching and usually raise them in rather heavy brooding cover. The fledging period is about six weeks, and females also begin to molt at about the time the young are fledged.

Suggested Viewing Opportunities: This species is widespread throughout the state during migration and nests on often rather small, grass-lined wetlands. It is Nebraska's third most common breeding duck. Migrants are found on generally shallow ponds, ditches, marshes, and the like and rarely occur in deep, open water. Breeding is typically done in marshes surrounded by native prairies and grassy sedge meadows. During the breeding season it is most often seen on habitat "patches" of 2–10 acres (Mollhoff 2016).

Selected References: BNA 625 (R. C. Rohwer, W. P. Johnson, and E. R. Loos 2002); Johnsgard 1965, 1975b, 2017a.

CINNAMON TEAL. *SPATULA CYANOPTERA*

Status: A local summer resident especially in western Nebraska and in somewhat saline wetlands. North American Breeding Bird Surveys between 1966 and 2015 indicate that the species underwent a survey-wide decline (2.97 percent annually) during that period (Sauer et al. 2017). This species is on the Nebraska Natural Legacy Project's Tier 2 list of threatened species in Nebraska (Schneider et al. 2018).

Habitats and Ecology: There is no obvious difference in habitat preferences between blue-winged and cinnamon teal, although the latter inhabits somewhat more alkaline marshes during the breeding season.

Breeding Biology: The social behavior and breeding biology of the cinnamon teal are extremely similar to those of the blue-winged teal. In a few areas they breed on the same marshes, nest at the same time, and use almost the same habitats. However, nesting densities of cinnamon teal in the middle of their breeding range are appreciably higher than those of blue-winged teal, and their home ranges tend to be smaller.

Suggested Viewing Opportunities: Persons wanting to see this beautiful little duck should visit Crescent Lake National Wildlife Refuge in June, when as many as six to eight males might be seen on a good day and where occasional breeding has been documented. Breeding records are scanty, but nearly all have occurred in small alkaline wetlands in Panhandle counties, such as at Kiowa Wildlife Management Area, Scotts Bluff County, and Chet and Jane Fleisbach Wildlife Management Area, Morrill County.

Selected References: BNA 209, rev. ed. (G. H. Gammonley 2012); Johnsgard 1965, 1975b, 2017a.

NORTHERN SHOVELER. *SPATULA CLYPEATA*

Status: A common summer resident essentially throughout the entire state; most common on wetlands of the Sandhills and decreasing southeastwardly. North American Breeding Bird Surveys between 1966 and 2015 indicate that the species underwent a survey-wide increase (2 percent annually) during that period (Sauer et al. 2017).

Habitats and Ecology: The specialized comb-like bill structure of this species allows for filter-feeding of surface organisms, and submerged plants also provide microhabitat organisms that can be filtered from the surface. Migrants utilize wetlands rich in plankton-sized foods, and during the nesting season the birds favor shallow prairie marshes rich in those food sources. Nonwooded shorelines are preferred over wooded ones, and shallow mud-bottom ponds are also apparently preferentially used over sandy-bottom ones.

Breeding Biology: Shovelers begin pair formation on their wintering grounds and continue it through their arrival on the breeding grounds. Most of their sexual displays are aquatic and resemble normal foraging behavior, but there are also noisy "jump flights" and aerial chases associated with courtship. The birds are seasonally monogamous, and at least in captivity some older birds remated with previous mates, while others chose new mates. Pairs spread out widely over breeding habitats and have been described as being territorial by some workers, while others have noted that they occupy overlapping home ranges of from 15 to 90 acres in area. The usual clutch size is 8–12 eggs. The females do all the incubation, and their mates abandon them during the incubation period, which lasts 26 days. The fledging period is about six or seven weeks, after which families break up and disperse.

Suggested Viewing Opportunities: Shovelers are widespread during migration but during the breeding season are closely associated with shallow Sandhills and Rainwater Basin wetlands.

Selected References: BNA 217 (P. J. Dubowy 1996); Johnsgard 1965, 1975b, 2017a.

GADWALL. *MARECA STREPERA*

Status: A common summer resident across the state, but most common on shallow, open-water prairie marshes. North American Breeding Bird Surveys between 1966 and 2015 indicate that the species underwent a survey-wide increase (2.66 percent annually) during that period (Sauer et al. 2017).

Habitats and Ecology: Migrants are normally found in shallow marshes and sloughs and sometimes occur on deeper waters such as lakes. Nesting occurs preferentially on shallow prairie marshes, especially those having grassy or weedy islands or surrounding weedy cover. During the breeding season gadwalls are most often seen on habitat "patches" of 101–1,000 acres and are the fourth most common breeding duck (Mollhoff 2016).

Breeding Biology: Gadwalls form their pair-bonding as early as autumn during a period of social courtship involving several aquatic displays, as well as aerial chases. Most birds are paired by the time they arrive on their nesting grounds, and pairs establish home ranges that sometimes exceed 50 acres, often overlapping with the home ranges of other pairs. Territorial behavior is not apparent, and nests are often close together, especially on grassy islands. The female constructs the nest alone and is usually abandoned by her mate about a week or two after incubation has begun. The usual clutch size is of 8–12 eggs, and incubation lasts 25–27 days. The hen raises her brood alone, usually on deeper marshes that are unlikely to dry up before fledging, which requires seven to eight weeks.

Suggested Viewing Opportunities: This species is widespread throughout the state during migration and often is seen in company with American wigeons.

Selected References: BNA 283 (C. R. Lechack, S. K. McKnight, and G. R. Hepp 1997); Johnsgard 1965, 1975b, 2017a.

AMERICAN WIGEON. *MARECA AMERICANA*

Status: A common summer resident, especially in the western Sandhills marshes. North American Breeding Bird Surveys between 1966 and 2015 indicate that the species underwent a survey-wide decline (2.13 percent annually) during that period (Sauer et al. 2017). This species is on the Nebraska Natural Legacy Project's Tier 2 list of threatened species in Nebraska (Schneider et al. 2018).

Habitats and Ecology: This species is associated with relatively open marshes and lakes having abundant aquatic vegetation at or near the surface and in the breeding season favors wetlands with sedge meadows or those with shrubby or partially wooded habitats nearby. Wigeons are strongly vegetarian and spend more time grazing grassy vegetation than do most ducks.

Breeding Biology: Wigeons form seasonally monogamous pair bonds after a period of prolonged social courtship in winter and spring. Males perform fairly simple displays, mainly uttering whistled calls, lifting their chins, and raising their folded wings high above their back. After pair formation, pairs establish a home range on marshes of from less than an acre to more than 20 acres in area. There is no territorial defense, although males evict other males from the vicinity of their mates. Nest sites are well hidden, and shortly after incubation begins males abandon their mates. The usual clutch size is 9–11 eggs, and incubation lasts 24–25 days. The female incubates and rears the brood alone. Broods are reared on relatively open marshes, and fledging occurs at about 45 days of age.

Suggested Viewing Opportunities: This species is widespread throughout the state during migration. Most breeding records have been in the central and western Sandhills and on habitat "patches" of 11–100 acres.

Selected References: BNA 401, rev. ed. (A. E. Mini, E. R. Harrington, B. D. Dugger, and T. Mowbray 2014); Johnsgard 1965, 1975b, 2017a.

MALLARD. *ANAS PLATYRHYNCHOS*

Status: An abundant resident throughout the state, breeding nearly everywhere wetlands ranging from ponds to large marshes occur. A total of 363 possible to confirmed breeding records was reported for all but a few counties during fieldwork for *The Second Nebraska Breeding Bird Atlas* (Mollhoff 2016). North American Breeding Bird Surveys between 1966 and 2015 indicate that the species underwent a slight population increase (0.3 percent annually) during that period (Sauer et al. 2017). During the 2018 Nebraska Christmas Bird Counts over 63,000 mallards were observed, mostly on the lower Platte River and DeSoto Bend and Boyer Chute Refuges.

Habitats and Ecology: This highly adaptable species occurs on nearly all aquatic habitats but prefers nonforested wetlands over forested ones. Breeding birds favor fairly shallow waters, either still or slowly flowing, and surrounding areas of grassy vegetation. Migrants are often found on large marshes, lakes, or reservoirs, especially where nearby grain fields provide food. Mallards quickly recognize and utilize protected areas, even when close to human activities, and thus have remained common in spite of intensive hunting pressures on them.

Breeding Biology: Mallards begin social display early in the fall, with many adults often renewing pair bonds with earlier mates and those hatched the previous summer beginning courtship for the first time. By spring, nearly all females have formed pair bonds, and on arrival at their breeding grounds paired birds spread out across the available wetlands. Home ranges of pairs vary greatly in size but at times exceed 700 acres; spacing is maintained when breeding males evict other males from the vicinity of their mates. Females choose their nest sites and are abandoned by their mates after incubation gets underway. The usual clutch size is of 10–12 eggs, and incubation lasts 24–25 days. Broods are quickly led to water, and the fledging period is about 50–55 days. Mallards often try to renest if their first attempt fails; the clutch sizes of renesting efforts tend to be slightly smaller than the original clutches.

Suggested Viewing Opportunities: This abundant species can be found on regional wetlands almost anywhere throughout the state; only a few southeastern counties lacked breeding records during *The Second Nebraska Breeding Bird Atlas* surveys. Most breeding records have been in the Sandhills and Rainwater Basin and on habitat "patches" of 2–10 acres.

Selected References: BNA 658, rev. ed. (N. Drilling, R. Tiitman, and F. McKinney 2018); Johnsgard 1965, 1975b, 2017a.

NORTHERN PINTAIL. *ANAS ACUTA*

Status: A common summer resident throughout the state. North American Breeding Bird Surveys between 1966 and 2015 indicate that the species underwent a survey-wide decline of 2.4 percent annually during that period and an even greater rate of decline (6.17 percent) in Nebraska (Sauer et al. 2017). This species is on the Nebraska Natural Legacy Project's Tier 2 list of threatened species in Nebraska (Schneider et al. 2018).

Habitats and Ecology: This is a tundra- and grassland-adapted breeding species, and it is rarely if ever found in wooded wetlands. It can breed on small and temporary ponds and on permanent marshes and frequently nests in exposed situations hundreds of yards from water. While on migration it uses nearly all aquatic habitats, ranging from flooded fields to large lakes and reservoirs. It frequently overwinters in considerable numbers where open water occurs and, along with mallards, is one of the earliest spring migrants. It is one of the northernmost breeders of the dabbling duck group, sometimes nesting in tundra habitats well beyond the Arctic Circle in northernmost Canada and Greenland.

Breeding Biology: Northern pintails form monogamous pair bonds during a prolonged period of social courtship, which continues as the birds migrate north in spring. Most or all females are paired by the time the birds arrive on their nesting grounds, and the pairs tend to become well spaced as they establish large home ranges. Females begin nesting shortly after hillsides are free of snow, and, like most ducks, they produce their clutches at the rate of one egg per day. The usual clutch size is of seven to nine eggs. Incubation lasts 25–26 days and begins with the laying of the last egg; by that time or shortly afterward the pair bond is broken. After the brood hatches the female leads them to water, sometimes shifting ponds and moving them up to nearly a mile from where they were hatched. The fledging period varies from 47 to 57 days in South Dakota but is usually 41–46 days for broods in Manitoba, where longer summer days allow more extended foraging times.

Suggested Viewing Opportunities: This species is widespread throughout the state during migration; great numbers of pintails migrate through the Rainwater Basin in early spring. Nearly all the recent breeding records have been from the Sandhills and western Rainwater Basin.

Selected References: BNA 163, rev. ed. (R. G. Clark, J. P. Fleske, K. Guyn, D. A. Haukos, J. E. Austin, and M. R. Miller 2014); Johnsgard, 1965, 1975b, 2017a.

GREEN-WINGED TEAL. *ANAS CRECCA*

Status: A common summer migrant over nearly the entire state, breeding locally. It is very common on the prairie marshes during migration and often overwinters. North American Breeding Bird Surveys between 1966 and 2015 indicate that the species underwent a slight population decline (0.14 percent annually) during that period (Sauer et al. 2017).

Habitats and Ecology: Migrants are associated with a wide variety of shallow-water aquatic habitats in Nebraska. Breeding normally occurs where ponds or sloughs are surrounded by a mixture of grasses, sedge meadows, and upland areas supporting shrubby or tall woody vegetation.

Breeding Biology: Green-winged teal are highly social and display over a long period of late winter and spring while forming their pair bonds. Pair-forming displays are numerous, elaborate, and highly animated. On reaching their breeding grounds, pairs spread out and establish home ranges that usually center on small ponds. Females select nest sites while accompanied by their mates, and pair bonds last until incubation is under way. The usual clutch size is 10–12 eggs, and incubation lasts 23–24 days. After the clutch has hatched, the tiny ducklings grow very rapidly. They fledge in no more than 44 days in the Great Plains and as little as 35 days in Alaska. Growth is unusually rapid at higher latitudes, where a 24-hour summer daylight period makes continuous feeding possible.

Suggested Viewing Opportunities: This species is widespread throughout the state during spring migration, which, in contrast to the blue-winged teal, is almost as early as the mallard's. Nearly all the recent breeding records have been from the Sandhills and western Rainwater Basin.

Selected References: BNA 193 (K. Johnson 1995); Johnsgard 1965, 1975b, 2017a.

CANVASBACK. *AYTHYA VALISINERIA*

Status: A local and uncommon summer resident in the western Sandhills and most abundant in shallow cattail- and rush-lined prairie marshes. North American Breeding Bird Surveys between 1966 and 2015 indicate that the species underwent a slight population increase (0.2 percent annually) during that period. This species is on the Nebraska Natural Legacy Project's Tier 2 list of threatened species in Nebraska (Schneider et al. 2018).

Habitats and Ecology: In the breeding season, canvasbacks are found on shallow prairie marshes with abundant growths of emergent vegetation and on open-water areas that frequently are rich in aquatic plants such as pondweeds.

Breeding Biology: Canvasbacks renew their pair bonds annually, and courtship is usually intense while the birds are returning to their nesting grounds. Several aquatic displays, including cooing calls and head-throw displays, are then conspicuous. After pairing, the birds seek out nesting areas in smaller and shallower ponds than those used for courting. In densely populated breeding areas, a substantial amount of "brood-parasitic" egg-laying occurs among canvasbacks and redheads: females of both species often deposit some of their eggs in the nests of other females. Although redhead hens are prone to lay their eggs in canvasback nests, canvasback hens usually lay eggs only in the nests of other canvasbacks. Thus, mixed-species clutches and broods sometimes occur, but parasitically laid eggs are less likely to hatch and be reared by the host parent than are nonparasitized ones, so such behavior is of limited reproductive benefit to parasitic birds. The canvasback's usual unparasitized clutch size is of seven to nine eggs, and incubation lasts 24–27 days. The fledging period is eight or nine weeks, a substantially longer period than that of most surface-feeding ducks.

Suggested Viewing Opportunities: Deeper and larger marshes, lakes, and reservoirs attract this species on migration. It is a local summer resident in some larger Sandhills marshes of Cherry, Sheridan, Garden, and Rock Counties, especially at Valentine and Crescent Lake National Wildlife Refuges.

Selected References: BNA 659 (T. B. Mowbray 2002); Johnsgard, 1965, 1975b, 2017c.

REDHEAD. *AYTHYA AMERICANA*

Status: A common summer resident over most of the state and a locally uncommon to rare breeder on prairie marshes. North American Breeding Bird Surveys between 1966 and 2015 indicate that the species underwent a slight population increase (0.74 percent annually) during that period (Sauer et al. 2017).

Habitats and Ecology: Breeding habitats of redheads consist of nonforested country with water areas sufficiently deep to provide permanent, fairly dense, emergent vegetation as nesting cover. Water areas at least an acre in area are preferred for nesting, where substantial areas of open water exist for taking off and landing. Migrants are found on large prairie marshes, lakes, and reservoirs, especially where submerged vegetation is abundant.

Breeding Biology: Redheads have seasonally monogamous pair bonds, established each winter and spring. Their displays and associated behavior are much like those of canvasbacks, and the two species often associate. On reaching their nesting grounds, pairs establish home ranges that typically include nest-site potholes and waiting-site potholes, often shared with other pairs. Nest parasitism by redheads is common: females frequently deposit some or even all of their eggs in the nests of various other marsh birds, although not all females are parasitic. The redhead's usual unparasitized clutch size is eight to ten eggs, but many parasitized nests have 12–15 eggs. Incubation lasts 24–25 days. Males abandon their mates early in incubation and often fly elsewhere to undergo their molt and associated flightless period. In Iowa the ducklings have been reported to fledge at 70–84 days of age, but shorter fledging periods have been reported for Canada.

Suggested Viewing Opportunities: This species is widespread on larger marshes and lakes on migration. Redheads are considerably more common breeders than canvasbacks in Nebraska and can usually be seen during summer at Crescent Lake and Valentine Wildlife Refuges and in some deeper Rainwater Basin wetlands.

Selected References: BNA 695, rev. ed. (M. C. Woodin and T. C. Michot 2012); Johnsgard 1965, 1975b, 2017c.

RING-NECKED DUCK. *AYTHYA COLLARIS*

Status: A common migrant throughout Nebraska. North American Breeding Bird Surveys between 1966 and 2015 indicate that the species underwent a survey-wide increase (1.77 percent annually) during that period (Sauer et al. 2017).

Habitats and Ecology: Unlike any of its near relatives, the ring-necked duck is strongly associated with beaver ponds and other forest wetlands, where it is often among the commonest of breeding ducks. Sedge-meadow marshes and boggy areas are preferred for nesting, and the presence of water lilies and surrounding heather (ericaceous) cover seems to be important parts of breeding habitats.

Suggested Viewing Opportunities: This species is widespread on deeper marshes and lakes throughout the state during migration, but nesting habitat is lacking in the state.

Selected References: BNA 329 (C. L. Roy, C. M. Herwig, W. L. Hohman, and R. T. Eberhart 2012); Johnsgard 1965, 1975b, 2017c.

LESSER SCAUP. *AYTHYA AFFINIS*

Status: A regular migrant throughout Nebraska that overwinters locally where open water is present. North American Breeding Bird Surveys between 1966 and 2015 indicate that the species underwent a survey-wide decline (1.84 percent annually) during that period (Sauer et al. 2017); wintering survey evidence also suggests a significant rate of long-term national decline. This species is on the Nebraska Natural Legacy Project's Tier 2 list of threatened species in Nebraska (Schneider et al. 2018).

Habitats and Ecology: This is largely a prairie-adapted breeder and also breeds on ponds in northern Canadian woodlands, especially those supporting good populations of amphipods (scuds) and other aquatic invertebrates. No definite evidence of state nesting exists.

Suggested Viewing Opportunities: This species is widespread on deeper marshes and lakes throughout the state during migration and is far more common than the greater scaup.

Selected References: BNA 338, rev. ed. (M. J. Anteau, J. Devink, D. N. Koons, J. Austin, C. M. Custer, and A. D. Afton 2014); Johnsgard 1965, 1975b, 2017c.

BUFFLEHEAD. *BUCEPHALA ALBEOLA*

Status: An uncommon migrant. North American Breeding Bird Surveys between 1966 and 2015 indicate that the species underwent a survey-wide increase (2.8 percent annually) during that period (Sauer et al. 2017).

Habitats and Ecology: This species is so small that females can use old nest holes of flickers (which are also used by bluebirds, starlings, and similar-sized cavity-nesters) for nesting. At other times the birds are generally found on large and deep waters; no state nesting has been reported.

Suggested Viewing Opportunities: This species occurs commonly on deeper marshes and lakes throughout the state during migration and during winter.

Selected References: BNA 67, rev. ed. (G. Gauthier 2014); Johnsgard 1965, 1975b, 2016d.

COMMON GOLDENEYE. *BUCEPHALA CLANGULA*

Status: A widespread migrant and wintering species throughout Nebraska. North American Breeding Bird Surveys between 1966 and 2015 indicate that the species underwent a slight population increase (0.21 percent annually) during that period (Sauer et al. 2017).

Habitats and Ecology: While in Nebraska these ducks occur on deeper and larger bodies of water, such as lakes and reservoirs.

Suggested Viewing Opportunities: This species occurs commonly on deeper marshes and lakes throughout the state during migration and on ice-free waters during winter.

Selected References: BNA 170 (J. M. Eadie, M. L. Mallory, and H. G. Lumsden 1995); Johnsgard, 1965, 1975b, 2016d.

COMMON MERGANSER. *MERGUS MERGANSER*

Status: A common overwintering migrant and rare breeder, with recent documented breeding limited to the middle Niobrara River Valley and at Victoria Springs State Recreation Area, Custer County. North American Breeding Bird Surveys between 1966 and 2015 indicate that the species underwent a slight rate of population decline (1.98 percent annually) during that period (Sauer et al. 2017).

Habitats and Ecology: This fish-eating species occurs in areas of clear water supporting large fish populations and is by far the commonest merganser of the state. Nesting occurs in tree cavities and rock crevices and sometimes under boulders or among dense shrubbery.

Breeding Biology: During fall and winter, these mergansers usually stay in small flocks that sometimes feed cooperatively, but as spring approaches much time is spent in social display and in establishing pair bonds, and flock sizes decrease. Females often nest close together, and probably some dump-nesting (the indiscriminate depositing of eggs in one nest by two or more females) occurs in locations where nest sites are limited. The clutch size varies from 7 to 14 eggs. The males usually leave their mates before hatching but on rare occasions have been seen with broods. Incubation lasts 28–32 days. The young are led to water a day or two after hatching, and the brood is usually raised on shallow, clear rivers. The female sometimes carries part of her brood on her back, especially when the brood is frightened. The fledging period is 60–70 days.

Suggested Viewing Opportunities: This species occurs commonly on most deeper marshes, rivers, and lakes throughout the state during migration and over the winter. During the 2018–19 Nebraska Christmas Bird Counts over 21,000 common mergansers were observed, with more than 90 percent of them on Harlan County Reservoir.

Selected References: BNA 442, rev. ed. (J. Pearse, M. Mallory, and K. Metz 2015); Johnsgard 1965, 1975b, 2016d.

RUDDY DUCK. *OXYURA JAMAICENSIS*

Status: An occasional to rare summer resident mainly in shallow mud-bottom Sandhills marshes and a migrant almost throughout the state. North American Breeding Bird Surveys between 1966 and 2015 indicate that the species underwent a slight population increase (0.94 percent annually) during that period (Sauer et al. 2017).

Habitats and Ecology: Nonbreeding birds are found on larger wetlands that have silt or muddy bottoms and that are the foraging strata of ruddy ducks. Breeding occurs on overgrown shallow marshes with abundant emergent vegetation and some open water.

Breeding Biology: Ruddy ducks apparently mature in their first year, although not all females breed as yearlings. Pair bonds are rather weak, and much of the male's display behavior is related to territorial advertisement rather than courtship. Females show little or no pair-forming or pair-maintaining behavior, although males might remain in the vicinity of their mates after they have begun nesting. The usual clutch size is six to ten eggs, but dump-nesting by multiple females might produce larger clutches. Incubation lasts 24 days. Some males persist in remaining with brood-tending females, although the males do not assist in rearing or defending broods. The young are highly precocial and very independent, so they often become scattered long before they fledge, at about six or seven weeks after hatching.

Suggested Viewing Opportunities: This species occurs commonly on deeper marshes and lakes throughout the state during migration. Most breeding records are from the western Sandhills, with scattered records from the Rainwater Basin and elsewhere.

Selected References: BNA 696 (R. B. Brua 2002); Johnsgard 1965, 1975b, 2017c; Joyner 1975.

Family Odontophoridae (New World Quails)
NORTHERN BOBWHITE. *COLINUS VIRGINIANUS*

Status: The bobwhite was once a common resident over most of eastern Nebraska, but the state and national population has declined greatly in the past several decades. Habitat losses and pesticide effects on the birds and their major foods, mostly seeds, leaves, and (especially for growing chicks) insects, are likely causes. National Breeding Bird Surveys from 1966 to 2015 indicate that this species' population underwent a survey-wide average annual decline of 3.48 percent (Sauer et al. 2017), reflecting a population reduction of more than 90 percent over the bobwhite's overall U.S. range during recent decades. They are currently most abundant from Lincoln, Hayes, and Hitchcock Counties, east along the counties bordering Kansas to Richardson County, north to Custer, Valley, and Greeley Counties, and in southeastern Nebraska north to Butler and Saunders Counties. However, they extend locally west to the Wyoming border in the North Platte Valley and to Cherry, Tomas, and Logan Counties in the central Sandhills.

Habitats and Ecology: The bobwhite is not so much a grassland species as one needing a grassy cover for nesting, brushy cover for escape, and a foraging source of herbaceous plants, especially legume seeds, weedy herbs, and grains. It also needs a nearby source of water.

Breeding Biology: Bobwhites have long been considered to be strongly monogamous, although recent radio telemetry research has shown that the pairing pattern of bobwhites is better described as ambisexual polygamy (both sexes changing partners), at least when two or more breedings efforts occur in a single season. Older birds typically reestablish pair bonds in early spring. They then seek out a nesting site, and both sexes participate in nest-building. A scraped area is filled with leafy materials, and grasses or other leaves are arched over the nest to conceal it from above. The female incubates the clutch of about 14 eggs for 23 days, sometimes with limited male participation. The male takes over incubation if his mate is killed, and both sexes normally participate in brood rearing. No more than about five or six chicks are likely to survive through the 14-day fledging period, and as the families mature they begin to merge with other families to form fall coveys. Covey sizes of about 10–11 birds are an ideal size, as that many birds are needed to form a tight circle of roosting and outward-facing birds. Thus the contact of their bodies probably helps retain body heat during cold weather, and the many pairs of eyes provide an entire 360-degree visual panorama.

Suggested Viewing Opportunities: Sites that have a combination of woodland edges bordered by pastures and fields and a local source of surface water are likely to support northern bobwhites. The male's distinctive spring *bob-white* advertisement call provides a good clue for locating the birds, and many other vocalizations are important in social interactions.

Selected References: BNA 397, rev. ed. (L. A. Brennan, F. Hernandez, and D. Williford 2014); Johnsgard 1973, 2017b.

Family Phasianidae (Pheasants, Grouse, and Turkeys)

RING-NECKED PHEASANT. *PHASIANUS COLCHICUS*

Status: A common permanent-resident, nonnative species, the ring-necked pheasant was successfully introduced into Nebraska in 1914. National Breeding Bird Surveys from 1966 to 2015 indicate that this species' population underwent a survey-wide average annual decline of 0.64 percent (Sauer et al. 2017). Nebraska surveys indicate a 1.78 percent annual decline, based on a sample of 51 routes. Their primary range extends from the Panhandle east through south-central Nebraska to Jefferson and Saline Counties and in northeastern Nebraska from Knox County south and east to Burt County.

Habitats and Ecology: Breeding occurs mainly in native grasslands, edges of woodlands and marshes, irrigated agricultural areas, and small patches with tall grass and weedy forbs.

Breeding Biology: Nesting in Nebraska extends from about mid-April to mid-June. Males are polygynous and by frequent calling and displaying attempt to attract several females into their harems. Although females nest within their male's large home range, he does not participate in nest-building nor in protecting the female, her nest, or young. The clutch size varies from 7 to 15 eggs, and incubation begins when the clutch is complete. Incubation lasts 23–27 days, and the chicks are highly precocial at hatching. They fledge at 12–14 days. Pheasants do not compete significantly with native grouse or quails; the recent declines of these native species are clearly the result of habitat losses and probably extensive pesticide use rather than from competition with pheasants (Johnsgard 2017b).

Suggested Viewing Opportunities: Pheasants can often be seen near dusk as they search for small stones along gravel roads. They swallow the stones, which are then lodged in the gizzard, where they serve for grinding hard seeds.

18. Sharp-tailed grouse, displaying male in alert posture.

Selected References: BNA 572 (J. H. Guidice and J. T. Ratti 2001); Johnsgard 2017b.

SHARP-TAILED GROUSE. *TYMPANUCHUS PHASIANELLUS*

Status: An uncommon resident of mixed-grass plains and the Sandhills region from the Wyoming border east through the northern Panhandle to the eastern and southern boundaries of the Sandhills. National Breeding Bird Surveys from 1966 to 2015 indicate that this species' population underwent a slight average annual increase of 0.37 percent (Sauer et al. 2017).

Habitats and Ecology: This species is associated with native grasslands and grassy shrub areas, expanding into cultivated fields during fall and winter. The birds eat some bush and tree twigs during winter, and during heavy snow periods the birds may tunnel into the snow for roosting.

Breeding Biology: By early spring onward, male sharp-tailed grouse begin to assemble and establish or reestablish territories in communal male display areas, or leks. Only older, more experienced males are able to obtain and defend the more central and desirable territories, and females selectively choose these more robust birds when they arrive for fertilization. Much of the males' elaborate display behavior (dancing) is thus directed toward the other males for establishing social dominance and largely consists of ritualized hostile behavior, although a few displays and calls are specifically reserved for females. Females visit the leks only long enough to be fertilized by the most fit males, who take no further part in reproduction. The usual clutch size is of 10–12 eggs, and incubation lasts 23–24 days. The young hatch simultaneously and soon leave the nest to begin feeding on small insects. They can fly short distances by the time they are ten days old and might move up to a quarter mile in a day, even before fledging. They are nearly independent by the time they are six to eight weeks old and often disperse considerable distances at that time.

Suggested Viewing Opportunities: This species' display grounds are nearly all situated on loess or sandy grasslands north of the Platte and are active from about mid-March to early May. One location with free-use sharp-tailed grouse viewing is at Crescent Lake National Wildlife Refuge, Garden County, where a blind is available by reservation (308-761-4893). Another, at the Oglala National Grassland, Nebraska National Forest, near Crawford, Sioux County, is available by reservation (308-432-0300) on Fridays to Sundays. On other days it is available on a first-come basis. See the following species profile for locations where leks of both sharp-tailed grouse and prairie-chickens exist and viewing blinds are available.

Selected References: BNA 354, rev. ed. (J. W. Connelly, M. W. Schroeder, and L. A. Robb 1998); Lumsden 1965; Twedt 1974; Johnsgard 2002a, 2016b.

19. Greater prairie-chicken, displaying male in alert posture.

GREATER PRAIRIE-CHICKEN. *TYMPANUCHUS CUPIDO*

Status: This grassland-dependent species bred nearly throughout the state from the late 1800s to the 1920s but is now localized and is most common in the Sandhills and in the scattered remnant prairies of central and southeastern Nebraska (Johnsgard and Wood 1968). National Breeding Bird Surveys from 1966 to 2015 indicate that this species' population underwent a survey-wide average annual decline of 2.8 percent (Sauer et al. 2017); the Nebraska population currently appears to be stable, but no serious efforts to estimate its size have been made for decades. This species is on the Nebraska Natural Legacy Project's Tier 2 list of threatened species in Nebraska (Schneider et al. 2018).

Habitats and Ecology: Greater prairie-chickens are associated with native tall grasslands and with combinations of native grasslands and grain croplands where the proportion of croplands to native grasslands is fairly low.

Breeding Biology: Male prairie grouse competitively establish individual territories in early spring on traditional communal display sites called "booming grounds" (prairie-chickens), "dancing grounds" (sharp-tailed grouse), or leks (both species). They perform their species-specific displays every day for several months, usually from about mid-March to early May, with activity peaking in mid-April. In some locations prairie-chickens and sharp-tailed grouse co-occur and occasionally display simultaneously on the same lek (Johnsgard and Wood 1968). Such mixed-species leks are rare and usually involve only a single male of one species intruding into a lek of the other, although hybrids occasionally result from these encounters (Johnsgard 2002a, 2007a, 2016b). Females are attracted to and mostly mate with those experienced and dominant males able to obtain and defend central territories, the "master cocks." After fertilization the female lays her clutch in a ground nest hidden among taller grasses. Prairie-chicken nests in Nebraska are typically in grassy, open habitats, such as ungrazed meadows or hayfields. From 9 to 14 eggs (averaging 11 or 12) are laid, incubation begins at about the time the last egg is laid, and the incubation period is 23–26 days. Until they are about a week old the chicks are brooded much of the time, but they are highly precocial and can fly in less than two weeks. Families gradually disintegrate when the young are about six to eight weeks old.

Suggested Viewing Opportunities: Nebraska is one of very few states that have populations of both prairie-chickens and sharp-tailed grouse that are high enough to make watching both species in a single day very feasible. Free-access prairie-chicken leks are available on a first-come basis at Burchard Lake Wildlife Management Area, Pawnee County, where two blinds are available (402-335-2534). One blind is available with reservations on Nebraska Cooperative Republican Platte Enhancement project (N-CORPE) land, near North Platte, from mid-March to early May (308-534-6752 or 308-414-2140). Locations having free-use blinds where either or both prairie grouse might be observed include Valentine National Wildlife Refuge, Cherry County, where two free-use blinds are available by reservation starting April 1 (402-376-3789 or 402-376-1889), and two blinds are available on a first-come basis starting April 1 at the Bessey Ranger District of Nebraska National Forest, Thomas County (308-533-2257). Calamus Outfitters (308-346-4697), a Sandhills ranch near Burwell, Garfield County, offers commercial sharp-tailed and prairie-chicken lek excursions and associated overnight accommodations in a ranch setting. Similar lek-viewing tours of either or both species, also with overnight motel accommodations, are offered at the Sandhills Motel in Mullen, Hooker County (308-546-2417 or 888-278-6167).

Selected References: BNA 36, rev. ed. (J. A. Johnson, M. A. Schroder, and L. A. Robb 2011); Zimmerman 1993; Johnsgard 2002a, 2016b.

WILD TURKEY. *MELEAGRIS GALLOPAVO*

Status: A common resident in the state as a result of introduction efforts involving several different races. National Breeding Bird Surveys from 1966 to 2015 indicate that this species' population underwent an amazing average annual increase of 7.51 percent (Sauer et al. 2017), which is the most rapid rate of increase documented among native Nebraska birds. Possible to confirmed breeding records were reported for all but two counties during fieldwork for *The Second Nebraska Breeding Bird Atlas* (Mollhoff 2016).

Habitats and Ecology: Breeding habitats vary greatly among the several introduced subspecies. In eastern parts of the state, turkeys of the originally native eastern race are found in floodplain forests having a variety of hardwood trees, especially those bearing acorns or other large and edible seeds. In the Pine Ridge region, birds from the western montane Merriam's race are closely associated with pines, junipers, running water, and a fairly rugged topography. Birds of the Rio Grande race favor the drier habitats of southwestern Nebraska. Breeding often occurs in woodlands or forested areas, with the nests being well concealed, often under a log or at the base of a tree.

Breeding Biology: Turkeys spend the winter in small flocks consisting of adult males or larger groups of hens and family units. When the spring "gobbling" season begins the males establish individual gobbling or strutting areas. Groups of closely related "brothers" typically associate, displaying in synchrony and allowing the most dominant of the males to fertilize any female that is attracted. Alternatively, a single highly aggressive male might be able to dominate an entire local population in a manner equivalent to the "master cock" situation of lekking grouse. Females have only brief contact with males until they are fertilized; then they leave the courting groups and establish their nests. The usual clutch size is 8–12 eggs, and incubation lasts 25–26 days. Within a week the chicks can make short flights, and they soon begin to roost in trees. After young males reach the age of six or seven months they often leave their family and begin to establish multimale associations that might persist for their entire lives.

Suggested Viewing Opportunities: Breeding records for turkeys indicate that the largest populations are along well-wooded rivers (Platte, Niobrara, and Republican) and in the Pine Ridge region, but few occur in the western Sandhills or the driest parts of the Panhandle.

Selected References: BNA 22, rev. ed. (J. T. McRoberts, M. C. Wallace, and S. W. Eaton 2014); Lewis 1973; Dickson 1992.

Family Podicipedidae (Grebes)

EARED GREBE. *PODICEPS NIGRICOLLIS*

Status: A local summer resident in the state, especially in the western Sandhills (Sheridan and Garden Counties), plus occasional nesting in the Rainwater Basin. National Breeding Bird Surveys from 1966 to 2015 indicate that this species' population underwent a survey-wide average annual increase of 1.11 percent (Sauer et al. 2017).

Habitats and Ecology: This species is associated in the breeding season with rather shallow marshes and lakes having extensive reed beds and submerged aquatic plants. It is generally found in larger and more open ponds than either pied-billed grebes or horned grebes use, and, unlike these species, colonial nesting typically occurs.

Breeding Biology: Pair-forming displays occur during spring migration while the birds are in flocks and continue after arrival on the breeding grounds. Courting occurs within semicolonial breeding areas; no territorial behavior is evident. Displays are mutual and include an advertising call by unpaired or separated birds, as well as head-shaking and ritualized habit-preening. They also perform a "penguin dance," when both members of a pair stand upright in the water while paddling and facing each other, and a "cat attitude," with withdrawn head and fluffed body feathers (McAllister 1958). The female builds the semifloating nest, and copulation occurs on the nest platform. The usual clutch size is of three or four eggs, and incubation lasts 20–22 days. Thereafter both parents tend the young, which often ride on their parents' backs. Young are relatively independent by their third week, and the fledging period is about 45 days.

Suggested Viewing Opportunities: During migration, marshes and lakes throughout the state attract this species. Marshes near Lakeside and at Goose Lake, Crescent Lake National Wildlife Refuge, usually have breeding colonies.

Selected References: BNA 433 (S. A. Cullen, J. R. Jehl Jr., and G. L. Nuechterlein 1999); McAllister 1958; Johnsgard 1997a.

CLARK'S GREBE. *AECHMOPHORUS CLARKII*

WESTERN GREBE. *AECHMOPHORUS OCCIDENTALIS*

Status: Both species are locally uncommon (western grebe) and highly localized (Clark's grebe) summer residents of reedy marshes and shallow Sandhills lakes. National Breeding Bird Surveys from 1966 to 2015 indicate that these species' populations underwent a collective average annual decline of 2.02 percent (Sauer et al. 2017). Both species are on the Nebraska Natural Legacy Project's Tier 2 list of threatened species in Nebraska (Schneider et al. 2018).

Habitats and Ecology: Breeding typically occurs on permanent ponds and shallow lakes that are often slightly brackish and have large areas of open water, as well as semiopen growths of emergent vegetation. The largest numbers of recent breeding records have occurred in habitat patches of 101–1,000 acres and in locations defined as open-water lakes (Mollhoff 2016).

Breeding Biology: Most display activity occurs before the start of nesting and apparently serves primarily for pair-bond formation and maintenance. Most courtship displays are performed mutually by both sexes. They include crest-raising while the birds swim together, with an associated whistling note. There are also high arch and low arch postures, with necks stretched and bills pointed downward, stereotyped habit-preening, and a spectacular race. During the race display two birds (but sometimes as many as six) call, then rise in the water and race side by side over the water surface with arched necks, bills pointed diagonally upward, and wings partially raised. Behavior prior to the race usually includes threat-pointing with the bill and mutual bill-dipping as the birds approach each other; diving often terminates the sequence (BNA 26b). When more than two birds perform the race the additional ones are always males. Copulation normally occurs on the semifloating nest. The usual clutch size is three to four eggs, and incubation by both sexes lasts 23 days. Both pair members tend the young, and the fledging period is approximately 70 days.

Suggested Viewing Opportunities: The fall populations of migrating western and Clark's grebes at Lake McConaughy often number in the tens of thousands during late September, with smaller numbers present during late April. Breeding records for western grebes have mostly come from Cherry, Sheridan, and Garden Counties but extend east locally to Holt County. A few possible or probable

breeding records for Clark's grebes have come from Lake McConaughy (Keith County), Willy Lake (Sheridan County), and Valentine National Wildlife Refuge (Cherry County).

Selected References: Western: BNA 26a (N. LaPorte, R. W. Storer,
and G. L. Nuechterlein 2013). Clark's: BNA 26b (R. W. Storer and
G. L. Nuechterlein 1992); Johnsgard 1997.

Family Columbidae (Pigeons and Doves)

EURASIAN COLLARED-DOVE. *STREPTOPELIA DECAOCTO*

Status: An invasive and rapidly expanding resident species that is
now common and occupies the entire state. National Breeding Bird
Surveys from 1966 to 2015 indicate that this species' populations
underwent a collective estimated average annual increase of 29.18
percent (Sauer et al. 2017), by far the most rapid rate of increase
estimated for any Nebraska species. The estimated annual rate of
increase in Nebraska was 40.66 percent, a rate probably well beyond
the species' biological capacity for reproductive increase.

Habitats and Ecology: This self-introduced species is usually found
in smaller towns and villages rather than in large cities. Its foraging
ecology is notably different from that of the mourning dove (Poling
and Hayslette 2006). During the 2018 Nebraska Christmas Bird
Counts over 2,400 collared-doves were observed, and more than
half were in the Scottsbluff survey area.

Breeding Biology: This species usually nests in urbanized or agricultural locations. The nest is often placed in a tree or rarely on
the ledge of a building. Like other doves, the nest is a frail, thin
structure of small twigs and as high as 40 feet above ground, but
often is under 20 feet. Two white eggs, similar to but larger than a
mourning dove's eggs, constitute the clutch. The nesting period is
long, from at least March to August, and from three to six broods
might be raised in a single season. The incubation period is 14
days, and the duties are shared by both sexes. The young are able
to fly by 18 days of age and leave the nest vicinity by 21 days of age,
when the adults begin another nesting cycle. Probably because of
their persistent repeated nesting behavior, this species has one of

20. Adult common poorwill (above) and male mourning dove cooing.

the highest rates of natural increase of any North American bird, which helps account for their explosive spread across the continent within about half a century.

Suggested Viewing Opportunities: This species is most likely to be seen near grain elevators in smaller towns and villages, especially around feedlots or granaries where waste grain is abundant.

Selected References: BNA 630, rev. ed. (C. M. Romagosa 2012); Poling and Hayslette 2006.

Status: In both volumes of *The Nebraska Breeding Bird Atlas* (Mollhoff 2001, 2016), the mourning dove ranked as the most frequently reported bird species, with a 99 percent occurrence rate in all the 557 blocks surveyed in the second survey. A total of 555 possible to confirmed breeding records was reported for all but a few counties during fieldwork for *The Second Nebraska Breeding Bird Atlas* (Mollhoff 2016). National Breeding Bird Surveys from 1966 to 2015 indicate that this species' population underwent a survey-wide average annual decline of 0.29 percent (Sauer et al. 2017). Nebraska surveys indicate a 0.22 percent annual population decline, based on a sample of 51 routes.

Habitats and Ecology: Mourning doves are the most numerous breeding bird species in the state. They breed in all of Nebraska's terrestrial habitats, from the driest arid plains of the Panhandle to the wet riparian forests of the Missouri Valley. The species ranked first in overall state breeding abundance during both *Nebraska Breeding Bird Atlas* surveys. Other species that ranked from the second to tenth most abundant species during both surveys were the western meadowlark, barn swallow, eastern kingbird, red-winged blackbird, common grackle, and killdeer (Mollhoff 2016).

Breeding Biology: Mourning doves begin to form pairs at the onset of the breeding season, when males that are dominant in winter flocks mate with high-ranking females; such pairs are the first to establish territories and appear to be the most successful in their reproductive efforts. They place their nests on the ground, in bushes, or in trees and are persistent renesters. The availability of choice nesting materials (usually twigs) is important in determining territorial boundaries in captive birds, whereas food and water sites are not defended. The two eggs are usually laid at about 24-hour intervals, and incubation is by both sexes, the male normally incubating during the day and the female at night. Typically, the eggs hatch on successive days; thus the first-thatched chick tends to be larger and more aggressive in food-begging than the other. By the time the young are 12 days old they are ready to leave the nest, and they normally fledge when they are 13–15 days old. By then the adults have generally begun a second clutch in a new nest. In subsequent nestings the two nests might be used alternately; in Texas as many as six nesting cycles by a single pair during one summer have been reported, using three different nest sites. Nesting in Nebraska extends over a nearly five-month nesting span, so raising four broods is theoretically possible, inasmuch as about a month is required for each nesting cycle. As the breeding season ends, fall flocks begin to form, and a variably extensive migration southward occurs, depending on the severity of the weather.

Suggested Viewing Opportunities: This is an extremely widespread species that is easily found throughout the region in both towns and countrysides.

Selected References: BNA 117 (R. E. Mirachi and T. S. Baskett 1994); Hanson and Kossack 1963; Johnsgard 1975b.

Family Cuculidae (Cuckoos)

YELLOW-BILLED CUCKOO. *COCCYZUS AMERICANUS*

Status: An uncommon to locally rare and declining summer resident in Nebraska. National Breeding Bird Surveys from 1966 to 2015 indicate that this species' population underwent a survey-wide average annual decline of 1.45 percent, similar to the estimate for Nebraska (1.83 percent) (Sauer et al. 2017). This species is on the Nebraska Natural Legacy Project's Tier 2 list of threatened species in Nebraska (Schneider et al. 2018).

Habitats and Ecology: This species is associated with thickets near water, second-growth woodlands, deserted farmlands, and brushy orchards. Dense woodlands are avoided. Caterpillars, such as tent caterpillars, are a favorite food.

Breeding Biology: Cuckoos are late spring migrants, arriving in Nebraska in late May or even early June and inconspicuously taking up breeding territories. Their distinctive clucking and repeated hollow notes of *kaw* or *kowp* are frequently uttered, especially on cloudy days or at night. The birds gather nesting materials from trees by breaking off small branches and carrying them back one at a time to the nest. The usual clutch size is three or four eggs. The eggs are laid at irregular intervals, and the incubation period of 10–11 days apparently begins during egg-laying, since the hatching sequence is staggered. Both sexes assist equally in incubation, and both also equally tend the young. The young birds remain in the nest about nine days but are still flightless when they leave it. At that time they are very agile in climbing about on branches. When they leave the nest most of the young are somewhat more than half the adult weight; the most recently hatched and smallest might be left behind in the nest, often to be neglected and starve.

Suggested Viewing Opportunities: Nearly all recent breeding records are from east of the Panhandle, with the greatest number in southeastern counties, from Thayer and Saline Counties east to the Missouri River.

Selected References: BNA 418, rev. ed. (J. M. Hughes 2015); Bent 1940; Bennett and Keinath 2001.

Family Caprimulgidae (Nightjars)

COMMON NIGHTHAWK. *CHORDEILES MINOR*

Status: An uncommon summer resident throughout the state but most common in the Panhandle, but with relatively few recent breeding records in the eastern quarter of Nebraska. National Breeding Bird Surveys from 1966 to 2015 indicate that this species' population underwent a survey-wide average annual decline of 1.93 percent (Sauer et al. 2017).

Habitats and Ecology: This species forages entirely in the air on flying insects. It is especially common over grassland and urban areas, sometimes extending to shrub and desert scrub. Nesting occurs on the ground, often in grasslands or at the edges of woods, and sometimes on the asphalt rooftops of buildings.

Breeding Biology: Nighthawks are fairly late arrivals on northern nesting areas, and the males soon announce their presence by aerial displays. The most conspicuous of these is the *peent* call, uttered during a series of four or five wingbeats and serving to announce territorial ownership. Males also perform steep dives with down-flexed wings, each dive ending with a rush of air that produces a booming noise. These dives are often performed almost directly over the nest site. The females deposit their two eggs on almost any flat surface and often move them about in the course of incubation. The female rolls the eggs in front of her as she settles on them for incubation, sometimes as far as five or six feet from their original position. Some investigators report that only the female incubates; other researchers have stated that one sex incubates at night and the other by day, which seems most likely, given the insect-dependent foraging behavior of the species. Incubation lasts 19 days, and both sexes are known to help care for the young. As the adults bring food to their offspring they apparently place their bills inside the gaping mouths of the chicks and regurgitate food with a strong pumping of the head. Feeding of the young is usually done at dusk, after sunset, and just before dawn. About three weeks are required for the fledging of the young, and they are independent at about 30 days of age.

Suggested Viewing Opportunities: This is a widespread species and is usually seen or heard coursing low over towns and cities near sundown.

Selected References: BNA 213, rev. ed. (R. Brigham, R. G. Poulin, and D. Grindel 2011); Bent 1940; Sutherland 1963; Holyoak 2001.

COMMON POORWILL. *PHALAENOPTILUS NUTTALLII*

Status: A common summer resident in western Nebraska mainly on dry, rocky habitats. Nearly all recent breeding records are from the Panhandle, with a few extending east along the Niobrara River to Keya Paha County and some possible breeding in the Republican Valley east to Frontier County. National Breeding Bird Surveys from 1966 to 2015 indicate that this species' population underwent a survey-wide average annual decline of 0.86 percent (Sauer et al. 2017).

Habitats and Ecology: Generally, this species is associated with rocky habitats having a cover of arid-adapted shrubs or low trees, such as juniper, sagebrush, and dry grasslands. Poorwills nest on the ground, often under scrub oaks, the leaves of which provide concealment for both adults and young.

Breeding Biology: Like other species in this family, poorwills are late spring migrants, and soon after arrival they begin to utter their distinctive *poor-will* or *poor-will-low* notes during the evening. Little is known of their courtship displays. Nests are extremely difficult to locate, and the well-camouflaged adult usually remains motionless on the nest until very closely approached. The clutch consists of two eggs. Although females perhaps do most of the incubating, males have been seen incubating at night and also have been observed brooding the young. When incubating or brooding, both adults and young keep their eyes almost completely shut, which adds to the effective camouflage of these already inconspicuous birds. When disturbed the adults often utter a loud hissing sound and maximally expand their body, which often has a startling and frightening effect on persons unfamiliar with the birds and might deter some predators. Incubation lasts 20–21 days. The young are hatched in a downy coat that is soon replaced with a juvenile plumage very similar to the adults'. The fledging period is 20–23 days. Two broods are sometimes reared. Most poorwills migrate south by early fall, but in a few locations, such as in California and New Mexico, individuals have been found torpid among rocks or vegetation during subfreezing temperatures. They can remain for weeks or months without eating, with a body temperature as low as 40° F and as few as ten heartbeats per minute. This adaptation was first reported in North Dakota by Lewis and Clark (Johnsgard 2003b) but was dismissed by biologists. It is a rare example of hibernation among birds, although hummingbirds also enter a nocturnal torpid state when exposed to cold overnight temperatures.

Suggested Viewing Opportunities: The Wildcat Hills and Pine Ridge regions are by far the best places to look (and listen) for this species. It is most easily found after dark by driving along dirt or gravel roads in pine woodlands and watching on the roadway for the birds' eye-shine when the car's headlights reflect them.

Selected References: BNA 32, rev. ed. (C. P. Woods, R. D. Csada, and R. M. Brigham 2005); Bent 1940; Holyoak 2001.

Family Apodidae (Swifts)

WHITE-THROATED SWIFT. *AERONAUTES SAXATALIS*

Status: A local summer resident in the western Panhandle, mainly in the western Pine Ridge, around Scotts Bluff, and in the Wildcat Hills. National Breeding Bird Surveys from 1966 to 2015 indicate that this species' population underwent a survey-wide average annual decline of 1.68 percent (Sauer et al. 2017). This species is on the Nebraska Natural Legacy Project's Tier 2 list of threatened species in Nebraska (Schneider et al. 2018).

Habitats and Ecology: This swift is associated with cliffs, canyons, and generally steep terrain, with associated nesting rock crevices.

Breeding Biology: Most white-throated swifts arrive at their nesting areas by mid-May. They soon begin to construct their unique nests, carrying individual feathers in their bills, sometimes apparently for miles. They also then begin their aerial courtship, which might even include copulation while in flight. To initiate copulation, the birds fly toward each other from opposite directions, meet, and begin to tumble downward while clinging together, sometimes falling several hundred feet. The usual clutch size is four to five eggs. It might be presumed that both sexes incubate, but nothing specific is known about incubation behavior. Incubation probably lasts about 20–27 days, but its average duration and the fledging period are still uncertain; the latter is 28 days in the chimney swift. During the nonbreeding period the birds often roost in communal quarters, much like chimney swifts. Observations on one such roosting site in California indicated that the birds, numbering 100–200, returned to their roosting crevice shortly after sunset. Within five minutes the entire flock had entered the crevice, passing through an entry only about 2–3 inches wide! In very cold weather, roosting birds might become torpid, although this is not known to be a regular adaptation of swifts for coping with cold periods.

Suggested Viewing Opportunities: The summit of Scotts Bluff

National Monument is an excellent place from which to watch these amazing birds in flight during summer months.

Selected References: BNA 526 (T. P. Ryan and C. T. Collins 2000); Bent 1940; Chantlier 1995.

Family Trochilidae (Hummingbirds)
RUBY-THROATED HUMMINGBIRD. *ARCHILOCHUS COLUBRIS*

Status: This hummingbird's breeding distribution is centered on the immediate Missouri Valley, with westward extensions into the Niobrara Valley possibly to Keya Paha County, the Elkhorn Valley to Stanton County, the Platte Valley to Platte County, and the Little Blue Valley possibly to Jefferson County. National Breeding Bird Surveys from 1966 to 2015 indicate that this species' population underwent a survey-wide average annual increase of 1.44 percent (Sauer et al. 2017). This species is on the Nebraska Natural Legacy Project's Tier 2 list of threatened species in Nebraska (Schneider et al. 2018).

Habitats and Ecology: A variety of wooded habitats are used by this species, from rather dense to open coniferous and hardwood woodlands and manmade environments (orchards, shade trees). Herbs or shrubs that provide tubular nectar-bearing flowers (honeysuckle, lantana, gilia, trumpet vine, etc.) are an important part of the habitat.

Breeding Biology: Hummingbirds return to their Great Plains breeding grounds in April to late May. Territorial males then advertise by flying back and forth along an arc of a wide circle, frequently passing within a few inches of a perched female at the lowest part of the arc. A male might spend up to two months attracting and mating with available females. Copulation occurs on the ground and is apparently preceded by a period of aerial display by both birds, which hover in the air facing each other as they ascend and descend vertically. The female constructs the nest over several days, finally attaching lichens to the outside and adding plant down for lining. Nests are usually 6–50 feet above the ground on fairly level or downward-slanting twigs or branches that are protected from above by larger branches or a leafy canopy. They are frequently in trees near water, probably because favored flowers often grow there, and are more often placed in hardwood trees than in conifers. Two eggs are laid, and the incubation period is about 16 days. By the time the young are ten days old they are nearly as large as their mother and are fed a combination of nectar and insects by regurgitation. Fledging time records range from 14 to 28 days, the duration probably varying with available food supplies.

Suggested Viewing Opportunities: Setting out hummingbird feeders annually in May and again in August and September is the easiest and most enjoyable way to see hummingbirds up close. Hummingbirds remember the locations of feeders from year to year, and it is not difficult to establish regular seasonal visits if the feeders are tended regularly.

Selected References: BNA 204, rev. ed. (S. Weidensaul, T. R. Robinson, R. R. Sargent, and M. B. Sargent 2013); Bent 1940; Johnsgard 1997b.

Family Rallidae (Rails and Coots)

COMMON GALLINULE. *GALLINULA GALEATA*

Status: This reclusive marshland bird is limited to the wetlands of eastern Nebraska, with recent breeding records extending west to the Rainwater Basin wetlands of Seward and Fillmore Counties (Mollhoff 2016). Nebraska is at the western edge of the species' breeding range; it has not yet colonized the Sandhills wetlands, where suitable habitat is abundant.

Habitats and Ecology: The favored habitat of this species consists of freshwater ponds and marshes with an abundance of emergent vegetation. It prefers small wetlands with abundant aquatic vegetation and dense peripheral cover where there are many weed and grass seeds, as well as aquatic invertebrates and small vertebrates.

Breeding Biology: Common gallinules are monogamous, highly territorial birds and in some areas maintain winter core areas that later expand to become breeding territories. Within the territories the birds build three kinds of structures: display platforms, egg nests, and brood nests. Up to five temporary display platforms are built early in the breeding season, and one or two functional nests are constructed a week or two before egg-laying begins. These might be built on the water surface, suspended above water, or on land surrounded by water. Deep-water nests usually have a ramp up the side, whereas those in shallow water or on land do not. In an Iowa study, 17 of 19 nests were in cattails, while the others were in bulrushes. From five to ten eggs, averaging about seven, are laid daily, and both sexes incubate. The incubation period is 21–22 days, starting (in first clutches) with the penultimate egg or (in later clutches) midway through the laying period. The young of the first brood hatch nearly synchronously and are fed by their parents within an hour after hatching. Up to five brood nests are built after the brood hatches. Both parents tend their young for varying periods; in one case, a pair began a new nest only 26 days after hatching their first brood. The chicks fledge at 60–65 days of age and tend to disperse soon afterward. At least in warmer latitudes the species is a regular renester and sometimes is double-brooded. Nebraska breeding records are few and scattered, from Clay County in the west, east to Otoe County, and north to Dakota County.

Suggested Viewing Opportunities: Even where common gallinules are common, they are difficult to find, although their many somewhat coot-like vocalizations often reveal their presence. Playbacks of their vocalizations are most likely to bring them into view, a technique that works with many rails.

Selected References: BNA 685 (B. K. Bannor and E. Kiviat 2002); Bent 1926; Frederickson 1971; Johnsgard 1975b.

AMERICAN COOT. *FULICA AMERICANA*

Status: Widespread and a common summer resident on wetlands throughout the state, with the greatest numbers in the Sandhills and the Rainwater Basin. National Breeding Bird Surveys from 1966 to 2015 indicate that this species' population underwent a slight average annual increase of 0.62 percent (Sauer et al. 2017).

Habitats and Ecology: This species is associated with ponds and

21. American coot (above) and sora.

marshes that have a combination of open water and emergent beds of cattails, reeds, or rushes, in which nesting occurs. Besides foraging on aquatic plants, the birds sometimes also graze on nearby shorelines and meadows.

Breeding Biology: Coots are monogamous, with potential lifelong pair bonds, and spend much of their summertime in advertising and defending territories. These are established soon after arrival on the breeding grounds, and although the male patrols the territory at first, later both members of the pair defend it. Pairs also construct display platforms for copulation and, as the egg-laying period approaches, construct one or more egg nests, as well as brood nests later on. Both sexes participate in incubation, with the male most often incubating at night. The usual clutch size is six to nine eggs, and incubation lasts 21–24 days. Hatching is typically staggered over several days. Apparently, the male assumes the major responsibility for brooding the young birds, although the female might take charge of the first-hatched chicks and leave the male to incubate and tend the later hatchlings. The chicks start begging shortly after hatching and soon begin to follow the adults during their foraging. After a month or so the young are nearly independent, but they beg occasionally almost to the time they become fully independent, at about eight weeks of age. If the adults begin a second clutch, they might expel the young of the first brood from the area while they are still fairly young.

Suggested Viewing Opportunities: Permanent shallow wetlands throughout the state provide easy viewing opportunities for this common species.

Selected References: BNA 697 (I. L. Brisbin Jr., H. D. Pratt, and T. B. Mowbray 2002); Bent 1926; Frederickson 1970; Johnsgard 1975b.

Family Gruidae (Cranes)

SANDHILL CRANE. *ANTIGONE CANADENSIS*

Status: Lesser sandhill cranes (*Antigone canadensis canadensis*) are abundant spring migrants in the Platte Valley, with numbers in recent years peaking between 500,000 and 800,000 in late March.

22. Adult sandhill crane, male uttering a mild alarm call.

Autumn number are far lower, with the great majority of migrants overflying the Platte Valley. Greater sandhill cranes (*A. c. tabida*) are local and rare summer residents in the more remote wetlands, pastures, and large meadows of the state.

Habitats and Ecology: Sandhill cranes are especially associated with remote marshy wetlands. The birds are highly territorial, and nests usually are well scattered. Their loud calls communicate over long distances and serve to advertise territories. They have bred in the Rainwater Basin since the mid-1990s, as well as in a few northern counties (Rock, Knox) and also some western counties (Sioux, Scotts Bluff, Morrill) since the early 2000s.

Breeding Biology: Cranes are monogamous, usually pairing for life after reaching reproductive maturity at about four years of age. Upon returning to their breeding areas, pairs establish territories as early as two to four weeks before nest-building gets under way. Nest-building is done by both sexes and might require from a day to a week or more. Two eggs are laid within three days. Both sexes participate in incubation, with the female apparently always doing the nighttime incubation. Incubation lasts 30–32 days. The eggs typically hatch 24 hours apart, and the "colts" begin to feed immediately, with the first hatched often taken away from the nest by one adult while the other remains to hatch the second. Perhaps because the young are usually aggressive toward each other, they are often brooded separately. Fledging occurs at 67–75 days of age, and the family soon migrates as a unit. Family bonds persist until the young are sexually mature, and multigeneration family groups often migrate together.

Hunting and Population: National Breeding Bird Surveys from 1966 to 2015 indicate that this species' population underwent a survey-wide average annual increase of 4.74 percent (Sauer et al. 2017), a figure that primarily reflects the flourishing status of the greater race. However, the breeding population of greater sandhill cranes in Nebraska might be fewer than ten pairs. This subspecies is on the Nebraska Natural Legacy Project's Tier 2 list of threatened species in Nebraska (Schneider et al. 2018). The Arctic-breeding lesser sandhill's population is essentially stable or increasing only slowly, probably because of extremely high hunting mortality. "Sport" hunting of the lesser sandhill crane in the United States and Canada annually killed about 50,000 birds in 2018 and has been increasing annually. This mortality nearly cancels out the lesser's annual production, which in recent decades has been excellent, probably owing to current unusually warm and long breeding seasons occurring in the High Arctic. Luckily, no crane hunting is allowed in Nebraska, and the annual income generated in the state each spring from 20,000 to 30,000 crane-watchers traveling from all over the world has been estimated to be some $14–20 million (about $400–1,000 per person). By comparison, over 85,000 small game hunting licenses (to hunt upland game birds, small game mammals, and waterfowl) were sold in Nebraska in 2018. Probably nearly all of these were bought by residents and likely involved smaller associated per-person expenditures than those of crane-watchers.

Suggested Viewing Opportunities: The central Platte Valley, from late February to early April, is simply the best place in the world to watch sandhill cranes, with a half-million birds concentrating mostly in a 40-mile stretch between Kearney and Grand Island.

Selected References: BNA 431, rev. ed. (B. Gerber, J. F. Dwyer, S. A. Nesbitt, R. C. Drewein, C. Littlefield, T. C. Tacha, and P. A. Vohs 2014); Niemeier 1979; Johnsgard 1981b, 1983, 1991, 2011d, 2015b; Ellis et al. 1998; Tacha 1998; Krapu et al. 2014.

WHOOPING CRANE. *GRUS AMERICANA*

Status: An occasional spring and fall migrant in Nebraska, more often seen in spring than in fall. It has been observed in at least 26 counties but has most commonly been seen in Buffalo and Kearney Counties. This species is on the Nebraska Natural Legacy Project's Tier 1 list of threatened species in Nebraska (Schneider et al. 2018) and is a federally listed endangered species.

Habitats and Ecology: While whooping cranes are in Nebraska, the Platte Valley is their primary habitat, and a wide and slow-flowing river, with its numerous sand bars and islands and adjacent wet meadows, grain fields, and marshlands, is evidently an important combination of habitat characteristics. The species migrates later in spring than does the sandhill crane and thus does not normally associate with it. It uses marshy wetlands and similar wet areas for foraging to a much larger degree than does the sandhill crane and rarely forages in cornfields.

Suggested Viewing Opportunities: The vast majority of Nebraska sightings have occurred within 30 miles of the Platte River. The birds forage and roost more often in large wetlands in the Rainwater Basin than along the Platte. The spring migration extends from early March to late May, with a peak during the period April 1–15. No doubt the entire flock of birds (500-plus in 2018) that migrates between Aransas National Wildlife Refuge (Texas) and Wood Buffalo National Park (Alberta) passes through Nebraska each spring and fall. Often individual whooping cranes arrive with early-arriving sandhill cranes in late February and associate with them. The fall migration extends from mid-September to early November, with a peak during the period October 11–25. Harassing whooping cranes is a federal offense; viewers must avoid disturbing them and approach no closer than about 75 yards.

Selected References: BNA 153, rev. ed. (R. P. Urbanek and J. C. Lewis 2015); Ellis et al. 1998; Johnsgard, 2003a, 2011d.

Family Recurvirostridae (Stilts and Avocets)

BLACK-NECKED STILT. *HIMANTOPUS MEXICANUS*

Status: A spring and fall migrant and an uncommon and local summer resident in alkaline wetlands of the Sandhills. National Breeding Bird Surveys from 1966 to 2015 indicate that this species' population underwent a survey-wide average annual increase of 1.75 percent (Sauer et al. 2017). This species is on the Nebraska Natural Legacy Project's Tier 2 list of threatened species in Nebraska (Schneider et al. 2018).

Habitats and Ecology: Breeding in this species usually occurs in the grassy shoreline areas of shallow freshwater or brackish pools of wetlands having extensive mudflats or sometimes along the shore-

23. Adults of black-necked stilt (above) and American avocet.

lines of salt lakes where vegetation is essentially lacking. Stilts are often found in company with American avocets, which use similar habitats. Breeding in Nebraska has most often occurred in Garden (Crescent Lake National Wildlife Refuge) and Sheridan (Lakeside) Counties, but there have also been records from Dawes and Scotts Bluff Counties and the Rainwater Basin (Clay, York, and Phelps Counties).

Breeding Biology: Like avocets, stilts form pair bonds gradually and without associated elaborate displays, mainly through the persistent association of a female with a particular male, in spite of initial aggressiveness by the male. Stilts defend territories on their breeding grounds and advertise them by aerial displays. Ritualized breast-preening by both sexes, apparently identical to that typical of avocets, precedes copulation by stilts. Nest-building is probably done by both sexes, and materials are added to the nest through incubation. During periods of rising water, materials added by the pair can raise the nest considerably, and both sexes apparently share incubation about equally. The clutch size is three to five eggs, usually four. Incubation begins when the penultimate or last egg is laid and lasts 25 days. The eggs hatch relatively synchronously, and the young remain in the nest no more than 24 hours. They are probably brooded for at least a week and become independent at about four weeks.

Suggested Viewing Opportunities: Breeding records in recent decades have occurred in the western Sandhills (Sheridan and Garden Counties) and in the Rainwater Basin (Phelps to Seward Counties).

Selected References: BNA 449 (J. A. Robinson, J. M. Reed, J. P. Skorupa, and L. W. Oring 1999); Bent 1927; Hamilton 1975; Johnsgard 1981a.

AMERICAN AVOCET. *RECURVIROSTRA AMERICANA*

Status: A locally common summer resident over much of the state, mainly on shallow marshes of the western Sandhills and, to a lesser extent, the Rainwater Basin. National Breeding Bird Surveys from 1966 to 2015 indicate that this species' population underwent a slight average annual decline of 0.31 percent (Sauer et al. 2017). This species is on the Nebraska Natural Legacy Project's Tier 2 list of threatened species in Nebraska (Schneider et al. 2018).

Habitats and Ecology: During breeding this species favors ponds or shallow lakes with exposed and sparsely vegetated shorelines and somewhat saline waters that have large populations of aquatic invertebrates, which the birds gather by performing scythe-like movements of the curved bill through the water and extracting surface-dwelling invertebrates.

Breeding Biology: In Oregon, avocets arrive on their breeding areas 15–20 days before egg-laying to establish territories and begin courtship. They apparently form pairs in late winter without associated elaborate posturing. Copulation is preceded by a rather simple breast-preening ceremony that might be initiated by either bird.

Pairs form close bonds and forage together, and they defend their territory as a unit. Both sexes develop incubation patches and begin to incubate their clutch as soon as it is completed. The clutch size is three to five eggs, usually four. Early in incubation the male spends more time on the nest than the female, but the female is more attentive later on. Incubation lasts 22–24 days. The eggs hatch over a period of one to two days, and the young soon become very active, feeding themselves almost from hatching. They fledge in four to five weeks, and thereafter the families begin to form flocks, which remain intact until the following breeding season.

Suggested Viewing Opportunities: Wetlands in southern Sheridan and northern Garden Counties have had the greatest number of avocet breeding records in recent decades. Wetlands in Clay County have also had a few breeding records, and nests have also been found in Cherry, Dawes, Dawson, Morrill, and Keith Counties.

Selected References: BNA 275, rev. ed. (J. Ackerman, C. A. Hartman, M. P. Harzog, J. V. Takekawa, J. A. Robinson, L. W. Oring, J. P. Skorupa, and R. Boettcher 2013); Bent 1927; Hamilton 1975; Johnsgard 1981a.

Family Charadriidae (Plovers)

KILLDEER. *CHARADRIUS VOCIFERUS*

Status: The killdeer is a common summer resident, remaining well into fall and arriving early in spring. A total of 510 possible to confirmed breeding records was reported for every county during fieldwork for *The Second Nebraska Breeding Bird Atlas,* and the species was judged to be the state's tenth most common breeding bird (Mollhoff 2016). National Breeding Bird Surveys from 1966 to 2015 indicate that this species' population underwent a survey-wide average annual decline of 1.09 percent (Sauer et al. 2017). Nebraska surveys indicate a similar 0.8 percent annual decline, based on a sample of 51 routes.

Habitats and Ecology: This species is widely distributed in open landscapes, including roadsides, reservoirs, ponds, gravel pits, golf courses, and suburban lawns. Gravelly areas with rocks about the size and color of the birds' eggs are favored nesting sites. Nesting on rooftops sometimes occurs where gravelly habitats are absent. Killdeers forage visually on surface-dwelling insects such as beetles rather than probing for invisible foods in the manner of sandpipers and snipes.

Breeding Biology: During pair formation the male makes a series of shallow scrapes on the ground, the last of which is used by the female for depositing her eggs. Stones that are 0.2–0.4 inches (5–10 mm) in diameter are selectively used for nest lining. Nearly all killdeer nests have clutches of four eggs, which typically are about the same size and color as the stones lining the nest. They are incubated by both sexes for 21–22 days, and initial flights occur about 30 days after hatching. Nesting in Nebraska extends from about early May to late June, only enough time for a single breeding effort. Killdeers remain north long into the fall and also are usually the first shorebirds to arrive in Nebraska in spring.

Suggested Viewing Opportunities: This abundant plover can be seen on grassy meadows or shorelines almost anywhere in the state.

Selected References: BNA 517 (B. J. Jackson and J. A. Jackson 2000); Bent 1929; Johnsgard 1981a.

PIPING PLOVER. *CHARADRIUS MELODUS*

Status: Breeds uncommonly and locally in eastern Nebraska along several river systems, including the Missouri, the lower Niobrara, the central and lower Platte, and the North Loup, Middle Loup, and Loup. Borrow-pit and sand-pit lakes are also used to a substantial degree. There were 21 possible to confirmed nesting records during the fieldwork for *The Second Nebraska Breeding Bird Atlas* (Mollhoff 2016). This species is on the Nebraska Natural Legacy Project's Tier 1 list of threatened species in Nebraska and is listed as state and federally classified as threatened (Schneider et al. 2018).

Habitats and Ecology: Piping plovers are associated with sparsely vegetated shorelines of shallow lakes and impoundments, especially those that have salt-encrusted areas of gravel, sand, or pebbly mud. Sand dunes with little or no vegetation are also used for nesting.

Breeding Biology: Piping plovers are monogamous, but mate-changing in successive breeding seasons is fairly frequent, even when the original mate is still available. Their nests are simple hollows in the sand, sometimes lined with pebbles, or scrapes in gravel or pebbly mud. The clutch size varies from two to four eggs (typically four in first clutches). Incubation ranges from 27 to 29 days, starting with either the third or last egg. Eggs are laid on alternate days, and incubation responsibilities are about equally divided by the two sexes. In most nests the eggs all hatch on the same day, and within two or three hours the young have dried off and are able to leave the nest. They are brooded by both adults until they are about 20 days old, and although they can run very well they tend to crouch and "freeze" when approached. Adults of both sexes feign injury when their brood is threatened. Until they fledge at 30–35 days of age, the young remain within 400–500 feet of the nest. The species is single-brooded, but renesting usually occurs if the clutch is lost in the first half of the breeding season.

Suggested Viewing Opportunities: Many breeding efforts in recent decades have occurred on the lower Niobrara and lower Platte, where the river is slow and wide and has many sandy bars and islands. The population at Lake McConaughy has exceeded 200 pairs during summers when the reservoir has been low and large sandy areas are exposed (Brown, Dinsmore, and Brown 2012). The Nebraska-nesting birds winter on the Gulf Coast.

Selected References: BNA 2, rev. ed. (E. Elliott-Smith and S. M. Haig 2004); Bent 1929; Wilcox 1959; Stout 1967; Johnsgard 1981a.

MOUNTAIN PLOVER. *CHARADRIUS MONTANUS*

Status: A local uncommon summer resident on the drier short-grass plains of southwestern Nebraska. National Breeding Bird Surveys from 1966 to 2015 indicate that this species' population underwent a survey-wide average annual decline of 3.41 percent (Sauer et al. 2017). This species is on the Nebraska Natural Legacy Project's Tier 1 list of threatened species in Nebraska (Schneider et al. 2018). There were three possible to confirmed nesting records during the fieldwork for *The Second Nebraska Breeding Bird Atlas* (Mollhoff 2016).

Habitats and Ecology: This species breeds exclusively in early spring on arid grasslands where grasses are usually no more than 3 inches in height and sometimes in semidesert areas with cacti and scattered shrubs far from water. During the nonbreeding seasons the birds are also found in relatively dry habitats.

Breeding Biology: In northeastern Colorado, mountain plovers arrive in late March and soon disperse over their shortgrass breeding grounds. Males commonly reestablish their old territories, whereas females also return to the same general area but might visit several territories before choosing mates. Territorial males advertise with calls and an aerial "falling-leaf" display and occasionally with a slow "butterfly flight." As in other plovers, scraping of the often sandy substrate is the most frequent courtship display of the male on several potential nest sites throughout his territory. The clutch size is usually three eggs, and incubation lasts 28–31 days. At least some females begin a second clutch with new mates within about two weeks of completing their first clutches, leaving their first mates to attend to their original nests. Evidently, the female often incubates the second clutch herself, as only one sex is involved in incubation and brooding duties for each clutch and brood.

Suggested Viewing Opportunities: Nearly all of Nebraska's breeding records have occurred on dry croplands in Kimball County.

Selected References: BNA 211, rev. ed. (F. L. Knopf and M. B. Wunder 2006); Bent 1929; Graul 1975; Johnsgard 1981a; Dinsmore 2001; Ellison and White 2001; Bly, Snyder, and VerCauteren 2008.

SNOWY PLOVER. *CHARADRIUS NIVOSUS*

Status: A local and very rare summer resident, the snowy plover is largely limited to saline flats and sandy riverbeds in several river valleys, including the Niobrara (Knox County), the Missouri (Cedar County), the North Platte (Lake McConaughy, Keith County), the Platte (Rowe Sanctuary, Buffalo County), and the Republican (Harlan County Lake and Swanson Reservoir, Hitchcock County). By far the greatest number of recent nestings has occurred along the shoreline of Lake McConaughy. There were four possible to confirmed nesting records during the fieldwork for *The Second Nebraska Breeding Bird Atlas* (Mollhoff 2016).

Habitats and Ecology: Barren salt plains represent prime breeding habitat for this arid-adapted species, and sandy riverbeds or bare shorelines of reservoirs are used secondarily.

Breeding Biology: After arriving on their breeding areas and establishing territories, males begin to advertise with various calls and displays, including scraping, a ritualized nest-building behavior involving scraping out a depression in the earth. One of the other male displays is a slow "butterfly flight," accompanied by a trilling call. Although the birds commonly breed around salt water, they cannot drink saline water and must obtain liquid by eating insects or other succulent foods. Thermal extremes are also common in their often vegetation-free and highly reflective environment. Thus, during hot weather parental activity increases. The birds spend most of their time standing over the eggs or chicks rather than sitting on them. The clutch size is two to four eggs, usually four. The eggs are laid about three days apart, and both sexes incubate. Incubation lasts 24 days, and hatching is synchronous. Both sexes also defend the eggs and young, performing effective broken-wing diversion behavior when threatened. The young fledge in 27–31 days.

Suggested Viewing Opportunities: Bare sandy shores at Lake McConaughy provide the best opportunities for seeing this rare species.

Selected References: BNA 154, rev. ed. (C. W. Page, L. E. Stenzel, J. S. Warriner, J. C. Warriner, and P. W. Paton 2009); Bent 1929; Johnsgard 1981a.

Family Scolopacidae (Sandpipers, Snipes, and Phalaropes)

UPLAND SANDPIPER. *BARTRAMIA LONGICAUDA*

Status: An uncommon summer resident nearly throughout the state but rarer in the easternmost counties and in southwestern Nebraska. The greatest numbers occur in the Sandhills grasslands (hayfields and pastures) in habitat patches of at least 100 acres.

24. Upland sandpiper, adult incubating.

National Breeding Bird Surveys from 1966 to 2015 indicate that this species' population underwent a slight average annual increase of 0.4 percent (Sauer, et al. 2017). There were 313 possible to confirmed nesting records during the fieldwork for *The Second Nebraska Breeding Bird Atlas* (Mollhoff 2016).

Habitats and Ecology: This species is generally associated with wet meadows, hayfields, mowed prairies, or midlength prairies, avoiding both shortgrass steppe areas and extremely tall grasses. It is often found far from water and rarely if ever wades for its foods.

Breeding Biology: In the Great Plains the first spring arrivals appear about two weeks before the start of nesting in May and are usually already paired. Territorial birds perform a flight display consisting of circling with quivering wingbeats while uttering a musical purring or "wolf-whistle" call and finally diving abruptly back to the earth. Nesting begins almost simultaneously with aerial display, and the eggs are laid at approximately daily intervals. The clutch size is usually four eggs, and incubation lasts 21 days. Both sexes incubate, and incubating adults typically feign injury when disturbed. There is a fairly long interval between the first-laid eggshell's initial cracks and the hatching of the last egg, which might vary from less than 24 hours to about three days. The chicks are brooded by both parents, and by the time they are 30 days old they appear to be fully grown and are virtually fledged.

Suggested Viewing Opportunities: Large native meadows and hayfield in the Sandhills are prime places to search for this iconic prairie species. It is usually present from early May until August, when it undertakes an approximate 8,000-mile migration to Argentine wintering areas.

Selected References: BNA 580, rev. ed. (C. S. Houston, C. Jackson, and D. E. Bowen Jr. 2011); Bent 1929; Higgins and Kirsch 1975; Johnsgard 1981a, 2001b.

LONG-BILLED CURLEW. *NUMENIUS AMERICANUS*

Status: An uncommon summer resident in grassland areas in the central and western Sandhills and on the grassy tablelands of the northern Panhandle. Nearly all recent breeding records are from the western half of the state, and grassland patches of 10,000 acres or more are favored nesting habitats. There were 61 possible to confirmed nesting records during the fieldwork for *The Second Nebraska Breeding Bird Atlas* (Mollhoff 2016). National Breeding Bird Surveys from 1966 to 2015 indicate that this species' population underwent a slight average annual increase of 0.17 percent (Sauer et al. 2017). This species is on the Nebraska Natural Legacy Project's Tier 1 list of threatened species in Nebraska (Schneider et al. 2018).

Habitats and Ecology: On the breeding grounds this species occurs in shortgrass sites, grazed taller grasslands, and overgrazed grasslands with scattered shrubs or cacti. Hilly or rolling areas seem to be favored over flat land, and the birds often nest rather far from standing water. However, migrating and wintering birds are usually found on sandy beaches and shorelines.

Breeding Biology: In the Nebraska Sandhills, long-billed curlews arrive by early April, usually in flocks of fewer than 12 birds. The rest of the month is spent in prenesting activities, including establishing core areas and foraging areas. Core areas typically consist of rolling sandhills and are advertised by extended flight displays and calling above the ultimate nest site. Meadows adjacent to nesting locations are used for foraging and are advertised by similar flight displays. The foraging area is a part of the defended territory, and other curlews are forcibly excluded from it. The clutch size is usually four eggs. Both sexes incubate, and incubation lasts 27–28 days. Both sexes care for the brood, and the fledging period is about 30 days. When a nest or brood is disturbed, the loud alarm calls of the resident pair will quickly attract nearby pairs to help in distraction behavior.

Suggested Viewing Opportunities: Very large, often hilly hayfields with nearby wet meadows are favored habitats in the Sandhills. The birds are present from July or August, leaving shortly after their young fledge. The Nebraska breeding population evidently winters along the Gulf Coast, based on a few radio-tracked birds.

Selected References: BNA 628 (B. D. Dugger and K. M. Dugger 2002); Bent 1929; Forsythe 1972; Bicak 1977; Fitzner 1978; Johnsgard 1981a, 2001b.

BAIRD'S SANDPIPER. *CALIDRIS BAIRDII*

Status: A common spring and fall migrant throughout the state; one of the most abundant of the small "peep" sandpipers in Nebraska.

Habitats and Ecology: Migrants are associated with wet meadows and shallow ponds. They often feed in grassy areas somewhat away from water but also forage along muddy shorelines, where they tend to peck at surface food sources rather than to probe for them.

Suggested Viewing Opportunities: The eastern Rainwater Basin is perhaps the best place to look for this and other small sandpipers, where spring counts of up to 6,400 Baird's sandpipers, 3,200 semipalmated sandpipers, and 3,200 white-rumped sandpipers have been made (Jorgensen 2012a). Most spring sightings have historically occurred between April 21 and May 13, and fall sighting records are mostly from August 12 to October 6.

Selected References: BNA 661 (W. Moskoff and R. Montgomerie 2002); Bent 1927; Johnsgard 1981a.

LEAST SANDPIPER. *CALIDRIS MINUTILLA*

Status: A common spring and fall migrant throughout the state, becoming less common westward.

Habitats and Ecology: While on migration these sandpipers are found on a variety of moist habitats, often in company with semi-palmated, Baird's, or western sandpipers and probably feeding on much the same small invertebrates as these species.

Suggested Viewing Opportunities: The Rainwater Basin and Sandhills wetlands are major habitats for migrating sandpipers. Most spring sightings records fall between May 2 and May 12, and fall sightings are centered from August 2 to September 18.

Selected References: BNA 115, rev. ed. (S. Nebel and J. M. Cooper 2008); Bent 1927; Johnsgard 1981a.

LONG-BILLED DOWITCHER. *LIMNODROMUS SCOLOPACEUS*

Status: A common spring and fall migrant throughout the state. It is the more common of the two dowitcher species.

Habitats and Ecology: Migrating birds use marshy habitats that provide for foraging at wading depths.

Suggested Viewing Opportunities: Spring counts of up to 1,500 long-billed dowitchers have been made in the eastern Rainwater Basin (Jorgensen 2012). Most spring sighting records fall between May 1 and May 11, and fall sightings are centered from August 8 to October 14.

Selected References: BNA 493 (J. Y. Takekawa and N. Warnock 2000); Bent 1927; Johnsgard 1981a.

SPOTTED SANDPIPER. *ACTITIS MACULARIA*

Status: A common summer resident throughout the state, but with lower populations in the western third of the state. National Breeding Bird Surveys from 1966 to 2015 indicate that this species' population underwent a survey-wide average annual decline of 1.35

25. Long-billed dowitchers, adults in flight.

percent (Sauer et al. 2017). There were 157 possible to confirmed nesting records during the fieldwork for *The Second Nebraska Breeding Bird Atlas* (Mollhoff 2016).

Habitats and Ecology: This species is associated with forest streams, pools, and rivers. It utilizes a wide diversity of terrains near water and rarely occurs in the absence of nearby water. Shaded watercourses are favored, and sometimes the birds are found along rather rapidly flowing streams.

Breeding Biology: Male and female spotted sandpipers arrive on their breeding grounds at about the same time. Pair bonds are formed extremely rapidly during a period of intense aggression, especially among females, which are larger and more aggressive than males. Females establish territories, and males form pair bonds by entering such territories and being either accepted or expelled by unmated females. The female might lay her first egg as soon as five days after the male's arrival. The clutch size is usually four eggs, which are laid at approximately daily intervals. By the time she lays her third egg the female begins to show a resurgence of sexual activity, with increased singing and territoriality. Although some females remain monogamous and assist with incubation, others leave their first mate to finish any remaining incubation duties and find a second mate. Successive mating with as many as four mates in a single season has been found, and typically the female helps incubate the final clutch. The young birds leave the nest as soon as their feathers dry and are able to fly as early as 13–16 days after hatching.

*Suggested Viewing Opportunities: T*his species is likely to be found along rocky or grassy shorelines of streams, with lakeshores and pond edges of seemingly lesser attraction.

Selected References: BNA 289, rev. ed. (J. M. Reed, L. W. Oring, and E. M. Gray 2013); Bent 1927; Oring and Knudson 1973; Johnsgard 1981a.

LESSER YELLOWLEGS. *TRINGA FLAVIPES*

Status: A common spring and fall migrant nearly throughout the state. National Breeding Bird Surveys from 1966 to 2015 indicate that this species' population underwent a survey-wide average annual decline of 1.87 percent (Sauer et al. 2017).

Habitats and Ecology: Outside the breeding season these birds occur along mudflats and shallow ponds, often with vegetated shorelines, and they sometimes also visit flooded fields. Breeding typically occurs in boreal habitats that have a combination of rather open and tall woodlands with low and sparse brushy undergrowth and are fairly close to grassy or marshy ponds.

Suggested Viewing Opportunities: Lesser yellowlegs occur on the same kinds of wetlands as are used by the greater, namely, marshy edges and shallow wetlands. Most spring sighting records fall between April 14 and May 13, and fall sightings are centered from August 16 to October 5, although yellowlegs are often the first of the northern breeding shorebirds to arrive in July.

Selected References: BNA 427 (T. L. Tibbits and W. Moskoff 1995); Bent 1927; Johnsgard 1981a.

GREATER YELLOWLEGS. *TRINGA MELANOLEUCA*

Status: A common spring and fall migrant throughout the state. National Breeding Bird Surveys from 1966 to 2015 indicate that this species' population underwent a survey-wide average annual increase of 3.17 percent (Sauer et al. 2017).

Habitats and Ecology: In migration these birds occupy the edges of marshes and slow-moving rivers, foraging along the shorelines and sometimes wading out belly-deep to probe in the mud or skim the surface for invertebrates. On their boreal breeding grounds the birds favor muskeg areas having a mixture of ponds, trees, and clearings.

Suggested Viewing Opportunities: Almost any shallow, fairly open wetland with a shoreline of sand or mud seems to attract this species during migration. Most spring sightings fall between April 13 and May 5, and fall sightings are centered from August 17 to October 8.

Selected References: BNA 355, rev. ed. (T. L. Tibbitts and W. Moskoff 2014); Bent 1927; Johnsgard 1981a.

BUFF-BREASTED SANDPIPER. *TRYNGITES SUBRUFICOLLIS*

Status: An uncommon spring and fall migrant in eastern Nebraska. This Arctic-breeding species is highly local in distribution during migration. Joel Jorgensen determined that Nebraska is probably the species' most important spring staging area, supporting 22,000–78,000 buff-breasted sandpipers that stage in agricultural fields in the eastern Rainwater Basin during May. They also occur in much smaller numbers during fall (late July to late September) (Jorgensen 2012).

Habitats and Ecology: Migrants are usually found on recently plowed fields, mowed or burned grasslands, meadows, heavily grazed pastures, and other rather dry habitats.

Suggested Viewing Opportunities: Jorgensen (2004, 2012) observed maximum numbers during the second and third weeks of May. He found this species to be regular in Seward and Fillmore Counties of the eastern Rainwater Basin, especially in soybean stubble. It has also been reported several times in York and Lancaster Counties.

Selected References: BNA 91, rev. ed. (J. P. McCarty, L. L. Wolfenbarger, C. D. Laredo, P. Pyle, and R. B. Lanctot 2017); Johnsgard 1981a; Jorgensen 2004, 2012.

WILSON'S PHALAROPE. *PHALAROPUS TRICOLOR*

Status: A common statewide spring and fall migrant and local summer resident over most of Nebraska's wetlands, especially the Sandhills. Breeding Bird Surveys from 1966 to 2015 indicate that this species' population underwent a slight average annual decline of 0.48 percent (Sauer et al. 2017). There were 64 possible to confirmed nesting records during the fieldwork for *The Second Nebraska Breeding Bird Atlas* (Mollhoff 2016).

Habitats and Ecology: Breeding habitats are typically wet meadows adjoining shallow marshes that range from fresh to highly saline. Ditches, river edges, and shallow lakes are sometimes also used for breeding. Migrating birds use similar habitats.

Breeding Biology: Although female phalaropes are appreciably larger and more brightly colored than males, recent studies have cast doubt on the idea that they are regularly polyandrous. Pair bonds apparently are formed after the birds arrive on the breeding areas during a period of behavior that is intensely aggressive but scarcely indicative of typical territoriality. The female probably makes the nest scrape after the pair is formed, but the male adds the nest lining. The clutch size is usually four eggs. Eggs are laid about 48 hours apart, and presumably the female plays no further role in parental care. Incubation lasts 20–21 days. The male alone incubates, and he leads his brood from the nest to foraging areas only a few hours after they hatch. The fledging period is probably less than three weeks.

Suggested Viewing Opportunities: Jorgensen (2004) found this species to be the second most abundant spring migrant shorebird in the eastern Rainwater Basin, where spring counts of up to 1,700 birds have been documented (Jorgensen 2012). Nesting there has occurred several times since 2000. It is a fairly common breeder in the Sandhills, with Cherry County having the largest number of recent breeding records. This species breeds commonly around Border Lake, a highly alkaline lake at the western edge of Crescent Lake National Wildlife Refuge, Garden County. It has also recently bred in Fillmore County (Sora Wildlife Management Area), Saunders County (Jack Sinn Memorial Wildlife Management Area), and probably in Lancaster County.

Selected References: BNA 83 (M. A. Colwell and J. R. Jehl Jr. 1994); Bent 1927; Kangarise 1979; Johnsgard 1981a, 2001b; Bomberger 1982.

RED-NECKED PHALAROPE. *PHALAROPUS LOBATUS*

Status: An uncommon to rare spring migrant in northern and western Nebraska; it is more common in western parts of the state. It is less common during fall in all regions.

Habitats and Ecology: Migrants use the same habitats as do Wilson's phalaropes, namely, open-water areas of marshes and shallow lakes where invertebrate life is abundant and can be captured by surface foraging.

Suggested Viewing Opportunities: Western Nebraska's alkaline marshes are the best places to search for this species during spring migration in mid-May.

Selected References: BNA 538 (M. A. Rubega, D. Schamel, and D. M. Tracey 2000); Bent 1927; Johnsgard 1981a.

Family Laridae (Gulls and Terns)

LEAST TERN, *STERNA ALBIFRONS*

Status: Breeds locally in the Missouri, Platte, and lower Niobrara Valleys, west in the Niobrara Valley to Brown County, and in the Platte Valley to Dawes County. Also breeds locally along the North and South Platte Rivers, especially at Lake McConaughy, and along the lower Elkhorn and lower Loup drainages. The Great Plains race, *Sterna albifrons athalassos,* is federally listed as endangered. This species is on the Nebraska Natural Legacy Project's Tier 1 list of threatened species in Nebraska (Schneider et al. 2018). There were 53 possible to confirmed nesting records during the fieldwork for *The Second Nebraska Breeding Bird Atlas* (Mollhoff 2016).

Habitats and Ecology: In our region nearly all nesting occurs on river sandbars or islands, sandy lake shorelines, and the shorelines of impounded gravel mines or borrow-pit lakes.

Breeding Biology: During courtship a bird may make aerial glides while carrying fish, then alight and offer the fish to another bird. Sex recognition may be achieved in this way; if a male is offered a fish, it responds by attacking. Incipient nest-building by the male may stimulate the female to begin the actual nest. Solitary nesting is frequent in Nebraska. Nests consist of a simple scrape in sand or gravel, with little or no lining. Nest sites are usually widely spaced, lessening antagonism between nesting pairs. From two to four eggs are laid, typically two. The eggs are laid on consecutive days or at two-day intervals, and incubation probably begins with the first egg. At first the female incubates alone, but gradually the male assumes part of this duty. The incubation period is 20–21 days. The eggs typically hatch on consecutive days, and the female does most of the brooding. Within a day the chick and parent have learned to recognize each other, and thus the parents feed no young other than their own. Within two days after hatching the young begin to wander away from the nest and usually do not return. They fledge on about the twentieth day after hatching, and the colony is gradually deserted. Although the least tern is single-brooded, renesting is typical early in the season.

Suggested Viewing Opportunities: Lake McConaughy provides the best viewing opportunities in spring and summer, but disturbance of nesting sites is prohibited.

Selected References: BNA 290, rev. ed. (B. C. Thompson, J. A. Jackson, J. Burger, L. A. Hill, E. M. Kirsch, and J. L. Atwood 1997); Bent 1921; Jenniges and Plettner 2008; Brown, Dinan, and Jorgensen 2016.

BLACK TERN. *CHLIDONIAS NIGER*

Status: A common, almost statewide migrant but increasingly rare summer resident in the state, mainly in Sandhills marshes. National Breeding Bird Surveys from 1966 to 2015 indicate that this species' population underwent a survey-wide average annual decline of 1.37 percent (Sauer et al. 2017). This species is on the Nebraska Natural Legacy Project's Tier 1 list of threatened species in Nebraska (Schneider et al. 2018). There were 51 possible to confirmed nesting records during the fieldwork for *The Second Nebraska Breeding Bird Atlas* (Mollhoff 2016).

Habitats and Ecology: Typical nesting habitat consists of small to large marshes with extensive stands of emergent vegetation and some open water. Fish populations are not necessary, as the birds feed mostly on insects while on the nesting grounds. Nests are more often placed among emergent vegetation than on solid substrates, although muskrat houses are sometimes used. Nesting has occurred in 2006 at Crescent Lake National Wildlife Refuge, and there is a 2007 breeding record for Funk Waterfowl Production Area, Phelps County, but more recent positive nesting records seem to be lacking.

Breeding Biology: Prenesting behavior in black terns is marked by two types of display flights, including "fish flights" (the birds usually carry insects rather than fish), normally performed by two birds, and "flock flights," involving most or all of the birds of an entire nesting area. In the courtship phase, one bird (probably the male) postures and calls while standing on a potential nest site. The two birds also make aerial glides downward from several hundred feet while maintaining a fixed position relative to each other. The nests are built from materials gathered in the immediate vicinity of the nest rather than carried in from a distance, but nesting sites of the previous year apparently are not reused. The clutch size is usually three eggs, and incubation lasts 20–22 days. Both sexes assist in incubation, and both brood the young for at least eight days after hatching. Little brooding is done thereafter, although the chicks are unable to fly until they are more than 20 days old. Nestlings are fed almost exclusively with insects and continue to feed on them after they fledge.

Suggested Viewing Opportunities: The Sandhills wetlands are the most likely, although increasingly uncertain, places to see black terns during the breeding season. Lake Ogallala once attracted migrating black terns by the thousands, but numbers are now much smaller. Migrants might also be seen anywhere in the Rainwater Basin or across central Nebraska.

Selected References: BNA 147 (E. H. Dunn and D. A. Agro 1995); Bent 1921; Goodwin 1960; Bergman, Swain, and Weller 1970.

FORSTER'S TERN. *STERNA FORSTERI*

Status: An uncommon statewide migrant and summer resident mainly on Sandhills marshes. Confirmed breeding records for the past several decades have been limited to sites in Rock, Cherry, Sheridan, Morrill, and Grant Counties. Breeding Bird Surveys from 1966 to 2015 indicate that this species' population underwent a survey-wide average annual decline of 1.6 percent (Sauer et al. 2017). This species is on the Tier 2 list of the Nebraska Natural Legacy Project's threatened species in Nebraska (Schneider et al. 2018). There were 14 possible to confirmed nesting records during the fieldwork for *The Second Nebraska Breeding Bird Atlas* (Mollhoff 2016).

Habitats and Ecology: Large marshes having extensive beds of rushes and cattails or muskrat houses for nest sites are the typical breeding habitats of this species, which breeds colonially in such locations, with as many as five nests sometimes situated on a single muskrat house. Such sites that are close to open-water areas and have surrounding wet meadows for foraging are especially favored nesting locations.

Breeding Biology: Shortly after they arrive on their nesting marshes, Forster's tern pairs begin to seek out nest sites. They are relatively colonial, and as many as five nests might be placed on a favorable dry site, such as atop a large muskrat house. All members of a colony initiate nest-building almost simultaneously. The clutch size is usually two or three eggs, incubation lasts 23–25 days, and both sexes incubate. Wind and wave action, house-building activities by muskrats, and possibly intraspecific hostility are likely major causes of egg losses, which seem to be relatively high in this species. Little information is available on the growth of the young, but presumably they fledge in less than a month, as is typical of the common tern.

Suggested Viewing Opportunities: In the western Sandhills, wetlands with extensive stands of emergent vegetation and some areas of open water are most likely to support Forster's terns. The largest populations are probably in the saline wetlands of northern Garden County and southern Sheridan County.

Selected References: BNA 595, rev. ed. (S. R. Heath, E. H. Dunn, and D. J. Agra 2009); Bent 1921; Bergman, Swain, and Weller 1970.

Family Gaviidae (Loons)

COMMON LOON. *GAVIA IMMER*

Status: An uncommon spring and fall migrant on large and deep wetlands, rivers, and reservoirs across the state.

Habitats and Ecology: Loons typically are found on clear reservoirs and lakes where fish are abundant and human disturbance is minimal. National Breeding Bird Surveys from 1966 to 2015 indicate that this species' population underwent a slight average annual increase of 0.66 percent (Sauer et al. 2017).

Suggested Viewing Opportunities: Loons can often be seen on Lake McConaughy and Lake Ogallala, especially below the Kingsley Dam spillway, mostly during spring and fall but even during summer. These individuals are usually in immature or winter plumage.

Selected References: BNA 313, rev. ed. (D. C. Evers, J. D. Paruk, J. W. McIntyre, and J. F. Barr 2010); Bent 1919; Olson and Marshall 1952; McIntyre 1988.

Family Phalacrocoracidae (Cormorants)

DOUBLE-CRESTED CORMORANT. *PHALACROCORAX AURITUS*

Status: A common spring and fall migrant and local summer resident in the Sandhills and North Platte Valley wetlands and locally common elsewhere. National Breeding Bird Surveys from 1966 to 2015 indicate that this species' population underwent a survey-wide average annual increase of 3.76 percent (Sauer et al. 2017).

Habitats and Ecology: This species is associated with lakes and rivers with good fish populations and often nests on islands or in flooded trees.

Breeding Biology: Cormorants are at least seasonally monogamous and usually breed initially when three years old. Courtship occurs on water and includes much chasing and diving. Males choose their small territories, which include a nest site and adjacent perching spot. Copulation occurs on the nest, mainly during the nest-building period. The clutch size is usually three to four eggs, and incubation lasts 25–29 days. Both sexes assist in incubation, which begins before the clutch is complete; thus, hatching is staggered over several days. The young leave the nest by about six weeks but continue to be fed by their parents until about nine weeks of age, when family bonds disintegrate.

Suggested Viewing Opportunities: During migration, marshes, reservoirs, and lakes throughout the state, especially those with good fish populations, attract this species. Breeding records in recent decades have occurred in Garden, Cherry, Lincoln, Keith, and Morrill Counties. Breedings at Goose Lake (Crescent Lake National Wildlife Refuge) and Valentine National Wildlife Refuge, Cherry County, have occurred frequently.

Selected References: BNA 441, rev. ed. (B. S. Dorr, J. J. Hatch, and D. V. Weseloh 2014); Johnsgard 1993.

26. Adults of American bittern (top left), pied-billed grebe (top right), double-crested cormorant (middle left), Neotropic cormorant (middle right), and American white pelican (bottom).

Family Pelecanidae (Pelicans)

AMERICAN WHITE PELICAN. *PELECANUS ERYTHRORYHNCHOS*

Status: A common spring and fall migrant and a local nonbreeding summer visitor. National Breeding Bird Surveys from 1966 to 2015 indicate that this species' population underwent a survey-wide average annual increase of 4.82 percent (Sauer et al. 2017).

Habitats and Ecology: This species is associated with lakes and rivers having large fish populations that can be reached by surface feeding. Gregarious, the birds typically forage and nest in groups and sometimes forage 10–20 miles away from their nesting grounds.

Suggested Viewing Opportunities: During migration, this species is attracted to marshes, reservoirs, and lakes throughout the state, especially those with good fish populations, such as Harlan County and Calamus Reservoirs, where it is not unusual to see 500–700 birds in May. During summer, immature or otherwise nonbreeding birds can usually be found at Lake McConaughy, Lake Ogallala, and other larger reservoirs.

Selected References: BNA 57, rev. ed. (F. L. Knopf and R. M. Evans 2004); Schaller 1964; Johnsgard 1993.

Family Ardeidae (Herons and Egrets)

AMERICAN BITTERN. *BOTAURUS LENTIGINOSUS*

Status: A widespread but uncommon and inconspicuous summer resident. Breeds locally, especially along overgrown edges of dense reedy marshes, in the Sandhills and Rainwater Basin. National Breeding Bird Surveys from 1966 to 2015 indicate that this species' population underwent a survey-wide average annual decline of 0.52 percent (Sauer et al. 2017). There were 46 possible to confirmed nesting records during the fieldwork for *The Second Nebraska Breeding Bird Atlas* (Mollhoff 2016).

Habitats and Ecology: Bitterns are associated with reed beds, cattails, and other emergent marsh vegetation and rarely observed feeding in open water in the manner of other herons. Foods include frogs, snakes, and other animal life in addition to fish, and thus the species is not limited to areas where fish occur.

Breeding Biology: Relatively little is known of the social behavior of this elusive bird, but males evidently establish and advertise territories with their distinctive "pumping" call, especially at dawn and dusk. Females are attracted to such territories and form possibly polygamous pair bonds. I once observed a copulation that occurred on open ground after the male had exposed a pair of normally hidden white and airy "shoulder" plumes that resembled two miniature wings. He then persistently advanced toward the female while repeatedly lowering and swaying his head from side to side. After overtaking the retreating female, he simply climbed on her back, grasped her nape, and copulated. About 40 years later I again observed this same amazing exhibition of the male's egret-like plumes, a display that has remained virtually undescribed in the ornithological literature (Johnsgard 1980a, 2016a). The female evidently chooses the nest location, which was about 50 yards from the copulation site in the case I observed, and apparently does all the nest-building and incubation. There are a few scattered recent nesting records in Cherry, Garden, Lincoln, and Clay Counties. The clutch size is usually four eggs. The male takes no part in defending the nest, but the female defends it fiercely. Incubation lasts 24–29 days, and the young remain in the nest for about two weeks. The fledging period is still uncertain but is 50–55 days in a closely related European species.

Suggested Viewing Opportunities: Large, shallow Sandhills lakes with extensive emergent vegetation are the best places to find this elusive species.

Selected References: BNA 18, rev. ed. (P. E. Lowther, A. F. Poole, J. P. Gibbs, S. Melvin, and F. A. Reid 2006); Palmer 1962; Hancock and Elliott 1978; Hancock and Kushlan 1984; Johnsgard 1980a, 2016a.

GREAT BLUE HERON. *ARDEA HERODIAS*

Status: A common statewide spring and fall migrant and summer resident, this heron breeds widely across the state, from eastern wooded waterways west to the Colorado and Wyoming borders. National Breeding Bird Surveys from 1966 to 2015 indicate that the species' population underwent a slight average annual increase of

0.44 percent (Sauer et al. 2017). Nebraska surveys indicate a 2.38 percent annual increase, based on a sample of 48 routes. There were 331 possible to confirmed nesting records during the fieldwork for *The Second Nebraska Breeding Bird Atlas* (Mollhoff 2016).

Habitats and Ecology: This species occurs along the riparian areas where there are fish and suitable nesting trees. Large cottonwoods near water are a favored location for colonial nesting colonies; the birds construct stick nests near the crowns of such trees. Sometimes the nests are situated a mile or more from water. Herons remain in southern Nebraska until fall freeze-up, sometimes well into December.

Breeding Biology: Great blue herons are seasonally monogamous, and both sexes arrive at the nesting ground about the same time. Birds probably breed initially when two years old. Older males return before the females and begin to refurbish old nests in preparation for their arrival. The male selects the breeding territory, which usually centers on an old tree nest. Several obviously hostile displays are associated with territorial defense. Additionally, numerous highly ritualized territorial advertising displays occur, including the "stretch," "snap," and others. These are predominantly male displays performed at the nest site, and they serve to attract females and aid in pair formation (Mock 1976). Mutual behavior between members of a pair includes twig-passing, feather-nibbling, bill-stroking, and similar activities. Pair-bonding involves a variety of complex postural displays, often performed mutually. After pairing, the male brings in more materials, and the female finishes the nest. Both sexes incubate, and mutual nest-relief ceremonies are performed. Nesting in Nebraska extends from May to late June. Typically, four eggs are laid, and incubation by both sexes lasts 25–29 days. The eggs usually hatch over an interval of five to eight days, and adults feed the young by regurgitating food onto the bottom of the nest. Although the youngsters can make short flights in the nest vicinity when only seven weeks old, they usually continue to use the nest and are fed by the adults until they are about 10–11 weeks old. They leave the nest at 64–90 days.

Suggested Viewing Opportunities: Nesting or attempted nesting has been reported in almost every county in recent decades, and foraging individuals are likely to be seen in any shallow wetland rich in fish and amphibians between early April and late October.

Selected References: BNA 25, rev. ed. (R. C. Vennesland and R. W. Butler 2011); Meyeriecks 1960; Palmer 1962; Mock 1976; Hancock and Elliott 1978; Hancock and Kushlan 1984.

SNOWY EGRET. *EGRETTA THULA*

Status: A regular and almost statewide spring and fall migrant but a rare and highly local summer resident; breeding has been reported in Hall and Garden Counties. National Breeding Bird Surveys from 1966 to 2015 indicate that this species' population underwent a slight average annual increase of 0.65 percent (Sauer et al. 2017). There was one possible nesting record during the fieldwork for *The Second Nebraska Breeding Bird Atlas* (Mollhoff 2016).

Habitats and Ecology: These birds occur in a wide range of aquatic habitats but seem to prefer somewhat sheltered locations for breeding and often occur in company with other larger heron species. When foraging the birds are fairly active and sometimes rush about in shallow water in an apparent attempt to flush out their prey.

Breeding Biology: After returning to their breeding grounds, males establish a territory that centers on a potential nest site but need not include an old nest. The birds are typically colonial but might nest singly at the edge of their range. The male performs hostile displays, as well as several sexual displays that include both a stationary and an aerial "stretch" as major advertisement displays. A single "circle flight" around a potential mate is also common, and a more spectacular flight is a towering circular flight from 50 to 150 yards above the female, followed by a spectacular tumbling downward to land beside her. A mutual display called the "jumping-over" display, in which one bird makes a short jump flight over the back of the other, is a probable indication that a pair bond has been formed. Copulation occurs on the nest site or on a limb close to it. The male gathers material, and the female constructs the nest.

The nests are rather flat and elliptical rather than round and are loosely constructed. They are usually in shrubs or low trees from 2 to 10 feet above the ground, but some nests up to 30 feet high have been recorded. The first egg might be laid before the nest is completed, and eggs are laid about two days apart. The clutch size varies from two to five eggs, and the incubation period is 22–23 days. Since incubation (by both sexes) begins before the clutch is complete, the first young typically hatches about 18 days after the last egg is laid. At 20–25 days of age the young are ready to leave the nest. There is one brood per season.

Suggested Viewing Opportunities: Most sightings have been made in counties bordering the Platte and Missouri Rivers. The largest number of Nebraska sightings have occurred in May, followed by April, August, and September.

Selected References: BNA 489 (K. C. Parsons and T. L. Master 2000); Palmer 1962; Jenni 1969; Meyeriecks 1960; Hancock and Elliott 1978; Hancock and Kushlan 1984.

Family Threskiornithidae (Ibises and Spoonbills)

WHITE-FACED IBIS. *PLEGADIS CHIHI*

Status: An uncommon spring and fall migrant and local summer resident in the Sandhills and Rainwater Basin. National Breeding Bird Surveys from 1966 to 2015 indicate that this species' population underwent a survey-wide average annual increase of 2.4 percent (Sauer et al. 2017). There were 29 possible to confirmed nesting records during the fieldwork for *The Second Nebraska Breeding Bird Atlas* (Mollhoff 2016).

Habitats and Ecology: Ibises are generally associated with freshwater or brackish marshes having an abundance of cattails, bulrushes, or phragmites.

Breeding Biology: Relatively little is known of the social behavior of this species. Monogamous pair bonds are formed, and both sexes help construct the nest, which takes about two days. It might be placed on floating vegetation or supported by emergent vegetation up to 6 feet above the water. The platform-shaped nest is built of old reeds, rushes, twigs, and grasses. Incubation begins with the last egg. The clutch size is usually three to four eggs. Both sexes also incubate, and during nest relief they do mutual billing and preening and utter guttural cooing notes. The adults continue to add material to the nest during the incubation period, which lasts 21–22 days, and into the fledging period, which lasts about six weeks. The adults feed the young by regurgitation, the youngster inserting its bill into that of the parent, or at times the adults will disgorge food into the nest to be picked up by the young. By the time the young are about seven weeks old they fly with their parents to the foraging grounds and return with them at night for roosting.

Suggested Viewing Opportunities: Crescent Lake National Wildlife Refuge has long been the most reliable place to find this ibis, but it is now widely dispersed across the Sandhills and in both the eastern and western parts of the Rainwater Basin.

Selected References: BNA 130 (R. A. Ryder and D. E. Manry 1994); Bent 1926; Palmer 1962; Burger and Miller 1977.

Family Cathartidae (New World Vultures)

TURKEY VULTURE. *CATHARTES AURA*

Status: A common spring and fall migrant statewide and a local summer resident. National Breeding Bird Surveys from 1966 to 2015 indicate that this species' population underwent a survey-wide average annual increase of 2.25 percent (Sauer et al. 2017). Nebraska surveys indicate a statistically questionable 10.23 percent annual increase, based on a sample of 40 routes. Probably this apparent rapid population increase is the result of turkey vultures adapting to "city life" by moving into the safety and warmth of villages, towns, and even large cities for nighttime roosting and finding carrion among road kills along nearby highways. There were 388 possible to confirmed nesting records during the fieldwork for *The Second Nebraska Breeding Bird Atlas* (Mollhoff 2016).

Habitats and Ecology: The turkey vulture is a scavenger species that consumes the carcasses of mostly larger animals, such as livestock and deer, which it finds both visually and by using its remarkable olfactory abilities. Vultures can often be seen soaring above on thermals, with few if any wingbeats, owing to their broad wingspan relative to their light body mass, resulting in a very low wing loading. Turkey vultures often travel up to 200 miles per day, and have been seen soaring at elevations as high as 20,000 feet. The nesting and brooding season in Nebraska extends from about late May to late August. Fall migration begins in late September, before hard fall frosts have begun. Turkey vultures winter in the southern United States and Mexico and arrive in Nebraska after most spring frosts.

Breeding Biology: Turkey vultures are monogamous, and during turkey vulture courtship, potential pairs perform a "follow flight" activity, during which one bird leads the other through twisting, turning, and flapping flights for a minute or more, a pattern sometimes repeated over periods as long as three hours. Their nests are well scattered, even where nest sites are restricted. Nesting often occurs in abandoned buildings, such as those that are common at old farmsteads. The nests are also sometimes located on or near the ground in hollow logs or large snags, and in the Sandhills they have been found in rock crevices. The pair often uses a cave or other potential nest site as a roost for some time before laying their eggs there. The clutch size is usually two eggs. Incubation lasts 31–37 days. Both sexes participate in incubation, and the incubating bird usually takes morning and afternoon breaks to preen and sit in the sunshine. Injury feigning at the nest by defensive parents has been reported, and young birds will often disgorge their food or bite when approached. The young are relatively precocial and soon move to the mouth of the nesting cavity to sun themselves. The fledging period is surprisingly long, at 70–80 days. They are also long-lived and are known to have survived in the wild for as long as 24 years. Recent genetic studies indicate that the turkey vulture and other New World vultures comprise an order (Cathartiformes) related to the typical hawks, eagles, and Old World vultures (Accipitriiformes) rather than to the storks, with which they had until recently been associated.

Suggested Viewing Opportunities: Many small towns in the Sandhills and cities such as Lincoln have large trees that support nocturnal roosts of turkey vultures. The birds leave their roosts every morning to search for carrion in the countryside and return at about sunset.

Selected References: BNA 339 (D. A. Kirk and M. J. Mossman 1998); Bent 1937; Brown and Amadon 1968; Glinski 1998; Ferguson-Lees and Christie 2001; Snyder and Snyder 2006.

Family Pandionidae (Ospreys)

OSPREY. PANDION HALIAETUS

Status: A common spring and fall migrant nearly statewide and a rare and local summer resident in areas near lakes or streams. Breeding or attempted breeding has occurred several times since 2008 on nesting platforms in western Nebraska (Scotts Bluff, Lincoln, and Keith Counties), with attempts also reported for Washington and Burt Counties. National Breeding Bird Surveys from 1966 to 2015 indicate that this species' population underwent a survey-wide average annual increase of 2.54 percent (Sauer et al. 2017).

Habitats and Ecology: This species is commonly seen along clear rivers and lakes. Ospreys often nest in tall trees near water, building large stick nests resembling those of eagles. The erection of artificial nesting platforms in areas lacking good natural sites has helped expand the species' breeding range in treeless regions.

Breeding Biology: Ospreys arrive as the ice is melting from their nesting grounds, and males soon begin courtship flights. These swooping and soaring flights might serve to attract females and continue for a time after pair bonds are established or reestablished. Nest-building or repair of the old nest starts very soon, the male bringing most of the larger sticks and the female bringing in the lining materials, as well as doing the final shaping of the nest. From the time she arrives until the young are nearly fledged, the female catches few if any fish and thus relies on the male for virtually all her food. Mating occurs on the nest site or a nearby branch and continues during the egg-laying period. The clutch size is usually three eggs. Both sexes incubate, but the female undertakes most of the responsibility and does all the nighttime incubation. Incubation lasts 32–33 days. The eggs hatch at intervals of up to five days, which results in considerable differences in the sizes of the young. For the first month of brooding the female rarely leaves the nest, and the male does all the hunting. As the young approach fledging at about 55 days of age, the female might also help in hunting. After fledging the young continue to use the nest for roosting and as a feeding platform, but they gradually attempt to catch fish on their own. Ospreys do not mature sexually until they are three years old.

Suggested Viewing Opportunities: Ospreys are likely to appear at almost any fish-rich wetland, river, or reservoir in the state during migration. Most older spring sightings fall between April 14 and May 18, and fall sightings are centered from September 6 to September 26.

Selected References: BNA 684, rev. ed. (R. O. Bierregaard, A. F. Poole, M. S. Martell, P. Pyle, and M. A. Patten 2016); Brown and Amadon 1968; Johnsgard 1990; Glinski 1998; Ferguson-Lees and Christie 2001; Snyder and Snyder 2006; Gessner 2008.

Family Accipitridae (Hawks and Eagles)

GOLDEN EAGLE. *AQUILA CHRYSAETOS*

Status: An uncommon resident in western Nebraska, with nearly all the breeding efforts of the past few decades occurring in the Panhandle. Sioux County has had the largest number of possible to confirmed nestings. There have also been peripheral nestings in Garden and Cheyenne Counties. National Breeding Bird Surveys from 1966 to 2015 indicate that this species' population underwent a slight average annual decline of 0.13 percent (Sauer et al. 2017). There were 29 possible to confirmed nesting records during the fieldwork for *The Second Nebraska Breeding Bird Atlas* (Mollhoff 2016). This species is on the Nebraska Natural Legacy Project's Tier 2 list of threatened species in Nebraska (Schneider et al. 2018).

Habitats and Ecology: This is a mountain- and plains-adapted species that often occurs in grasslands, semideserts, juniper woodlands, and the ponderosa pine zone of coniferous forests. In Nebraska it nests on dry cliffs, rocky outcrops, and areas of scattered large

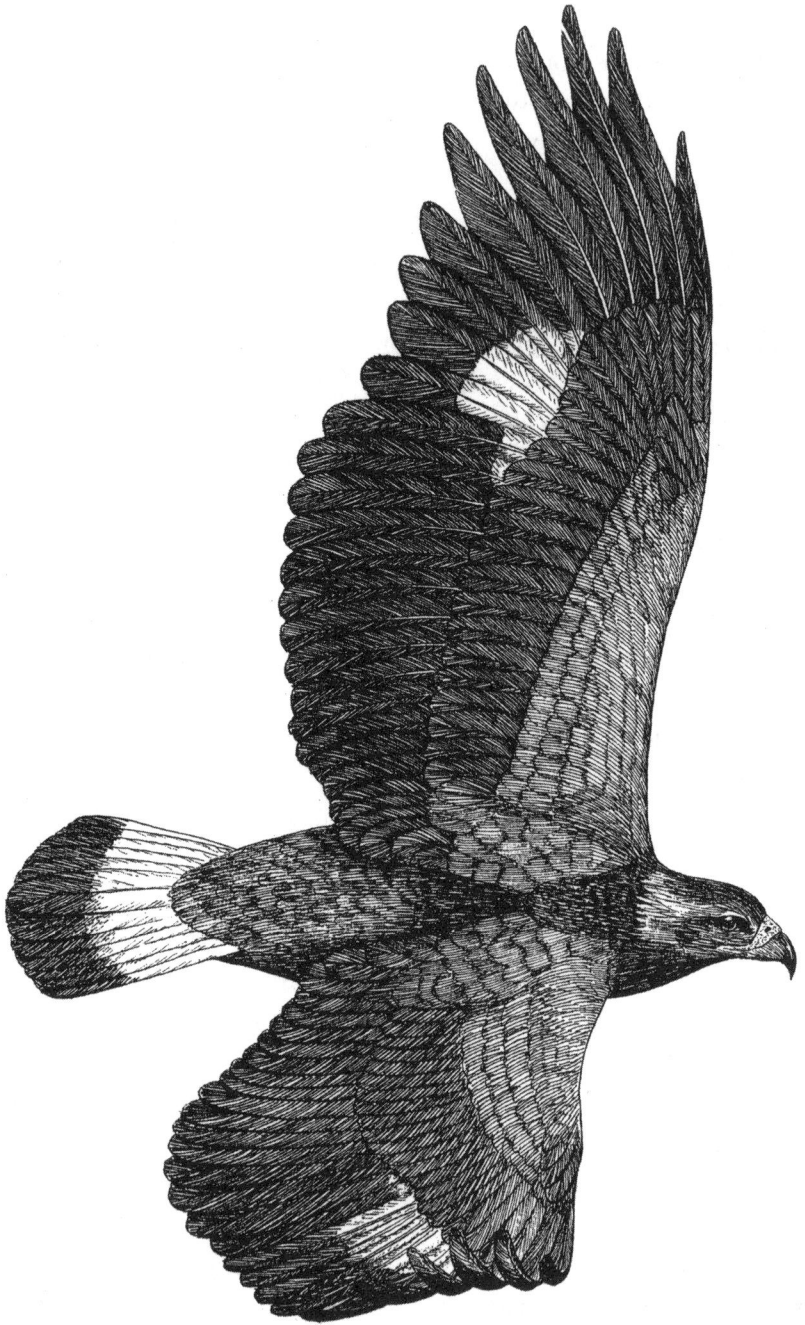

27. Golden eagle, adult in flight.

trees. In mountainous regions it nests on cliffs or in trees, rarely on the ground.

Breeding Biology: Golden eagles are monogamous, and pairs occupy large home ranges that averaged about 35 square miles in a California study. Aerial displays are most common before the nesting season but might occur at other times too. They consist of soaring and swooping by one or both members of the pair. Both work on the massive nests, and several alternate nests might be maintained. The clutch size is usually two eggs. They are laid at intervals of three to four days, and incubation begins almost immediately with the laying of the first egg. Incubation lasts 43–45 days. The female does most of the incubation, but the male begins to assist in brooding soon after the young have hatched. By about 50 days of age the young are fully feathered, and they fledge at about 65–70 days. However, they remain dependent upon their parents for at least some food for as long as three months after fledging. The adult plumage is not reached until the sixth year (Ferguson-Lees and Christie 2001).

Suggested Viewing Opportunities: Bluffs and dry canyon country provide viewing possibilities year-round, such as in the badlands region of Sioux County and elsewhere in the Pine Ridge.

Selected References: BNA 684 (M. N. Kochert, N. K. Steenhof, C. L. McIntyre, and E. H. Craig 2002); Brown and Amadon 1968; Snow 1973a; Ohlendorf 1975; Johnsgard 1990; Glinski 1998; Ferguson-Lees and Christie 2001; Penny 2001; Snyder and Snyder 2006.

BALD EAGLE. *HALIAEETUS LEUCOCEPHALUS*

Status: An uncommon summer resident and a common overwintering migrant. The species was removed from its prolonged status (since the early 1970s) as a federally threatened and endangered species in 2007. It is still on the Nebraska Natural Legacy Project's Tier 2 list of threatened species (Schneider et al. 2018), and individuals are not allowed to possess eagle feathers or other eagle parts without federal permits. Nationally, bald eagles increased at an estimated rate of 5.18 percent annually between 1966 and 2015 (Sauer et al. 2017). By 2006 there were more than 10,000 pairs in North America, excluding Alaska and Canada. Judging from the recent rate of increase, there were possibly 20,000 breeding pairs south of Canada by 2019. During the five Nebraska Christmas Bird Counts ending in 2017–18, an average of 436 bald eagles were observed.

Habitats and Ecology: During summer, breeding eagles are widely dispersed along rivers and around lakes. This species opportunistically feeds on carrion, such as dead livestock and road-killed deer, but it is primarily a fish-eating species. Immature eagles tend to be scavengers at carcasses or other easy sources of meat until they become proficient predators. Bald eagles are likely to be seen along any of the fairly clear rivers, lakes, and reservoirs in Nebraska, especially where fish populations are high or where local fish kills provide a sudden source of abundant food. Some eagles remain in the state year-round, and some northern migrants overwinter in Nebraska and add to local resident populations.

Breeding Biology: After maturing and acquiring the adult plumage at four to five years of age, eagles pair monogamously and remain paired permanently. They perform aerial displays, one of which involves locking talons and tumbling downward through the sky for several hundred feet, ending just in time to avoid crashing. Additionally, aerial fights between males over territorial rights may include similar talon-locking behavior that might result in serious injuries or even deaths for one or both combatants. These spectacular flights occur during the nest-building period. Copulation occurs at the same time, and egg-laying soon follows. The usual clutch size is two eggs, but three- and four-egg clutches have been reported, and incubation lasts 35–45 days. Both sexes assist in incubation, and the young hatch at intervals of several days. The female and young are brought food by the male. As the eaglets grow, both parents gather food for them, but rarely do more than two survive to fledging. This occurs at about 70 days of age, but the young eagles follow their parents for some time afterward, until the adults evict them from the area.

Suggested Viewing Opportunities: Eagle-watching sites have been established at Kingsley Dam in Keith County and at the J-2 power plant in Lincoln County, but eagles can be seen at many reservoirs, such as Calamus, where during a recent spring more than 800 gathered as the ice broke up and exposed countless dead fish. As many as 576 birds have been seen on recent Nebraska Christmas Bird Counts, most of which were centered along the Missouri River. In 2017 a total of 209 nests were documented in the state, a 29 percent increase above the previous year's total and an extension of a long-term population increase. The great majority of the nests were east of the Panhandle and north of the Platte River, with a concentration in the Loup and middle Niobrara drainage basins (Jorgensen and Dinan 2018).

Selected References: BNA 506 (D. A. Buehler 2000); Brown and Amadon 1968; Snow 1973b; Johnsgard 1990; Glinski 1998; Ferguson-Lees and Christie 2001; Snyder and Snyder 2006.

NORTHERN HARRIER. *CIRCUS HUDSONICUS*

Status: An uncommon summer resident throughout the state, nesting locally, especially in nonforested habitats such as grasslands, croplands, and meadows. National Breeding Bird Surveys from 1966 to 2015 indicate that this species' population underwent a survey-wide average annual decline of 1.21 percent (Sauer et al. 2017).

Habitats and Ecology: Unlike most hawks, harriers nest on the ground rather than in trees and favor extensive grassy or marshy habitats of at least 100 acres or more. The English word "harrier" refers to the birds' persistent harrying behavior toward potential prey rather than a hare-based diet.

Breeding Biology: Males migrate separately from females and arrive on the nesting grounds first. They display aerially by performing a series of spectacular dives and swoops (the generic name *Circus* refers to these acrobatic maneuvers), especially in the presence of females when they arrive. Later the pair might display in this same way and also by locking talons in flight, with both tumbling toward the earth. Primarily the female undertakes the construction of the nest, although the male might help gather materials. Frequently, the birds are semicolonial, with up to six nests reported within a square mile. The eggs are laid at intervals of several days, and the female might begin to incubate at almost any time during the egg-laying period. The clutch size is usually four to six eggs, and incubation lasts 29–39 days, averaging about 35 days. Males feed their incubating mates and, on the basis of a group of six nests studied in Manitoba, sometimes provide food for two females. The young hatch at staggered intervals, and while they are very small they are brooded continuously by the female while the male brings in food. Later the female also hunts, but she more often receives the food the male brings in by aerial talon-to-talon transfer. She is the only parent that feeds the young directly. When males are tending two nests the females must do more hunting alone, and starvation of nestlings is frequent. The young fledge at about five weeks, the smaller males a few days sooner than females, and they are independent by about 65–70 days.

Suggested Viewing Opportunities: This grassland-adapted hawk is probably most common in the Sandhills and the Rainwater Basin. Most historic initial spring sightings have occurred in mid-March, and last fall sightings have usually been in mid-December. Adult males return in spring a week or two before females and immatures and remain later in the fall.

Selected References: BNA 210 (R. B. Macwhirter and K. L. Bildstein 1996); Brown and Amadon 1968; Watson 1977; Johnsgard 1990; Glinski 1998; Ferguson-Lees and Christie 2001; Snyder and Snyder 2006.

SHARP-SHINNED HAWK. *ACCIPITER STRIATUS*

Status: An uncommon winter resident statewide and a rare or local breeder in northern areas, with only one confirmed breeding during the two *Nebraska Breeding Bird Atlas* surveys. National Breeding Bird Surveys from 1966 to 2015 indicate that this species' population underwent a survey-wide average annual decline of 0.92 percent (Sauer et al. 2017).

Habitats and Ecology: Fairly dense forests, either mixed or coniferous, are the preferred habitats of this species, which is swift and elusive and usually nests in dense groves of trees. Aspens, riparian woodlands, and coniferous forests are all used for breeding.

Breeding Biology: Like other hawks, these birds are monogamous, no doubt because of the need for a male to be present to provide food for the incubating and brooding female. The female is appreciably larger than her mate and forages on somewhat larger prey, primarily small birds. In Utah the birds appear at their nest sites as long as a month before egg-laying begins. They probably spend much of that time constructing a new nest, since old ones are rarely used, even if they are still intact. However, a crow or squirrel nest is sometimes modified for use. The clutch size is usually four to five eggs, and incubation lasts 30–35 days. After the clutch is complete both sexes assist in incubation, and the young hatch almost simultaneously. They grow rapidly, with the males reportedly fledging at only 24 days of age and the somewhat larger females at 27 days.

Suggested Viewing Opportunities: This widespread species might occur anywhere in wooded areas from fall to spring but is most likely to be seen visiting city or suburban bird feeders. Early fall sightings center in mid-September, and a second apparent movement of birds occurs in March as birds that had wintered farther south pass northward.

Selected References: BNA 482, rev. ed. (K. G. Smith, S. R Wittenberg, R. B. Macwhirter, and K. L. Bildstein 2011); Brown and Amadon 1968; Platt 1976; Johnsgard 1990; Glinski 1998; Ferguson-Lees and Christie 2001; Snyder and Snyder 2006.

COOPER'S HAWK. *ACCIPITER COOPERII*

Status: Uncommon permanent resident. National Breeding Bird Surveys from 1966 to 2015 indicate that this species' population underwent a survey-wide average annual increase of 2.24 percent (Sauer et al. 2017). Nebraska surveys indicate a statistically weak and unbelievable 14.58 percent annual population increase, which was based on a limited sample of only ten survey routes. However, Christmas Bird Counts also indicate a recent increase in Nebraska's

Cooper's hawks, especially relative to sharp-shinned hawks, and a marked increase in city nesting. During Lincoln's five Christmas Bird Counts ending in 2017–18, almost twice as many sharp-shinned as Cooper's hawks were seen (22 sharp-shinned, 13 Cooper's), whereas during Lincoln's Christmas Bird Counts from 1946 through 1997, the sharp-shinned outnumbered the Cooper's by more than three to one (Johnsgard 1998). This trend may reflect both more northerly wintering on the part of the sharp-shinned and an actual increase in the Cooper's hawk population. There were 104 possible to confirmed nesting records during the fieldwork for *The Second Nebraska Breeding Bird Atlas* (Mollhoff 2016).

Habitats and Ecology: This hawk is associated with mature forests. Like other accipiter hawks, it is a highly effective predator of birds up to about the size of a quail, and at times it attacks prey as large as domestic chickens, grouse, and pheasants. Mammals and reptiles are less frequent prey and include mammals as large as skunks, opossums, rabbits, and hares (Johnsgard 1990).

Breeding Biology: In New York, Cooper's hawks arrive in their nesting areas in March, and the male establishes a territory about 100 yards in diameter. From this area he calls and feeds any female that might appear. While a pair bond is being formed the birds perform courtship flights either singly or together. Such flights might be seen for a month or more. During that time the male selects a nest site; rarely, he uses an old nest, but more frequently, he chooses a new location. Nests are usually placed in tall trees, up to 60 feet above the ground. The male gathers most of the materials and does most of the actual nest-building, and he also continues to feed his mate during this period. Typically, the female lays four eggs on an alternate-day basis, and she often begins incubation with the laying of the third egg. The clutch size is usually four eggs, and incubation lasts 35–36 days. The female incubates while the male provides food for her, and he guards the nest briefly while she is eating. At the time of hatching the female carries the eggshells away from the nest and might even help the young birds out of the shell. For the first three weeks after hatching the female rarely leaves the nest, and the male does all foraging. Fledging by the substantially smaller males occurs at about 30 days, and females fledge by about 34 days. Fledglings become independent by about two months of age.

Suggested Viewing Opportunities: This is a widely distributed species and might occur anywhere in wooded areas, especially in denser forests, throughout the year.

Selected References: BNA 75, rev. ed. (O. E. Curtis, R. N. Rosenfield, and J. Bielfeldt 2006); Meng 1951; Brown and Amadon 1968; Johnsgard 1990; Glinski 1998; Ferguson-Lees and Christie 2001; Snyder and Snyder 2006.

NORTHERN GOSHAWK. *ACCIPITER GENTILIS*

Status: An occasional to rare winter visitor statewide.

Habitats and Ecology: This species favors dense conifers or deciduous trees near water for breeding and ranges into low woodlands, riparian woods, and sage areas at other times. National Breeding Bird Surveys from 1966 to 2015 indicate that this species' population underwent a survey-wide average annual decline of 0.39 percent (Sauer et al. 2017).

Suggested Viewing Opportunities: Most recent sightings have been from eastern and central Nebraska, but possible sightings could occur in almost any well-wooded locations. About half the historic fall records of this species occurred between early October and late December, and half of the spring records occurred between early January and late April. The nearest breeding area might be northern Minnesota; no recent proof of breeding has been observed in the Black Hills.

Selected References: BNA 198 (J. R. Squires and R. T. Reynolds 1997); Schnell 1958; Brown and Amadon 1968; Johnsgard 1990; Glinski 1998; Ferguson-Lees and Christie 2001; Snyder and Snyder 2006.

SWAINSON'S HAWK. *BUTEO SWAINSONI*

Status: A common summer resident over most of western Nebraska. National Breeding Bird Surveys from 1966 to 2015 indicate that this species' population underwent a survey-wide average annual increase of 0.72 percent (Sauer et al. 2017). There were 165 possible to confirmed nesting records during the fieldwork for *The Second Nebraska Breeding Bird Atlas* (Mollhoff 2016).

Habitats and Ecology: This species is associated with open grasslands, sagebrush, and agricultural lands and rarely with riparian areas, typically nesting in isolated trees but sometimes on man-made structures. During *The Second Nebraska Breeding Bird Atlas* studies, nearly half of all breeding encounters occurred on habitat patches of 101–10,000+ acres (Mollhoff 2016).

Breeding Biology: Swainson's hawks are monogamous and arrive on their breeding ground in western Nebraska about a month before egg-laying begins. They soon begin nest-building and sometimes use old magpie nests for a base; infrequently, they use their own old nests. The clutch size is usually two or three eggs. Although males rarely assist in incubation, they do bring prey to the incubating female. Incubation lasts 28 days. The female broods the young during the first 20 days after hatching but thereafter spends considerable time hunting to feed the young, which fledge in 28–35 days.

Suggested Viewing Opportunities: Swainson's hawks are probably most often seen standing on telephone poles or soaring over open grasslands. They are especially numerous in late September and October, when fall migration is under way and large flocks (kettles) are common.

Selected References: BNA 265, rev. ed. (M. J. Bechard, C. S. Houston, J. H. Sarasota, and A. S. England 2010); Dunkle 1977; Brown and Amadon 1968; Johnsgard 1990, 2001b; Glinski 1998; Ferguson-Lees and Christie 2001; Snyder and Snyder 2006.

RED-TAILED HAWK. *BUTEO JAMAICENSIS*

Status: Common summer resident and migrant. A total of 495 possible to confirmed breeding records was reported for every county during fieldwork for *The Second Nebraska Breeding Bird Atlas*, and the species was judged to be the state's twelfth most common breeding bird (Mollhoff 2016). Many red-tailed hawks migrate to the southern United States for the winter, although some winter in the area, and others enter Nebraska from farther north to winter here, including the subarctic-breeding melanistic form known as the Harlan's hawk (*Buteo jamaicensis harlani*). Numbers vary, with migration peaks of the breeding Great Plains race in early April and October. National Breeding Bird Surveys from 1966 to 2015 indicate that this species' population underwent a survey-wide average annual increase of 1.42 percent (Sauer et al. 2017). Nebraska surveys indicate a 3.58 percent annual population increase, based on a sample of 50 routes. Judging from the two *Nebraska Breeding Bird Atlas* surveys, red-tails have expanded west in recent decades, with now about as many breeding records in the western half of the state as in the eastern half (Mollhoff 2016).

Habitats and Ecology: This is a common buteo hawk that occupies a broad range of habitats, extending to open country. However, trees, especially large cottonwoods in shelterbelts and woodlands, are favored sites. Red-tailed hawks are extremely beneficial and are the most common and widespread buteo in North America. Their foods are highly diverse, but in a survey of 11 published studies, mammals ranged from 37 to 88 percent of prey taken, averaging 68 percent; birds from 4 to 58 percent, averaging 17.5 percent; reptiles (mostly snakes) and amphibians from 0 to 41 percent, averaging 7 percent; and invertebrates from 0 to 21 percent, averaging 3.2 percent. When food biomass is taken into consideration the primary importance of mammals in the overall diet becomes much more evident, with rabbit and rodents being major prey items (Johnsgard 1990).

Breeding Biology: Red-tailed hawks pair monogamously and arrive at their nesting areas already mated. Nonetheless, courtship flights are common in early nesting phases, with the birds dramatically soaring and swooping together and occasionally locking talons in flight. Copulation often follows such flights. Both birds build the nest well before egg-laying begins, and after it is completed the female stays near the nest while the male feeds her and brings nest-lining materials. The clutch size is usually two to three eggs, and incubation lasts 28–32 days. Both sexes help incubate, but the female assumes most of the responsibility and is fed by her mate during this period. The young are hatched at intervals of several days and grow rapidly. By the time they are a month old they might climb out onto adjoining branches, and they fledge at about 45 days. After leaving the nest they are fed progressively less by their parents and become relatively independent in about a month.

Suggested Viewing Opportunities: This nearly ubiquitous buteo is likely to be found anywhere in open country. Telephone poles or tall trees in shelterbelts are typical perching and lookout sites. Unusually pale-plumaged birds (Krider's hawks) sometimes occur in Nebraska and are variously regarded as a distinct Great Plains race (*Buteo jamaicensis krideri*) or simply as leucistic variants. Melanistic Harlan's hawks also comprise a small percentage (ca. 2–5 percent) of the wintering red-tailed population and are most apparent from October through March.

Selected References: BNA 52, rev. ed. (C. R. Preston and R. D. Beane 2009); Austin 1963; Brown and Amadon 1968; Johnsgard 1990; Glinski 1998; Ferguson-Lees and Christie 2001; Snyder and Snyder 2006.

FERRUGINOUS HAWK. *BUTEO REGALIS*

Status: An uncommon resident primarily in open-country habitats in the Panhandle. National Breeding Bird Surveys from 1966 to 2015 indicate that this species' population underwent a survey-wide average annual increase of 0.61 percent (Sauer et al. 2017). This species is on the Tier 1 list of the Nebraska Natural Legacy Project's Tier 1 list of threatened species in Nebraska (Schneider et al. 2018). There were 15 possible to confirmed nesting records during the fieldwork for *The Second Nebraska Breeding Bird Atlas* (Mollhoff 2016).

Habitats and Ecology: These hawks are found during the breeding season in grasslands, brushlands, and badlands. They nest in small conifers or on cliff ledges, rock outcrops, and sometimes human-made structures such as windmills. During *The Second Nebraska Breeding Bird Atlas* studies, most breeding encounters occurred on habitat "patches" of 1,000–10,000+ acres, with Sioux County accounting for the largest number of occurrences (Mollhoff 2016).

28. Ferruginous hawk, adult.

Breeding Biology: Pairs return to their breeding territory each year and usually use the same nest, so it gradually increases in size. Both sexes bring nesting material in the form of sticks and nest lining, which the female molds to fit her body. The clutch size is usually three to four eggs, and incubation lasts 32–36 days. Evidently the female does most of the incubation. After hatching, the male did most of the brooding in a nest observed in Washington. The young might leave the nest when only about a month old but do not fledge until they are about 44–48 days of age. They start catching live prey only a few days after fledging.

Suggested Viewing Opportunities: Ferruginous hawks are strongly attracted to prairie dog towns, a primary prey species, and more generally favor broad expanses of open landscapes, such as buttes, badlands, and rimrock topography.

Selected References: BNA 172, rev. ed. (J. Ng, M. D. Giovanni, M. J. Bechard, J. K. Schmutz, and P. Pyle 2017); Brown and Amadon 1968; Angell 1969; Snow 1974a; Ohlendorf 1975; Johnsgard 1990, 2001b; Zelenak, Rotella, and Harmata 1997; Glinski 1998; Ferguson-Lees and Christie 2001; Snyder and Snyder 2006.

ROUGH-LEGGED HAWK. *BUTEO LAGOPUS*

Status: A common winter visitor throughout the state, especially in open habitats.

Habitats and Ecology: This species is usually found hunting in grasslands and sagebrush or sometimes over marshes. This open-country hawk forages mostly on fairly small rodent prey such as lemmings during summer, which range up to 3 ounces in mass, and similar-sized voles during winter.

Suggested Viewing Opportunities: During winter this species might be found in open country across the state. During that season its prey tends to consist of small rodents, especially those of the widespread vole genus *Microtus,* and various New World (cricetid) mice. Larger prey, such as rabbits and even birds as large as the ring-necked pheasant, have been reported; birds represented 12 percent numerically of the total prey analyzed in one study (Johnsgard 1990).

Selected References: BNA 641 (M. J. Bechard and T. R. Swem 2002); Brown and Amadon 1968; Johnsgard 1990; Glinski 1998; Ferguson-Lees and Christie 2001; Snyder and Snyder 2006.

Family Strigidae (Typical Owls)
EASTERN SCREECH-OWL. *MEGASCOPS ASIO*

Status: Screech-owls are uncommon residents statewide. National Breeding Bird Surveys from 1966 to 2015 indicate that this species' population underwent a survey-wide average annual decline of 1.27 percent (Sauer et al. 2017).

Habitats and Ecology: Screech-owls are associated with a variety of wooded habitats, including farmyards, cities, orchards, wooded parks, and riparian woodlands. There were 93 possible to confirmed nesting records during the fieldwork for *The Second Nebraska Breeding Bird Atlas* (Mollhoff 2016). About 90 percent of the screech-owls in Nebraska are of the gray plumage morph; the rest are rufous or intermediate.

Breeding Biology: Screech-owls are so small and inconspicuous that they might nest in an urban backyard without the owner ever being aware of their presence. Most often the species can be detected by their distinctive wailing call or a series of short whistled notes that often speed up and become a trill, similar to the noise of a ball bouncing to a standstill. Eight to nine days are needed to complete a clutch of four eggs. From the time incubation begins the male probably hunts for both members of the pair, but he does not incubate the eggs. Incubation lasts 26 days. When all the young have hatched over a period of about three days, both parents feed them about equally. Classic studies by Arthur A. Allen (1924) indicated that a surprising variety of prey is brought to the nestlings, including numerous adult songbirds such as sparrows, warblers, phoebes, and tanagers. Over a 45-day period, 77 birds of 18 species were brought to the young, as well as insects, mammals, salamanders, crayfish, and other prey. The young begin to fly when they are about 28–30 days of age but continue to be fed for some time thereafter.

Suggested Viewing Opportunities: Screech-owls can often be seen or heard in city parks, woodlots, or other areas that are sometimes fairly close to humans. The owls are most easily seen by luring them close while imitating their calls or by using playbacks of recordings.

Selected References: BNA 165, rev. ed. (G. Richison, F. R. Gehlbach, P. Pyle, and M. A. Patten 2017); Allen 1924; Glinski 1998; König, Weick, and Becking 1999; Johnsgard 2002b; Lynch 2007; Mikkola 2012; Weidensaul 2015.

GREAT HORNED OWL. *BUBO VIRGINIANUS*

Status: A common permanent resident statewide. National Breeding Bird Surveys from 1966 to 2015 indicate that this species' population underwent a survey-wide average annual decline of 0.81 percent (Sauer et al. 2017). Nebraska surveys indicate a 0.69 percent annual population increase, based on a sample of 48 routes. Great horned owls are very uniformly distributed across the state, an indication of the species' high degree of adaptability. During *The Second Nebraska Breeding Bird Atlas* studies, nearly half of all breeding encounters occurred on habitat patches of 2–10 acres, There were 173 possible to confirmed nesting records during associated fieldwork (Mollhoff 2016).

Habitats and Ecology: A powerful and adaptable owl, this species occurs everywhere from riparian woodlands to the nearly treeless Sandhills and to wooded suburbs. Thirteen percent of 258 nesting reports by Mollhoff (2016) were in human-related habitats. Nesting sites are highly variable, but nesting often occurs in abandoned bird or squirrel nests, on tree crotches, or rarely even on the ground. Great horned owls nest very early in the season; as a result, the young are hatched at the time that small rodent populations are increasing rapidly. As a counterpart raptor to the red-tailed hawk, the great horned owl is the most widespread of North American owls and no doubt is the most common. Its prey is likewise highly variable but is strongly skewed toward mammals. Rabbits and hares are favorite foods in most areas. In a comparison of ten studies, lagomorphs (rabbit and hares) comprised 54–70 percent of all foods taken by biomass, followed by larger rodents at 9–39 percent, mice and voles at 7–23 percent, ducks and gallinaceous birds at 4–8 percent, and passerine birds at 1.5–4 percent (Sauer et al. 2017).

Breeding Biology: Great horned owls are strongly monogamous, and pairs keep in contact by using their familiar hooting calls, *who who who, whoo-whooo.* The male's call is appreciably lower in pitch than the female's. The breeding season begins amazingly early in the year, the pairs usually nesting in the same area and sometimes in the same nest that they used the previous year. Incubation begins as soon as the first egg is laid, perhaps partly to keep the eggs from freezing but also to ensure staggered hatching of the young. The clutch size is usually two to three eggs, and incubation lasts 36–35 days. Both sexes reportedly incubate, but the female probably does most, while the smaller male hunts for the pair. The young

29. Adults of barn owl (top left), short-eared owl (top right), and barred owl (middle), and burrowing owl adult and two owlets (bottom).

are hatched in a scanty down coating and do not open their eyes for a week or more. They are brooded by their parents for nearly a month and cannot fly until they are about nine or ten weeks old. Even after they fledge the owlets continue to beg for food until they are driven away from the area by their parents and must forage for themselves.

Suggested Viewing Opportunities: Like screech-owls, great horned owls sometimes live quite close to humans, such as in well-wooded city parks, and can be detected by using playbacks or imitations of their calls.

Selected References: BNA 372, rev. ed. (C. Artuso, C. S. Houston, D. G. Smith, and C. Rohner 2013); Errington, Hamerstrom, and Hamerstrom 1940; Glinski 1998; König, Weick, and Becking 1999; Johnsgard 2002b; Lynch 2007; Mikkola 2012; Weidensaul 2015.

BURROWING OWL. *ATHENE CUNICULARIA*

Status: An uncommon and increasingly local summer resident on the plains over much of western Nebraska; there have been no breeding records east of Grand Island for more than two decades. National Breeding Bird Surveys from 1966 to 2015 indicate that this species' population underwent a survey-wide average annual decline of 0.98 percent (Sauer et al. 2017). This species is on the Tier 1 list of the Nebraska Natural Legacy Project's threatened species in Nebraska (Schneider et al. 2018).

Habitats and Ecology: This is the only North American owl closely associated with grassland rodents such as prairie dogs, and as the range and abundance of these mammals have decreased, so too has the status of the burrowing owl. During summer it is largely an insectivorous species, often eating large beetles, but it also takes many small mice.

Breeding Biology: Based on studies in New Mexico, burrowing owls arrive on their nesting areas either singly or paired, with males returning to the same burrows they had occupied previously. Unpaired males display from their burrow locations by bowing and uttering a dove-like double-noted *coo-coooo* "song" through the night. Pair formation might occur in a single evening. The clutch size is usually five or six eggs, and incubation lasts 27–30 days. Evidently only females incubate, and males feed their mates during pair formation, incubation, and brooding. When the young are three or four weeks old the female begins to forage for herself and her brood, and at about this time the young birds become capable of flight and begin to find food independently. Not long after fledging the fall migration begins.

Suggested Viewing Opportunities: Prairie dog towns are rapidly declining both statewide and nationally as burrowing populations also decline. Prairie Dog Waterfowl Production Area (Kearney County) is now one of the easternmost Nebraska locations that has regularly reported burrowing owls in recent years. Other waterfowl production areas with recent burrowing owl sightings include three waterfowl production areas, Massie, Harvard, and Hultine. Some Panhandle counties (Sioux, Box Butte, Scotts Bluff, Banner, Morrill, and Garden) also have reported burrowing owl nestings in recent years. There was a total of 38 confirmed breedings during *The Second Nebraska Breeding Bird Atlas* field studies, only 4 of which were from the eastern half of the state, and only 12 were east of the Panhandle.

Selected References: BNA 61, rev. ed. (R. G. Poulin, L. D. Todd, E. A. Haug, B. A. Millsap, and M. S. Martell 2011); Lincer and Steenhof 1997; Glinski 1998; König, Weick, and Becking 1999; Johnsgard 2001b, 2002b; Lynch 2007; Mikkola 2012; Weidensaul 2015.

LONG-EARED OWL. *ASIO OTUS*

Status: An uncommon resident over much of the state, especially where tall trees are present.

Habitats and Ecology: A widespread but highly inconspicuous species, long-eared owls are often associated with coniferous or deciduous forests but also found in woodlots, orchards, large wooded parks, and even juniper woodlands during the breeding season. Trees surrounded by open country seem to be favored for nesting.

There were six possible to confirmed nesting records during the fieldwork for *The Second Nebraska Breeding Bird Atlas* (Mollhoff 2016).

Breeding Biology: A few weeks before egg-laying, courtship calling begins, marked by a series of short three-noted calls similar to mourning dove calls and uttered at intervals of about three seconds. Aerial display flights include wing-clapping noises and acrobatic flying maneuvers. Clutches are usually laid in old nests of other species and are often four to five eggs. The eggs are laid at irregular intervals of one to five days, and a clutch might be completed in ten to eleven days. Incubation lasts 25–30 days. Only the female incubates, and the young are usually hatched over a period of about 7–12 days. For the first 15 days of brooding the female does not leave the nest area and is fed by the male. By the time the young are 25–26 days old they are sufficiently developed to leave the nest and flutter to the ground, but they are not capable of full flight until they are about 30–32 days of age, and they are not independent of their parents until sometime thereafter.

Suggested Viewing Opportunities: These elusive owls might be found almost anywhere in woodlands, ranging from riparian shrubs to mature and dense forests. They especially favor roosting in isolated stands of conifers during winter.

Selected References: BNA 133 (J. S. Marks, D. L. Evans, and D. W. Holt 1994); Glinski 1998; König, Weick, and Becking 1999; Johnsgard 2002b; Lynch 2007; Mikkola 2012; Weidensaul 2015.

SHORT-EARED OWL. *ASIO FLAMMEUS*

Status: An uncommon local resident in grasslands nearly statewide. Associated with open meadows and marshes. National Breeding Bird Surveys from 1966 to 2015 indicate that this species' population underwent a survey-wide average annual decline of 0.89 percent (Sauer et al. 2017). This species is on the Nebraska Natural Legacy Project's Tier 1 list of threatened species in Nebraska (Schneider et al. 2018). There were six possible to confirmed nesting records during the fieldwork for *The Second Nebraska Breeding Bird Atlas* (Mollhoff 2016).

Habitats and Ecology: This is a prairie-adapted species, breeding in areas of grassland, marshes, and low brushland. Nests are usually on the ground but sometimes are in burrows. More diurnal than most owls, these owls are often seen hunting during waning daylight hours.

Breeding Biology: The short-eared owl is one of the most diurnal of the grassland owls, and during spring it can sometimes be seen performing acrobatic courtship flights high above the prairies, marked by strong wing-clapping, swooping, diving, and somersaulting maneuvers, and by uttering a quavering, chattering cry as the bird plummets toward the ground. Copulation sometimes follows such aerial displays or might occur in their absence. Eggs are laid over a considerable period, at intervals of two to seven days. The clutch size is usually four to eight eggs, and incubation lasts 24–28 days. The female incubates alone, but her mate brings food to her during this period. The eggs usually hatch at intervals of about three days, and about two weeks after hatching the young begin to move some distance away from the nest. When they are about six weeks old they begin to catch some of their own food, such as insects and amphibians, but even after the young are well fledged at the age of two months the adults might continue to care for them. About 90 percent of this owl's food consists of rodents, which makes the species extremely valuable from a farmer's standpoint.

Suggested Viewing Opportunities: This species occurs widely across the state, mainly in meadows, grasslands, and marshes. Most recent nestings have occurred in fallow grain stubble in the Panhandle.

Selected References: BNA 62, rev. ed. (D. A. Wiggins, D. W. Holt, and S. M. Leasure 1993); Clark 1975; Glinski 1998; König, Weick, and Becking 1999; Johnsgard 2001b, 2002b; Lynch 2007; Mikkola 2012; Weidensaul 2015.

NORTHERN SAW-WHET OWL. *AEGOLIUS ACADICUS*

Status: Widespread if inconspicuous wintering migrant throughout the state and a local breeder in at least Scotts Bluff, Dawes, Garden, Cherry, and Antelope Counties. This species is on the Nebraska

Natural Legacy Project's Tier 2 list of threatened species in Nebraska (Schneider et al. 2018).

Habitats and Ecology: This tiny owl occurs widely, from deciduous riparian woodlands to coniferous forests. Ponderosa pine stands are probably the owls' favored habitats, where they nest in old woodpecker holes. Woodlands with open understories are favored for foraging.

Breeding Biology: The weak voice and relatively quiet nature of this species make its nesting easily overlooked; the courtship call consists of a monotonous series of uniform whistled notes that resemble more the dripping of water than the sound of a saw being whetted (the basis for the species' English name) and is primarily heard during the nesting period in March and April. Males court females by flying around them and landing nearby, often presenting a small prey item. As soon as egg-laying begins the female becomes very reluctant to leave the nest, and the combination of a large clutch size and an egg-laying interval of 24–74 hours results in a highly staggered period of hatching. The clutch size is usually five to six eggs. During the incubation and early brooding periods of 26–28 days the male is occupied with getting food, which often consists of small mice, frogs, and occasionally birds. Voles (*Microtus*) and other native mice (*Peromyscus* and *Reithrodontomys*) are major foods in Nebraska. The young remain in the nest about four weeks and by the end of this period are able to fly moderately well. However, parental care continues until late summer, when the distinctive juvenile plumage is lost and the first adult-like plumage is assumed.

Suggested Viewing Opportunities: This owl is so inconspicuous that any sighting is likely to be serendipitous, but most sightings have occurred from early November to late February, when the birds might be found roosting in dense junipers.

Selected References: BNA 42, rev. ed. (J. L. Rasmussen, S. G. Sealey, and R. J. Cannings 2008); Santee and Granfield 1939; König, Weick, and Becking 1999; Johnsgard 2002b; Lynch 2007; Glinski 1998; Rashid 2010; Mikkola 2012; Weidensaul 2015.

Family Alcedinidae (Kingfishers)

BELTED KINGFISHER. *MEGACERYLE ALCYON*

Status: Uncommon summer resident statewide, remaining well into late fall, until waters freeze over. National Breeding Bird Surveys from 1966 to 2015 indicate that this species' population underwent a survey-wide average annual decline of 1.37 percent (Sauer et al. 2017). Nebraska surveys indicate a 0.77 percent annual decline, based on a sample of 29 routes. Their distribution in Nebraska is closely tied to river systems, especially rivers with steep banks. There were 212 possible to confirmed nesting records during the fieldwork for *The Second Nebraska Breeding Bird Atlas* (Mollhoff 2016).

Habitats and Ecology: This species is found near flowing or still waters that are rich in fish populations and usually where nearby road cuts, eroded banks, gravel pits, or other steep earthen exposures provide opportunities for excavating earthen tunnel nests and where nearby tree branches provide convenient perching and observation sites. The birds typically choose nesting sites along streams where there are shallow riffles that provide habitat for the fish that they can easily see and catch. Small fish averaging about 3–4 inches long compose more than half their diet. When fish supplies decline, the birds eat crayfish, frogs, toads, salamanders, snakes, insects, and small mammals or young birds. Belted kingfishers take up residence in habitats that allow for large home ranges. At times they might forage up to 5 miles from the nest site; a population density of about one pair per 1.8 square miles of habitat has been estimated in Minnesota.

Breeding Biology: Nesting tunnels are dug into steep banks of clay or silt and up to 10–15 feet long. Both sexes participate in nest excavation, which might require up to three weeks. Sometimes the male digs a second short tunnel to use for resting and sleeping. The usual clutch size is six to eight eggs, and both sexes participate incubating over the 23–24-day period. The young are relatively helpless and spend much time clinging to one another, apparently to maintain body warmth. Fed by both parents, the young are fully feathered at 30–35 days. They leave the nest as soon as they are able to fly, then stay near it for the next few days while their parents teach them how to catch fish. An adult will capture a fish, beat it until it is nearly senseless, and drop it back into the water. The young are encouraged to capture such easy prey and gradually learn to catch unhindered fish. Within ten days after fledging they are relatively independent and soon leave the vicinity of the adult pair. If the first breeding is a failure, a second nesting is attempted, and the pair sometimes even excavates a new nest tunnel.

Suggested Viewing Opportunities: This widespread species can be seen along clear-water rivers or lakes that have good fish populations, nearby perching sites, and steep clay banks for nesting.

Selected References: BNA 84, rev. ed. (J. F. Kelley, E. S. Bridge, and M. J. Hamas 2009); Bent 1940; White 1953; Cornwell 1963.

Family Picidae (Woodpeckers)

LEWIS'S WOODPECKER. *MELANERPES LEWIS*

Status: A local and uncommon summer resident in forested areas of the Panhandle, especially in the Pine Ridge of Sioux (Fort Robinson and Sowbelly Canyon) and Dawes (Dead Horse and West Ash Canyons) Counties. There were two confirmed nesting records during the fieldwork for *The Second Nebraska Breeding Bird Atlas* (Mollhoff 2016). A nationally rapidly declining species; National Breeding Bird Surveys from 1966 to 2015 indicate that this species' population underwent a survey-wide average annual decline of 3.42 percent (Sauer et al. 2017). This species is on the Nebraska Natural Legacy Project's Tier 2 list of threatened species in Nebraska (Schneider et al. 2018).

Habitats and Ecology: Lewis's woodpeckers are mainly adapted to catching free-living insects rather than excavating for insects in wood burrows. This woodpecker is especially associated with pine forests that are rather open, with burned-over or otherwise dead trees with abundant snags or stumps and good insect populations, which means that its distribution tends to be labile and adaptable to local conditions. Streamside cottonwood groves are also favored foraging sites, and old, dying cottonwoods are favorite nesting trees.

Breeding Biology: Unlike other North American woodpeckers, this species is adapted to feeding on free-living insects and is remarkably adept at aerial fly-catching. As the breeding season approaches the male begins to utter his harsh *churr* breeding call, which serves to attract mates and defend or announce nest sites. Males also drum, but mutual wood-tapping and female drumming have apparently not been reported. Males take the predominant role in selecting the nest site and defending the nest. Since old nest cavities are usually used, little excavation is needed. The clutch size is usually six or seven eggs, and the incubation period lasts 12–14 days. Males incubate and brood at night, and both sexes share these responsibilities during the day. The fledging period is probably 31 days. A few days before the young fledge they move out of the nest cavity and begin to climb about. As the brood leaves the nest vicinity each pair member takes part of the brood and continues to feed them occasionally until they are able to catch insects on their own.

Suggested Viewing Opportunities: The only regional breeding population of this species is in Sheridan and Dawes Counties. Sowbelly Canyon, northeast of Harrison, and Fort Robinson State Park near Chadron have been reliable places to find this species. This woodpecker is attracted to areas of burned pines, but it usually appears only several years after a fire and is mostly attracted to larger burned areas.

Selected References: BNA 294, rev. ed. (K. T. Vierling, V. A. Saab, and B. W. Tobalske 2014); Bent 1939; Bock 1970; Short 1982; Winkler, Christie, and Nurney 1995; Shunk 2016.

RED-HEADED WOODPECKER. *MELANERPES ERYTHROCEPHALUS*

Status: A local and uncommon summer resident in the eastern two-thirds of the state, but uncommon to rare in the Panhandle. National Breeding Bird Surveys from 1966 to 2015 indicate that this species' population underwent a survey-wide average annual decline of 2.35 percent (Sauer et al. 2017). There were 472 possible to confirmed nesting records during the fieldwork for *The Second Nebraska Breeding Bird Atlas* (Mollhoff 2016).

Habitats and Ecology: Associated with open deciduous forests, woodlots, and riparian areas, red-headed woodpeckers sometimes extend into the ponderosa pine zone. Aspen stands and riparian cottonwood forests are the species' major habitats in this state. Like the Lewis's woodpecker, this species tends to nest in dead trees or the dead portions of living trees. It does less excavating for insects in wood than do most woodpeckers; instead, it often forages in open habitats at ground level.

Breeding Biology: After males return to their nesting areas in spring, they call and drum from their roosting and prospective nesting holes, apparently to attract mates to their excavations. When a female approaches, the male begins tapping from within the cavity, then typically flies away to allow the female to inspect the hole. He might also solicit copulation by inviting reverse mounting by the female while he is perched near the nest cavity. Mutual tapping at the nest hole seems to indicate that a pair bond is formed and that the female accepts the nesting cavity. The clutch size is usually five eggs, and the incubation period lasts 12–14 days. Both sexes assist in incubation and brooding, with the male performing these activities at night. In one Illinois study, 3 of 15 pairs nested a second time in one season, sometimes starting while still feeding their first fledglings. The young birds tend to follow their parents for some time after leaving the nest, until they are chased away when they are about 25 days old.

Suggested Viewing Opportunities: The densest breeding populations are along river valleys on the state's eastern boundaries in riparian cottonwood-dominated woods.

Selected References: BNA 518, rev. ed. (B. Frei, K. G. Smith, J. Withgott, P. Rodewald, P. Pyle, and M. A. Patten 2017); Bent 1939; Reller 1972; Short 1982; Winkler, Christie, and Nurney 1995; Shunk 2016.

DOWNY WOODPECKER. *DRYOBATES PUBESCENS*

Status: A common resident in wooded habitats throughout the state. The species is broadly distributed across the state but is rare in the

Sandhills and the arid southwestern Panhandle grasslands. National Breeding Bird Surveys from 1966 to 2015 indicate that this species' population underwent a survey-wide average annual increase of 0.03 percent (Sauer et al. 2017). Nebraska surveys indicate a 0.27 percent annual increase, based on a sample of 45 routes. There were 351 possible to confirmed nesting records during the fieldwork for *The Second Nebraska Breeding Bird Atlas* (Mollhoff 2016).

Habitats and Ecology: A wide variety of wooded habitats are used by this species, but it has a preference for open deciduous riparian woodlands. Downy woodpeckers favor nesting in cottonwoods. Both males and females participate in excavating a nest cavity with a round entrance about 1 inch wide and a cavity 6–12 inches deep. Because of their ability to excavate nest holes in trees, old nest holes are valuable to several other hole-nesting birds, including the house wren, black-capped chickadee, tufted titmouse, tree swallow, and bluebirds. Ants, larvae of beetles, and caterpillars are common summer foods. Seeds and fruits probably are important during winter.

Breeding Biology: Like hairy woodpeckers, this species is nonmigratory, and birds spend the entire year near their breeding areas. Toward late winter territorial drumming and associated conflicts begin, and mates often drum at dawn to locate each other when their roosting holes are widely separated. In spring both sexes begin to seek out suitable nest sites, and when one has located a potential site it begins drumming and tapping to attract the other. Short courtship flights occur near the nest and might strengthen site attachment or stimulate copulation, which takes place near the nest. Both sexes help excavate the nest, but usually the female is more active. The clutch size is typically four to five eggs, and the incubation period lasts 12 days. The male assists with incubation and spends the night in the nest. For the first week after hatching, one or the other adult remains at the nest at all times, but as the young birds develop, both adults spend much time foraging. By the time the young are two weeks old they can climb to the nest entrance to be fed, and they are ready to fly less than four weeks after hatching. Pair bonds break down after the breeding season, and each bird excavates fresh roosting holes for use in winter, when each forages independently of its mate.

Suggested Viewing Opportunities: These woodpeckers are nearly ubiquitous throughout the state and are easy to find around bird feeders in winter or in river-bottom woodlands rich in cottonwood trees.

Selected References: BNA 613, rev. ed. (J. A. Jackson and H. R. Quellet 2018); Bent 1939; Kilham 1974; Short 1982; Winkler, Christie, and Nurney 1995; Shunk 2016.

HAIRY WOODPECKER. *DRYOBATES VILLOSUS*

Status: A common resident in wooded areas throughout the state. National Breeding Bird Surveys from 1966 to 2015 indicate that this species' population underwent a survey-wide average annual increase of 0.81 percent (Sauer et al. 2017). There were 163 possible to confirmed nesting records during the fieldwork for *The Second Nebraska Breeding Bird Atlas* (Mollhoff 2016).

Habitats and Ecology: Optimum breeding habitat for hairy woodpeckers consists of fairly extensive areas of woodlands of conifers or hardwoods, but nesting occurs in riparian hardwood forests, aspen groves, and coniferous forests. Generally, aspens and other relatively soft-wooded deciduous trees are preferred over conifers for breeding.

Breeding Biology: Hairy woodpeckers are largely nonmigratory and begin to form pairs in midwinter, about three months before the start of nesting. Typically, males are attracted at this time to territories that the females established the previous fall. During this time the pair performs drumming duets, and both sexes use drumming to locate a mate when they are visually separated. The male also uses drumming as a territorial display. When searching for a suitable nest site, the birds perform a slow tapping, which tends to attract the mate. At about this time the female begins to solicit copulation, but copulation frequency reaches a peak during actual excavation. The male does most of the excavation, excavating throughout the day and sometimes sleeping in the cavity at night. The clutch size is usually four eggs, and the incubation period lasts 11–15 days. During incubation the male continues to be most attentive to the eggs, even during daylight. Both sexes brood the nestlings for more than two weeks. After about 17 days the adults begin to feed the young from outside the nest rather than going inside with food. When the young are 28–30 days old they emerge from the nesting hole and are soon able to fly strongly.

Suggested Viewing Opportunities: This widespread species is often found among cottonwood groves in winter.

Selected References: BNA 702, rev. ed. (J. A. Jackson, H. R. Quellet, and B. J. Jackson 2018); Bent 1939; Short 1982; Winkler, Christie, and Nurney 1995; Shunk 2016.

NORTHERN FLICKER. *COLAPTES AURATUS*

Status: This woodpecker is a common summer resident statewide but is occasionally present during winter months. National Breeding Bird Surveys from 1966 to 2015 indicate that this species' population underwent a survey-wide average annual decline of 1.33 percent (Sauer et al. 2017). Nebraska surveys indicate a 2.01 percent annual decline, based on a sample of 49 routes. There were 496 possible to confirmed nesting records during the fieldwork for *The Second Nebraska Breeding Bird Atlas* (Mollhoff 2016).

Habitats and Ecology: Broadly distributed, flickers are unusual among Nebraska's woodpeckers because much of their food consists of insects such as ants and beetles that are obtained by probing in the ground. Flickers are often found nesting in cottonwood stands and riparian woodlands, especially where snags of fairly soft-wooded trees such as cottonwoods are available. There they excavate cavity nests that later are used by many other hole-nesting songbirds and by small tree-dwelling mammals such as fox squirrels and deer mice. There are typically six to eight eggs in a clutch, and incubation requires 11–13 days. Fledging occurs at 25–28 days. Where both the red- and yellow-shafted races occur together, as in central Nebraska, interracial matings are frequent, and intermediate plumages are common, as well as probable backcross plumages (Short 1965). These intermediate individuals evidently survive as well as do apparently "pure" phenotypes and are fully fertile. The zone of overlap and hybridization in Nebraska is very wide, and birds showing varying traits of red-shafted flickers often can be found in eastern Nebraska, just as intergrade phenotypes that variably resemble yellow-shafted flickers often appear in western Nebraska.

Breeding Biology: This species is relatively migratory over most of the state, and when returning to the nesting area both sexes seek out their old territories and nest sites. Males tend to arrive a few days before females and soon begin uttering location calls and drumming as a territorial advertisement. Recognition of previous mates is apparently site induced, and sex recognition is based on the "moustache" markings of the male. Courtship displays include exposing the undersides of wing and tail, bobbing, and billing ceremonies. Males apparently select the nest site, often using a previous year's nest or starting to excavate a new one. The male does most of the excavation, and copulation typically occurs just before the nest is finished. The clutch size is usually six to eight eggs, and the incubation period lasts 11–13 days. The eggs are laid at daily intervals, and both sexes share incubation, with the male assuming most of the nocturnal responsibilities. Both sexes care for and brood the young, feeding them by regurgitation, and the males again take most of the responsibility. The nestling period is about 26 days, but the parents continue to feed their offspring for some time after they leave the nest.

Suggested Viewing Opportunities: This widespread species is sometimes found among cottonwoods in winter (Dorn and Dorn 1990) and is generally found where large dead trees such as cottonwoods and elms are abundant.

Selected References: BNA 166, rev. ed. (K. L. Weibe and W. S. Moore 2017); Bent 1939; Kilham 1961; Short 1982; Winkler, Christie, and Nurney 1995; Shunk 2016.

Family Falconidae (Falcons)

AMERICAN KESTREL. *FALCO SPARVERIUS*

Status: The kestrel is an uncommon summer resident, and some birds overwinter during mild years. National Breeding Bird Surveys from 1966 to 2015 indicate that this species' population underwent a survey-wide average annual decline of 1.39 percent (Sauer et al. 2017). Nebraska surveys indicate a 2.01 percent annual increase, based on a sample of 48 routes. During *The Second Nebraska Breeding Bird Atlas* surveys, there were possible to confirmed breeding records from nearly every Nebraska county (Mollhoff 2016).

Habitats and Ecology: This is an open-country falcon that nests in tree cavities previously excavated by woodpeckers or in natural cavities of large trees. It will also nest in rocky crevices or similar cavities. It is the only Nebraska hawk that will nest in nest boxes or woodpecker holes. Its usual clutch size is four or five eggs, laid at intervals of two or three days. The female performs incubation, with the male providing her and the nestling young with food. The eggs hatch in 29–30 days, and fledging occurs at about 30 days. The kestrel is the smallest and most insectivorous of Nebraska's falcons. However, on a biomass basis, birds are generally the most important single prey component, with mammals second. Other prey types such as reptiles and insects are of less importance, although on a numerical basis, insects and other invertebrates might constitute almost 100 percent of the entire prey base (Johnsgard 1990).

Breeding Biology: American kestrels are perhaps the most sociable of the falcons and until pair formation might associate in small groups. During the courtship period, males perform aerial dive displays, whining vocalizations, and courtship-feeding. Courtship-feeding serves to maintain the pair bond and also provides food for the female and her young. Sometimes the female begs for food in flight by performing a distinctive "flutter-glide" display. Copulations reach a peak just before egg-laying, but courtship-feeding peaks during the egg-laying period and continues through the brooding period. The clutch size is usually four to five eggs, and the incubation period lasts 29–30 days. The female does nearly all the incubating, and the young often hatch at intervals of about a day in the same sequence as the egg-laying. During the fledging period of approximately 30 days, the male continues to do most of the food-gathering while the female broods and directly feeds the young.

After about 20 days the adults bring in entire prey animals and place them in the nest for the young birds to tear apart and feed themselves. The family typically remains together for some time after fledging, although the young must eventually catch their own food.

Suggested Viewing Opportunities: For most of the year this species can be found in any fairly open-country habitat, especially where there are some trees with woodpecker holes for nesting and perching sites such as overhead lines.

Selected References: BNA 602 (J. A. Smallwood and D. M. Bird 2002); Bent 1938; Balgooyen 1976; Glinski 1998; Johnsgard 1990; Ferguson-Lees and Christie 2001.

MERLIN. *FALCO COLUMBARIUS*

Status: An uncommon resident in the Pine Ridge and a wintering migrant statewide. National Breeding Bird Surveys from 1966 to 2015 indicate that this species' population underwent a survey-wide average annual increase of 2.63 percent (Sauer et al. 2017). This species is on the Nebraska Natural Legacy Project's Tier 2 list of threatened species in Nebraska (Schneider et al. 2018). There was a possible and a probable nesting record during the fieldwork for *The Second Nebraska Breeding Bird Atlas* (Mollhoff 2016).

Habitats and Ecology: This is a forest- and woodland-adapted falcon, usually breeding in clumps of open woodlands, often in bottomlands or valleys. During the nonbreeding season they also forage in grasslands, agricultural lands, desert scrub, and marshes or shorelines.

Breeding Biology: Typically, males return to the breeding ground prior to females and begin calling while flying from one perch to another. Little actual nest-building is done, since the birds usually take over an already constructed nest of another species. The clutch size is usually five or six eggs. The eggs are laid at two-day intervals, and the female begins to incubate when the clutch nears completion. Probably the only times the male takes over incubation are during the short periods the female is off the nest eating food the male has brought. The incubation period lasts 28–32 days. The eggs hatch at intervals of about two days, resulting in marked size differences among the young, which develop rapidly. Early in the brooding period the male brings all the food and passes it on to the female, who then tears it up and divides it among the young. Later the female assists in hunting and bringing in food, which mostly consists of small birds. The young fledge in 25–30 days but remain in the vicinity of the nest for some time. They initially begin to hunt by catching insects but soon learn to chase and capture young birds.

Suggested Viewing Opportunities: This species is mostly an overwintering migrant in Nebraska, with initial historic fall sightings usually occurring in late October, and final spring sightings in late April. It is an open-country predator and often forages around wetlands.

Selected References: BNA 44, rev. ed. (I. G. Warkentin, N. S. Sodhi, R. H. M. Espie, A. F. Poole, W. Oliphant, and P. C. James 2005); Bent 1938; Johnsgard 1990; Glinski 1998; Ferguson-Lees and Christie 2001.

PEREGRINE FALCON. *FALCO PEREGRINUS*

Status: Extirpated in the wild but locally reestablished in Omaha and Lincoln directly or indirectly through rearing and release programs (hacking) of young birds (Johnsgard 2010b). National Breeding Bird Surveys from 1966 to 2015 indicate that this species' population underwent a survey-wide average annual increase of 2.77 percent (Sauer et al. 2017). This species is on the Nebraska Natural Legacy Project's Tier 2 list of threatened species in Nebraska (Schneider et al. 2018).

Habitats and Ecology: This species is largely a cliff-nesting species but has adapted to city-nesting on very tall buildings (usually at least 11 stories high). Nonbreeders occur over a wide habitat range, from mountain meadows to grasslands, marshes, and riparian habitats.

Breeding Biology: After a pair has returned to its nesting area, the male or both birds perform courtship flights consisting of diving and swooping and sometimes passing food in the air. Mating is frequent during this period and the period of egg-laying, with the eggs laid at intervals of about two or three days. The clutch

30. Peregrine falcon, stooping on a bufflehead.

size is usually three or four eggs, and the incubation period lasts 28–29 days. The female does most of the incubation, with the male bringing prey to his mate and also occasionally relieving her. The female typically eats away from the nest on a nearby "plucking post." For the first two weeks after hatching, the female does nearly all the brooding and feeding of the young, but later both adults hunt extensively and simply drop their prey into the nest, letting the young birds compete for it and tear it up. Usually only two or three young fledge from each brood; fledging occurs about 35–42 days after hatching. With the advent of modern pesticides and before they were outlawed, virtually no young were fledged at most nests, since the thin-shelled eggs laid by pesticide-poisoned females failed to hatch, or else the young did not survive to fledging.

Suggested Viewing Opportunities: The easiest way to see peregrine falcons in Nebraska is to survey the state capitol area in Lincoln or the Woodmen Tower in Omaha during the breeding season. Wintering migrants often arrive in September and leave by early May. They frequently forage around wetlands for waterfowl and larger shorebirds.

Selected References: BNA 660, rev. ed. (C. M. White, N. J. Clum, T. J. Cade, and W. C. Hunt 2002); Bent 1938; Baker 1967; Johnsgard 1990, 2010b; Glinski 1998; Ferguson-Lees and Christie 2001.

PRAIRIE FALCON. *FALCO MEXICANUS*

Status: A rare summer or year-round resident in the Panhandle, mainly in rimrock areas offering open country for hunting, and a seasonal wintering migrant elsewhere. National Breeding Bird Surveys from 1966 to 2015 indicate that this species' population underwent a survey-wide average annual increase of 1.05 percent (Sauer et al. 2017). There were 17 possible to confirmed nesting records during the fieldwork for *The Second Nebraska Breeding Bird Atlas* (Mollhoff 2016).

Habitats and Ecology: Prairie falcons are largely associated with plains, sagebrush, or other scrub habitats with steep cliffs nearby for nesting. In montane regions, tundra habitats also support breeders, and foraging might be done on mountain meadows or similar alpine habitats. Breedings have been reported most regularly in the Pine Ridge and Wildcat Hills. Scattered nestings or nesting attempts have also been made in Cherry, Garden, Keith, and Hitchcock Counties.

Breeding Biology: Prairie falcons probably initially nest when they are two years old; yearlings normally wander during the breeding season. The birds arrive on Nebraskan nesting grounds in late February or early March, and the male engages in aerial courtship for about a month while the pair examines potential nest sites. Frequently, nest sites of the previous year are used, even if the female is mated to a new male. The male does most of the hunting for the pair during the courtship period, and the female later undertakes nearly all the incubation. The clutch size is usually four or five eggs, and the incubation period lasts 29–31 days. Only when the female is eating food brought by her mate does he incubate, but the male performs the major role in nest defense. The young begin to acquire their flight feathers at about 30 days and fledge at about 40 days of age. Evidently a fairly high percentage of the young survive their first autumn, but there is an overall mortality rate of about 80 percent by the end of the first year of life.

Suggested Viewing Opportunities: Prairie falcons might occur anywhere in the state, with little seasonal pattern evident, but sightings mostly fall between September and April, with a possible peak in December.

Selected References: BNA 346, rev. ed. (K. Steenhof 2013); Bent 1938; Enderson 1964; Johnsgard 1990, 2001b; Anderson and Squires 1997; Glinski 1998; Ferguson-Lees and Christie 2001.

Order Passeriformes (Passerine Birds)

Family Tyrannidae (New World Flycatchers)

WESTERN WOOD-PEWEE. *CONTOPUS SORDIDULUS*

Status: A common spring and fall migrant and summer resident in the Panhandle, extending east in the Niobrara Valley to Cherry

County and probably Rock County, and in the North Platte Valley east at least to Garden County and possibly to Lincoln County. National Breeding Bird Surveys from 1966 to 2015 indicate that this species' population underwent a survey-wide average annual decline of 1.37 percent (Sauer et al. 2017). There were 68 possible to confirmed nesting records during the fieldwork for *The Second Nebraska Breeding Bird Atlas* (Mollhoff 2016).

Habitats and Ecology: Wood-pewees breed in most coniferous forest types and also to a varying extent in aspens, riparian forests, and various open deciduous or mixed woodland habitats. Open forests are favored, especially those dominated by conifers.

Breeding Biology: Rather late arrivals among the flycatchers, wood-pewees usually reach Nebraska in late May, near the end of the spring migration period. The nest is a surprisingly small, shallow, cuplike structure lined with fine grass or hairs. It is usually placed 8–20 feet above the ground, sometimes up to 45 feet high. The clutch size is usually three eggs, and the incubation period lasts 12–13 days. The female incubates, but the male occasionally feeds her and remains near the nest to help feed the young when they hatch. Like the nest, the juveniles closely resemble the surrounding bark and lichens. By about 14–18 days after hatching they are ready to leave the nest. They are probably dependent on the parents for food for some time after fledging, until they have become skilled in catching flying insects. They produce a single brood per season.

Suggested Viewing Opportunities: The best chances for seeing this species are in the pine forests of Sioux, Dawes, and Sheridan Counties between about May 13 and September 12.

Selected References: BNA 451 (C. Bemis and J. D. Rising 1999); Bent 1942; Eckhardt 1976.

EASTERN WOOD-PEWEE. *CONTOPUS VIRENS*

Status: A common spring and fall migrant and summer resident in eastern Nebraska, extending west in the Niobrara Valley to at least Cherry County, in the Platte Valley to at least Lincoln County and possibly to Keith County, and in the Republican Valley probably at least to Hitchcock and Hayes Counties. National Breeding Bird Surveys from 1966 to 2015 indicate that this species' population underwent a survey-wide average annual decline of 1.4 percent (Sauer et al. 2017). There were 189 possible to confirmed nesting records during the fieldwork for *The Second Nebraska Breeding Bird Atlas* (Mollhoff 2016).

Habitats and Ecology: The species is generally associated with deciduous forests, including floodplain and river-bluff forests at the western edge of its range. It is also found in woodlots, orchards, and suburban areas planted to trees. The eastern and western wood-pewees evidently come into local contact in the central Niobrara Valley and the eastern North Platte Valley, but there seem to be no proven cases of hybridization.

Breeding Biology: Like the western species, eastern wood-pewees usually reach the northern states near the end of the spring migration period. In Minnesota the species favors oak woodlands for nesting and continues to sing its distinctive three-noted song through nearly the entire summer. As is typical of the Tyrannidae, only the female constructs the nest, which is usually on the same branch year after year. In one case, a fork of an elm tree was used as a nest location by this species every year for 35 years. Nests are built on horizontal, often dead tree limbs, usually well out from the trunk, and 15–65 feet above the ground. They are placed on the tree bark, often but not necessarily in a crotch, and are well camouflaged by spider webs and lichens and might be easily overlooked. From two to four eggs are laid, usually three. The incubation period is 12–13 days, and the young are brooded by the female for most of the first four posthatching days. The chicks are feathered by seven days, and they leave the nest at 15–18 days.

Suggested Viewing Opportunities: The best chances for seeing this species are in the deciduous floodplain forests of Missouri Valley counties between about May 10 and September 10.

Selected References: BNA 245, rev. ed. (D. J. Watt, J. P. McCarty, S. W. Kendrick, F. L. Newell, and P. Pyle 2017); Bent 1942.

Status: A common spring and fall migrant and summer resident in wetlands of eastern Nebraska, extending west in the Niobrara Valley to at least Cherry County and possibly to Sioux County, in the Platte Valley to at least Keith and probably Garden County, and local in the eastern Republican Valley to Harlan County. National Breeding Bird Surveys from 1966 to 2015 indicate that this species' population underwent a survey-wide average annual decline of 1.48 percent (Sauer et al. 2017). There were 135 possible to confirmed nesting records during the fieldwork for *The Second Nebraska Breeding Bird Atlas* (Mollhoff 2016).

Habitats and Ecology: Willow flycatchers are specially associated with riparian or wetland habitats, including willow thickets, low gallery forests along streams, and prairie coulees.

Breeding Biology: In southern Michigan, males arrive at the nesting areas somewhat before females and begin to establish territories that average about 2 acres, always including shrubs and small trees, as well as clearings. Water is always present either on the territory or very close to it. Males usually sing from the highest point on the territory at a speed of up to 30 songs per minute. The female builds the nest, usually in upright crotches of shrubs that the returning birds can fly to directly. The clutch size is usually three to four eggs, and the incubation period lasts 12–13 days. Only the female is known to incubate. The young are fledged in 12–16 days. Fledglings continue to beg for food until they are about 24–25 days old, after they have become fairly adept at catching insects, and they remain in their parents' territory until fall.

Suggested Viewing Opportunities: This species is widespread and abundant throughout eastern Nebraska in thickets and wetland edges.

Selected References: BNA 533 (J. A. Sedgwick 2000); Bent 1942; Holcomb 1972.

LEAST FLYCATCHER. *EMPIDONAX MINIMUS*

Status: A common spring and fall migrant and possible summer resident in northern Nebraska. No definite nesting was discovered during fieldwork for *The Nebraska Breeding Bird Atlas* project, but possible nestings were reported from the Pine Ridge and the mouth of the Niobrara River, Knox County. There were eight possible and probable nesting records during the fieldwork for *The Second Nebraska Breeding Bird Atlas* (Mollhoff 2016). Other possible breeding records include pairs seen during June in Keya Paha and Brown Counties. National Breeding Bird Surveys from 1966 to 2015 indicate that this species' population underwent a survey-wide average annual decline of 1.71 percent (Sauer et al. 2017).

Habitats and Ecology: Least flycatchers are associated with open and edge-dominated habitats, such as mature deciduous floodplain forests with shrubby understories in prairie areas, scattered prairie grovelands, shelterbelts, woody lake margins, and urban parks or gardens.

Breeding Biology: Shortly after returning to their nesting areas, male least flycatchers establish breeding territories that are surprisingly small, averaging only 0.18 acre in one study. The territory usually also includes exclusive foraging areas, as well as a nest site; sometimes communal foraging areas are shared. Territories are advertised by the males' simple *cha-beck* songs and are primarily defended by males. Females defend only a small area around the nest. The female builds a nest in six to eight days, and the clutch size is usually four but might range from three to six. The incubation period lasts 14–16 days. The female also does all the incubating and brooding, although the male remains near the nest and occasionally feeds her. He also feeds the chicks when they hatch and at least initially provides most of their food. The juveniles leave the nest at 23–25 days of age and might leave the territory a few days after fledging, but they are dependent on their parents until about three weeks of age.

Suggested Viewing Opportunities: There are no proven breeding sites of this species in Nebraska, but well-forested locations in northeastern Nebraska such as Ponca State Park during May offer the best viewing possibilities.

Selected References: BNA 99, rev. ed. (S. Tarov and J. V. Briskie 2008); Nice and Collias 1961; Bent 1942; Davis 1959.

Status: A common spring and fall migrant and summer resident in the Pine Ridge (Sioux, Dawes, and Sheridan Counties) and a local breeder in the Wildcat Hills (Scotts Bluff County). There were six confirmed nesting records during the fieldwork for *The Second Nebraska Breeding Bird Atlas* (Mollhoff 2016). National Breeding Bird Surveys from 1966 to 2015 indicate that this species' population (including the Pacific Slope population, now classified as a separate species) underwent a survey-wide average annual decline of 0.3 percent (Sauer et al. 2017). This species is on the Nebraska Natural Legacy Project's Tier 2 list of threatened species in Nebraska (Schneider et al. 2018).

Habitats and Ecology: A widespread and adaptable small flycatcher, thus species ranges from riparian woodlands through aspens into the coniferous forest zone, but it favors riparian areas with shrubs and cottonwoods. Nests are often placed on ledges, including bridges and vacant buildings, in ravines with rock outcrops, on steep banks, and in similar sites.

Breeding Biology: These flycatchers are highly aggressive, and their territorial aggression is directed not only toward their own species but also toward other species of similar size. Unmated males sing their advertising *ps-seet, ptsick see* notes throughout most of the day, whereas mated males sing only at dawn. Nests are constructed by one of the pair, probably the female, over a period of four or five days, and the first egg is laid within a day or two of the completion of the nest. The clutch size is usually four eggs, and the incubation period lasts 14–15 days. The presumed female performs the incubation, while the mate occasionally feeds her on the nest. During the first few posthatching days, the presumed female again does all the brooding and most of the feeding, but soon both parents are kept busy feeding the growing brood. The nestling period lasts from about 14 to 18 days, and for a period after the young fledge, the adults continue to feed them at a rate even greater than when they were in the nest. After about four days the female might stop feeding the brood and begin her second nest. As the young birds grow they become more independent and gradually drift away from the territory.

Suggested Viewing Opportunities: This is a locally common species in Pine Ridge woodlands and is usually found along watercourses.

Selected References: BNA 556, rev. ed. (P. E. Lowther, P. Pyle, and M. A. Patten 2016); Bent 1942; Davis, Fisher, and Davis 1963.

SAY'S PHOEBE. *SAYORNIS SAYA*

Status: A common spring and fall migrant and summer resident over most of western Nebraska, especially the Panhandle east to Sheridan County, and southwestern Nebraska east to Gosper and Furnas Counties. Scattered nestings or possible nestings have occurred farther east. National Breeding Bird Surveys from 1966 to 2015 indicate that this species' population underwent a survey-wide average annual increase of 1.14 percent (Sauer et al. 2017). There were 91 possible to confirmed nesting records during the fieldwork for *The Second Nebraska Breeding Bird Atlas* (Mollhoff 2016).

Habitats and Ecology: Say's phoebes are generally associated with grasslands, brushlands, and agricultural areas in the state, especially prairie coulees and steep, eroded riverbanks.

Breeding Biology: Male phoebes arrive on the nesting grounds before females, and when the females return, pairs form or re-form rather rapidly. Nest-building or the repair of a previous nest might begin only a week or two after the birds arrive and is presumably done by the female. Often the same nest is used in subsequent years or for successive clutches. At favored nest sites the loss of one or both members of a pair brings a rapid replacement. The clutch size is usually four or five eggs, and the incubation period lasts 12–14 days. Males remain nearby on a convenient lookout post. However, the males guard their nests and feed their mates, as well as feeding the brood virtually alone during the first week or so of their lives. When a brood is ready to leave the nest at about two weeks of age, the male takes over and teaches them to capture insects, while the female prepares to produce a second clutch of eggs. Apparently she assumes the entire job of feeding herself and her second clutch, although reportedly the male might again appear to take care of the brood when it fledges, freeing the female for a possible third brood.

Suggested Viewing Opportunities: This is a typical species of open country and is often found perching on low fences along dry country roads. It often breeds where human construction such as bridges or abandoned buildings offers nesting sites or where rocky outcrops provide cavities with overhead cover.

Selected References: BNA 374 (J. M. Shukman and B. O. Wolf 1998); Bent 1942; Schukman 1974; Ohlendorf 1976.

WESTERN KINGBIRD. *TYRANNUS VERTICALIS*

Status: A common spring and fall migrant and summer resident in most of the state, but infrequent in the easternmost counties. National Breeding Bird Surveys from 1966 to 2015 indicate that this species' population underwent a slight average annual decline of 0.06 percent (Sauer et al. 2017). There were 474 possible to confirmed nesting records during the fieldwork for *The Second Nebraska Breeding Bird Atlas* (Mollhoff 2016).

Habitats and Ecology: This species is always associated with edge habitats near open country, such as shelterbelts, hedgerows, margins of forests, tree-lined residential districts, and riparian forests. It occupies more open country than the eastern or Cassin's kingbird.

Breeding Biology: Like the eastern kingbird, this species is highly territorial and generally is extremely intolerant of larger birds such as crows and hawks in the vicinity of its nest. The birds are also at least as noisy as the eastern kingbird, and during the early stages of territorial establishment and pair formation, they are particularly conspicuous for their calling and singing. Yet in spite of their overt aggressiveness, there are reported cases of several pairs occupying the same tree or sharing a small grove for nesting. A traditional return to a previous year's nesting territory is common, as it is in eastern kingbirds. The clutch size is usually four eggs, and the incubation period lasts 12–14 days. The female does most of the incubating, but males have been seen on the nest as well. Likewise, both parents actively feed the young, which remain in the nest for about two weeks. In one observed case in Oklahoma, a pair began a new nest only four days after its first brood fledged, but presumably the young birds remain at least partially dependent on their parents for some weeks after fledging.

Suggested Viewing Opportunities: This is a nearly ubiquitous breeder in open country throughout the entire state, commonly along riparian woodlands in otherwise tree-free habitats. In Nebraska the birds occur statewide, especially where flying insects are abundant.

Selected References: BNA 227, rev. ed. (L. R. Gamble and T. M. Bergen 2012); Bent 1942; Hespenheide 1964; Bergen 1987.

EASTERN KINGBIRD. *TYRANNUS TYRANNUS*

Status: A common spring and fall migrant and summer resident throughout the state, becoming slightly less common than the western kingbird in the Panhandle. A total of 538 possible to confirmed breeding records was reported for every county during fieldwork for *The Second Nebraska Breeding Bird Atlas* (Mollhoff 2016). National Breeding Bird Surveys from 1966 to 2015 indicate that this species' population underwent a survey-wide average annual decline of 1.28 percent (Sauer et al. 2017).

Habitats and Ecology: Eastern kingbirds are associated with open areas with scattered trees or tall shrubs, such as forest edges, fencerows, riparian woods, and farmsteads. In Nebraska the birds occur statewide at lower elevations, especially where scattered taller trees occur in otherwise open landscapes. In Colorado they occupy many of the same habitats as western kingbirds but are more likely to occur in woodlands adjacent to open water. Cottonwoods and elms are favored nesting trees there (Kingery 1998). In western Nebraska it is common to see an oriole pair nesting in the same tree where an active kingbird nest is present; the orioles apparently gain some protection from the presence of the highly territorial kingbirds.

Breeding Biology: Eastern kingbirds arrive on their breeding grounds when insect populations begin to become noticeable and soon become extremely conspicuous as the males begin territorial behavior and associated courtship. Aerial displays are common then, with the bird flying erratically in a series of swoops and dives not far above the ground and uttering harsh screams. Chases and fights between birds on adjacent territories are also prevalent at this time. Kingbirds have a strong tendency to return to the same nesting territory in subsequent years, although the specific nest site varies from year to year. Males help build the nest, but the female typically does the incubating. The clutch size is usually three to four eggs, and the incubation period lasts 12–14 days. After the eggs hatch both sexes are kept constantly busy bringing food to the young, the female generally being more active in feeding and brooding than her mate. The young are in the nest for approximately two weeks. After fledging the young remain as a group, often perched on a wire or an exposed tree branch, waiting for their parents to come and feed them. The young birds begin to catch flying insects after they are about eight days out of the nest, and they continue to improve in their flight abilities over the next month. The adults stop feeding them by about 35 days after fledging.

Suggested Viewing Opportunities: This is a widespread and common species throughout the state, with a somewhat more easterly (more woodland-adapted) distribution than the western kingbird.

Selected References: BNA 253, rev. ed. (M. T. Murphy and P. Pyle 2018); Bent 1942; Bergen 1987.

Family Laniidae (Shrikes)

LOGGERHEAD SHRIKE. *LANIUS LUDOVICIANUS*

Status: An increasingly uncommon migrant and infrequent summer resident nearly throughout the state, but most common in the Panhandle and the Sandhills. National Breeding Bird Surveys from 1966 to 2015 indicate that this species' population underwent a survey-wide average annual decline of 2.78 percent (Sauer et al. 2017). There were 181 possible to confirmed nesting records during the fieldwork for *The Second Nebraska Breeding Bird Atlas* (Mollhoff

2016). This species is on the Nebraska Natural Legacy Project's Tier 1 list of threatened species in Nebraska (Schneider et al. 2018).

Habitats and Ecology: Like the northern shrike, this species is associated with open habitats having scattered perching sites and ranges altitudinally from agricultural lands on the prairies to montane meadows. Sagebrush, desert scrub, and pinyon-juniper woodlands offer ideal nesting and foraging areas, but some nesting also occurs in woodland-edge situations, farmlands, and similar habitats.

Breeding Biology: On a study area in north-central Colorado, nests were separated by at least 400 meters (1,300 feet), and nest sites of previous years were often used. The clutch size is usually four to five eggs, and the incubation period lasts 14–16 days. The female incubates, while her mate feeds her, and later both sexes bring food to the young. The fledging period is normally 17 days but might require up to 20 days if food is limited. The young birds gradually learn how to wedge food items in forks or impale them on thorns or other sharp objects. Impaling is apparently a method of short-term food storage and is thus most likely to occur in shrikes that are not hungry.

Suggested Viewing Opportunities: This species is usually found in the same open-country habitats as the migrant northern shrike and is often seen perching on telephone or fence wires along country roads. It has become quite rare in recent years, even in the Sandhills, its prime Nebraska habitat.

Selected References: BNA 231 (R. Yosef 1996); Bent 1950; Porter et al. 1975.

Family Vireonidae (Vireos)

WARBLING VIREO. *VIREO GILVUS*

Status: A very common spring and fall migrant and summer resident in forests throughout the entire state. It is the most common and widespread vireo of Nebraska, although the Bell's vireo is more common in shrubby habitats. It is least common in drier parts of the Sandhills and the southern Panhandle. National Breeding Bird

Surveys from 1966 to 2015 indicate that this species' population underwent a survey-wide average annual increase of 0.85 percent (Sauer et al. 2017). There were 395 possible to confirmed nesting records during the fieldwork for *The Second Nebraska Breeding Bird Atlas* (Mollhoff 2016).

Habitats and Ecology: Fairly open woodlands, especially of deciduous trees, are favored by this species. It is probably most common along riparian forests supporting tall trees but also occurs in aspen groves and well-wooded residential or park areas, especially where tall cottonwoods are present. In coniferous forest areas, the birds favor areas where single or clumped broad-leaved trees such as aspens or birches occur. Foraging is done near the crowns of fairly densely leaved trees, and nests are sometimes located as high as 90 feet above the ground in very tall forests. In Colorado, most breeders occupy aspen woodlands (Kingery 1998). Relatively few nest in the eastern riparian woodlands of prairie rivers, whereas in Nebraska these constitute a primary nesting habitat.

Breeding Biology: In spite of its widespread occurrence, this species has not been extensively studied, perhaps because many of its breeding activities take place high in trees. J. J. Audubon reported that both sexes helped build the nest, which is unusual in vireos, and that eight days were required to complete it. The clutch size is usually four eggs, and the incubation period lasts 12–13 days. Both sexes incubate, and males often sing while on the nest. The fledging period is about 16 days. Adult birds continue to sing well into the summer, and young males learn to sing fairly well before they leave in fall. In this unusual respect, this species resembles the red-eyed and yellow-throated vireos.

Suggested Viewing Opportunities: Riparian willow and cottonwood stands are highly attractive to this most common Nebraska vireo. The male's loud and distinctively syncopated songs, uttered even while he is sitting on the nest, are distinctive.

Selected References: BNA 551 (T. Gardali and G. Ballard 2000); Bent 1949; Sutton 1949; Dunham 1964.

Family Corvidae (Crows, Jays, and Magpies)

PINYON JAY. *GYMNORHINUS CYANOCEPHALUS*

Status: A local and rare resident in pine woodlands of the Pine Ridge (Sioux and Dawes Counties) and Wildcat Hills (Banner and Morrill Counties). National Breeding Bird Surveys from 1966 to 2015 indicate that this species' population underwent a survey-wide average annual decline of 3.69 percent (Sauer et al. 2017). This species is on the Tier 1 list of the Nebraska Natural Legacy Project's highly threatened species in Nebraska (Schneider et al. 2018). There were four confirmed nesting records during the fieldwork for *The Second Nebraska Breeding Bird Atlas* (Mollhoff 2016).

Habitats and Ecology: Pinyon jays are generally associated with pine forests growing on dry substrates, but they extend during the nonbreeding period into mountain mahogany, sagebrush, and arid scrub habitats. Where pinyon pines are lacking, as in Nebraska, they use juniper woodlands and consume juniper berries (Kingery 1998).

Breeding Biology: Pinyon jays are highly gregarious, usually gathering in flocks of up to 50 birds for much of the year. Pair bonds are probably rather permanent in these flocks, and as early as mid-November males begin feeding their mates by transferring pine seeds or other morsels. First-year birds also perform this behavior, although initial breeding might not occur until they are at least another year older. Later, females actively solicit feeding by courtship begging, a display that continues through nesting and stimulates the male to feed his incubating mate. In late stages of courtship, a male might pick up a bit of vegetation, present it to his mate, and then fly up into a nearby tree, as if to lure her away from the flock. In this way, specific courtship crotches or branches are established, although the actual nest is often built in another location. Nests are usually placed on the south side of trees, probably for warmth. Both members of a pair build them, usually over about a week, and the first egg is laid about two days later. Most birds in a colony begin and complete their nests at nearly the same time; the colony's location is related to the caches of pine seeds from the previous fall. The clutch size is usually three or four eggs, and the incubation period lasts 17 days. Fledging occurs about three weeks after hatching, and the parents remain with their young for a prolonged period, continuing to feed them well after they are fledged.

Suggested Viewing Opportunities: Ponderosa pine woodlands in Sioux and Dawes Counties might offer the best opportunities for seeing this very locally distributed arid-land species; no pinyon pines occur in Nebraska.

Selected References: BNA 605 (R. P. Balda 2002); Bent 1946; Madge 1993; Savage 1997.

BLACK-BILLED MAGPIE. *PICA HUDSONIA*

Status: A local and increasingly rare resident in western Nebraska. National Breeding Bird Surveys from 1966 to 2015 indicate that this species' population underwent a survey-wide average annual decline of 0.49 percent, although in Nebraska the rate of decline is estimated at 9.43 percent annually (Sauer et al. 2017), and since the advent of West Nile disease in the early 2000s, the species has all but disappeared over much of central Nebraska and even the Panhandle. There were 53 possible to confirmed nesting records during the fieldwork for *The Second Nebraska Breeding Bird Atlas* (Mollhoff 2016). This species is on the Nebraska Natural Legacy Project's Tier 1 list of threatened species in Nebraska (Schneider et al. 2018).

Habitats and Ecology: Magpies use diverse habitats but are especially common in riparian areas with thickets, agricultural areas with scattered trees, sagebrush, cottonwood groves, and the lower levels of the coniferous forest zones. Small, thorny trees are especially favored nest sites, but junipers and similar densely vegetated trees are also used.

Breeding Biology: At least in some areas, pairs often remain in the general vicinity of their breeding areas through the winter period, and many use old nests for nighttime brooding during cold weather. Rarely, however, are old nests used again for nesting; new ones are typically built each year and for each breeding attempt.

In Nebraska, birds often begin carrying mud to anchor nest bases in late February, but intensive nest-building does not occur until mid-March or April. Both sexes gather materials, the male bringing more sticks than the female, and rarely each partner will begin a nest at a different location. A surprisingly long average period of 43 days is required to complete a nest, and during the latter part of this time intensive displaying also occurs, especially courtship-feeding of the female. The clutch size ranges from five to nine eggs. The female does all the incubating, but her mate feeds her throughout the incubation period of 17–22 days. Both sexes feed the nestlings equally, and they remain in the nest for an average of nearly four weeks. After the young are able to fly well, the family gradually wanders out of its nesting area. Although it is known that birds sometimes acquire territories and breed when a year old, it is likely that most initial breeding occurs during the second year.

Suggested Viewing Opportunities: This declining species might be seen in dry habitats of the western Panhandle, especially in places with scattered small trees and that are fairly close to water. Finding the large, globular stick nests of the species provides a good clue to the presence of the birds.

Selected References: BNA 389 (C. H. Trost 1999); Bent 1946; Balda and Bateman 1973; Madge 1993; Savage 1997.

AMERICAN CROW. *CORVUS BRACHYRHYNCHOS*

Status: A common summer or year-round resident in wooded habitats throughout the state. National Breeding Bird Surveys from 1966 to 2015 indicate that this species' population underwent a slight average annual increase of 0.07 percent (Sauer et al. 2017). The Nebraska crow population now appears to be stable, following a slow recovery from West Nile disease. There were 363 possible to confirmed nesting records during the fieldwork for *The Second Nebraska Breeding Bird Atlas* (Mollhoff 2016).

Habitats and Ecology: Crows breed in all forested areas, including wooded river bottoms, orchards, farms, woodlots, large parks, and suburban areas.

Breeding Biology: In Nebraska, crows begin to flock after the breeding season, and at least in northern areas, they tend to migrate some distance southward, where they develop massive roosting flocks. It is likely, however, that pairs are maintained within these large flocks, and shortly after returning to the breeding areas the birds typically become well spaced and territorial. Crows utter a surprisingly broad range of notes, including more than a dozen distinct calls, and in addition they commonly mimic other species. Both sexes help build the nest, and although it has been reported that both sexes incubate, this seems unlikely in light of what is known of related species. The clutch size is usually four to five eggs, and the incubation period lasts 18 days. The young birds fledge in about 36 days but remain with their parents for a protracted period. Like jays, crows often move about in family groups, whose members quickly alert each other when they find food or observe possible danger.

Suggested Viewing Opportunities: Crows are widespread in Nebraska at all seasons and didn't suffer from the prolonged effects of West Nile disease to the degree of blue jays and magpies.

Selected References: BNA 647 (N. A. M. Verbeek and C. Caffrey 2002); Bent 1946; Madge 1993; Savage 1997.

Family Alaudidae (Larks)

HORNED LARK. *EREMOPHILA ALPESTRIS*

Status: This species has one of the broadest ranges of all North American songbirds, extending from high Arctic tundra to hot barren deserts. In Nebraska it is most common in treeless areas of the Sandhills and Panhandle and is present year-round. National Breeding Bird Surveys from 1966 to 2015 indicate that this species' population underwent a survey-wide average annual decline of 2.46 percent (Sauer et al. 2017). Nebraska surveys indicate a 0.47 percent annual decline, based on a sample of 51 routes. During the more recent *Second Nebraska Breeding Bird Atlas* surveys there were 463 possible to confirmed breeding records from nearly all of Nebraska's counties (Mollhoff 2016).

31. Grassland birds: McCown's longspur (top left), chestnut-collared longspur (top right), lark bunting (middle left), horned lark (middle right), mountain plover (bottom).

Habitats and Ecology: Horned larks are notably cold-tolerant, and they are often the only sign of life one might see when driving Nebraska's snow-rimmed roads during the middle of winter. In the midwestern states, horned larks begin to establish and defend territories in January and February, while pairs are being formed. Territories are large (averaging about 4 acres in two studies) and are defended by males, but only against other males.

Breeding Biology: Two advertisement songs are uttered either on the ground or in the air. These songs seem to be related to courtship rather than to territorial defense and are most common after losing a mate or fledging a brood. Courtship-feeding and other displays are also performed at this time. The female selects a nest site almost anywhere in the territory and constructs the nest alone. She digs a cavity with her bill and feet, often "paving" it on one side with various objects, for still uncertain reasons. The paving might cover and hide the fresh dirt that has been dug out, or it might help keep the nest lining from blowing away during early nest-building stages. Horned lark nests are typically in a small hollow sheltered by a clump of grass or beside a rock. The clutch size is usually four eggs, and the incubation period is about 10–14 days. The female sometimes begins incubation slightly before the clutch is completed but more often begins when the last egg is laid. There is a relatively short nestling period of 9–12 days, averaging about 10, and the young birds leave the nest when their flight feathers are only about one-third to one-half grown. Until they are at least 15 days old, the young are able to fly only a few yards. The young birds begin to flock soon after leaving the nest, and in many areas the female shortly thereafter begins a second nesting.

Suggested Viewing Opportunities: Horned larks occur statewide and might easily be found during almost any season in open, grassy country. They seem to be most abundant in the Sandhills, where they often forage along roadsides and perch on low fence posts.

Selected References: BNA 195 (R. C. Beason 1995); Bent 1942; Verbeek 1967; Beason and Franks 1974; Johnsgard 2001b.

Family Hirundinidae (Swallows)

BANK SWALLOW. *RIPARIA RIPARIA*

Status: An uncommon spring and fall migrant and summer resident in suitable habitats throughout the state, mainly at lower elevations. National Breeding Bird Surveys from 1966 to 2015 indicate that this species' population underwent a survey-wide average annual decline of 5.33 percent (Sauer et al. 2017), the steepest rate of decline of any Nebraska swallow. There were 145 possible to confirmed nesting records during the fieldwork for *The Second Nebraska Breeding Bird Atlas* (Mollhoff 2016).

Habitats and Ecology: Breeding by bank swallows almost always occurs near water, such as in steep banks along rivers, road cuts near lakes, gravel pits, and similar areas with steep slopes of silt, clay, or sand. Outside the breeding season the birds are of broader distribution, sometimes foraging over agricultural lands.

Breeding Biology: Shortly after bank swallows arrive in a nesting area, they begin to gather near the breeding site. Unpaired birds apparently select a burrow site, which might be the same burrow they used the previous year. Thereafter they defend the area from intrusion, although potential mates continue to return to a defended spot until one is eventually tolerated and accepted. Sexual chases of the female by the male are a common feature of pair formation, accompanied by male song. Both members of a pair utter another vocalization, the mating song, as they sit side by side or face each other in the burrow opening. This behavior might be a preliminary to copulation, which probably occurs in the nest chamber. When a burrow needs to be dug or deepened, both sexes share equally in the task. Then they gather materials such as feathers and grass for nest lining. The clutch size is usually four or five eggs, and the incubation period lasts 12–15 days. Incubation is performed by the female and might begin before the clutch is completed. Thus some eggs might hatch as early as 13 days after the clutch has been completed. Both parents alternate at brooding the young, and both feed the young and keep the nest clean. Birds as young as 20 days of age might be able to fly but often do not leave the nest for some time thereafter.

Suggested Viewing Opportunities: This is a widespread species in summer, when it nests along road cuts in steep earthen banks near water. Where these slopes are unstable the colonies might last for only a year or two.

Selected References: BNA 414 (B. A. Garrison 1999); Bent 1942; Peterson 1955; Turner and Rose 1989.

TREE SWALLOW. *TACHYCINETA BICOLOR*

Status: A common spring and fall migrant and summer resident throughout the state, but absent from treeless areas. National Breeding Bird Surveys from 1966 to 2015 indicate that this species' population underwent a survey-wide average annual decline of 1.38 percent (Sauer et al. 2017). There were 302 possible to confirmed nesting records during the fieldwork for *The Second Nebraska Breeding Bird Atlas* (Mollhoff 2016).

Habitats and Ecology: Breeding by tree swallows extends from riparian woodlands to ponderosa pine forests. Nesting is especially prevalent where old woodpecker holes are available but also occurs at times in birdhouses erected for bluebirds. In some areas such as along plains rivers, cottonwoods serve for nesting; an easy access to open foraging areas is important (Kingery 1998).

Breeding Biology: One of the earliest spring migrants of the state's swallows, these birds reach the northern plains in late April, at least a month before nesting gets under way. Much of the courtship apparently occurs in the air, and it includes synchronized flying by a pair. In one reported case, a male grasped the female's breast in midair, and the two birds tumbled downward until they almost reached the ground. The female then flew to the vicinity of the nest and perched, whereupon the male glided above her and landed on her back. The female constructs the nest with little or no assistance from the male, at times carrying in more than 100 feathers for nest lining. Evidently the male brings food to his incubating mate only rarely; instead, she leaves the nest a few times during daylight to forage for herself. The males often spend the evening perched near the nest, leaving it for their own roosting sites only after dark. The clutch size is usually four to six eggs, and the incubation period lasts 13–16 days. The nestling period varies considerably, depending on brood size and thus on the rate of feeding, so that the young might spend as few as 16 days or as many as 24 days in the nesting cavity. After the young leave the nest, however, they rarely return to it.

Suggested Viewing Opportunities: Tree swallows are often associated with tree stands having good woodpecker populations. They are also very common among the riparian woodlands of Nebraska, such as along the Platte River.

Selected References: BNA 11, rev. ed. (D. W. Winkler, K. K. Halinger, D. R. Ardia, R. J. Robertson, B. J. Stutchbury, and R. R. Cohen 2011); Bent 1942; Stocek 1970; Turner and Rose 1989.

VIOLET-GREEN SWALLOW. *TACHYCINETA THALASSINA*

Status: A common spring and fall migrant and summer resident in mountainous areas, breeding locally in the Pine Ridge (Sioux and Dawes Counties) and Wildcat Hills (Scotts Bluff, Banner, and Morrill Counties). National Breeding Bird Surveys from 1966 to 2015 indicate that this species' population underwent a survey-wide average annual decline of 0.66 percent (Sauer et al. 2017). This species is on the Nebraska Natural Legacy Project's Tier 2 list of threatened species in Nebraska (Schneider et al. 2018). There were 24 possible to confirmed nesting records during the fieldwork for *The Second Nebraska Breeding Bird Atlas* (Mollhoff 2016).

Habitats and Ecology: This species is generally associated with open coniferous forests, such as ponderosa pines, but it also breeds in riparian woods and sometimes in urbanized areas. Nesting sites are rather variable and include old woodpecker holes, natural tree or cliff cavities, and occasionally also nests in birdhouses. Cliff-nest sites in Colorado are second only to upland deciduous forest habitats in nest-choice frequency (Kingery 1998).

Breeding Biology: The violet-green swallow is an unusually early migrant, usually arriving before other swallow species. It thus might not begin nesting for nearly a month after arrival. It spends some time seeking out a suitable cavity, and apparently the female makes the choice, with the male playing a minor role. But once the site is chosen, both sexes begin to bring in nesting materials, the female doing most of the carrying. About six days are spent in building the nest, and the female roosts on it at night. The clutch size is usually four or five eggs, and the incubation period lasts 13–15 days. Eggs are laid at daily intervals, and the female might begin incubating before the clutch is completed, although this does not happen often. Thus, the period of hatching is sometimes rather staggered and has been observed to require as long as five days. The female does most of the feeding of young and also broods during the first ten days or so after hatching. The fledging period is somewhat variable but averages about 23–25 days. After leaving the nest, neither the adults nor the young return to it.

Suggested Viewing Opportunities: This species occupies deep canyons during the breeding season. In Nebraska, the top of Scotts Bluff is a marvelous scenic viewing location.

Selected References: BNA 14, rev. ed. (C. R. Brown, A. M. Knoff, and E. Damrose 2011); Bent 1942; Edson 1943; Combellack 1954; Turner and Rose 1989.

NORTHERN ROUGH-WINGED SWALLOW.

STELGIDOPTERYX SERRIPENNIS

Status: A common spring and fall migrant and summer resident throughout the state, mainly in open habitats, with breeding reports from almost every county and with population numbers behind only the barn swallow and cliff swallow. National Breeding Bird Surveys from 1966 to 2015 indicate that this species' population underwent a survey-wide average annual decline of 0.53 percent (Sauer et al. 2017). There were 396 possible to confirmed nesting records during the fieldwork for *The Second Nebraska Breeding Bird Atlas* (Mollhoff 2016).

Habitats and Ecology: Associated with open areas, including agricultural lands, rivers and lakes, and grasslands near water, and breeding almost exclusively in cavities dug in earthen banks of clay, sand, or gravel. Ready-made cavities in rock are also used.

Breeding Biology: Almost as soon as they arrive on their nesting grounds, these swallows begin to show interest in suitable nesting sites, and they might seek out old kingfisher or bank swallow excavations that are still usable. Males establish a limited territory around a potential nest site, perching near it and pursuing females from it. Females carrying nesting materials are especially pursued, although this behavior might be associated more with copulation than with courtship. Copulation has not been described and presumably occurs in the nesting cavity. Evidently only the female gathers and carries nest-lining material; apparently neither bird does any excavating. About six days are needed to construct the rather bulky nest, but it might rarely take as long as 20 days. The clutch size is usually six or seven eggs, and the incubation period lasts 15–16 days. The female usually starts incubating with the laying of the penultimate egg, and hatching might extend a few hours or as long as several days. Brooding is done primarily if not exclusively by the female, but both sexes feed the young. The young birds are able to fly a few days before they leave the nest, which usually occurs at 18–21 days of age. Young birds rarely return to the nest after they leave it, and there is no evidence on how long the young remain dependent on their parents for food after fledging.

Suggested Viewing Opportunities: This swallow is found throughout Nebraska and can be seen along almost every stream with steep dirt banks or with rocky banks having deep crevices.

Selected References: BNA 234 (M. J. DeJong 1996); Bent 1942; Turner and Rose 1989.

BARN SWALLOW. *HIRUNDO RUSTICA*

Status: A common spring and fall migrant and summer resident throughout the state. A total of 540 possible to confirmed breeding records was reported for every county during fieldwork for *The Second Nebraska Breeding Bird Atlas* (Mollhoff 2016). National

Breeding Bird Surveys from 1966 to 2015 indicate that this species' population underwent a survey-wide average annual decline of 1.19 percent (Sauer et al. 2017).

Habitats and Ecology: Except for the urban-adapted purple martin, this species is the swallow that is most closely associated with humans. Although it might still occasionally nest on cliff or cave walls, its normal current nesting sites are the horizontal beams or upright walls of abandoned buildings, as well as around bridges and culverts.

Breeding Biology: Within about two weeks after their arrival in nesting areas, most barn swallows have formed pair bonds. Pair formation takes place on fences and utility lines near nesting areas; unpaired birds perch alone and sing and perch or fly between paired ones. Courtship also includes "flight songs" by flocks of swallows flying high and chasing each other. Both sexes gather mud for nests; when available, hair from livestock is added to the mud cup, and feathers are added later for lining. Many times an old nest is used, with new materials added as necessary. An average of about six days is needed to build a nest, and eggs are not laid until the nest is completed. Five eggs are a typical clutch, and the incubation period lasts 14–16 days. Only the female incubates in most nests, but in some cases males also participate. The nestling period in this species has been reported to average 21 days, with a range of 18–27 days. Partners are apparently not changed between broods, and the same nest is usually used again, often with more mud and feathers added to it. There is a gap of about a month between nesting cycles, and only about a third of the swallows in one New York study raised second broods. Second clutches most often have four rather than five eggs, but in one study, egg and nesting mortality rates were similar during the two nesting cycles.

Suggested Viewing Opportunities: This species is abundant, especially around farmyards or other human habitations.

Selected References: BNA 452, rev. ed. (M. B. Brown and C. R. Brown 2009); Bent 1942; Samuel 1971; Turner and Rose 1989.

CLIFF SWALLOW. *PETROCHELIDON PYRRHONOTA*

Status: A common to abundant spring and fall migrant and sum-mer resident throughout the state. National Breeding Bird Surveys from 1966 to 2015 indicate that this species' population underwent a survey-wide average annual increase of 0.72 percent (Sauer et al. 2017). There were 355 possible to confirmed nesting records during the fieldwork for *The Second Nebraska Breeding Bird Atlas* (Mollhoff 2016).

Habitats and Ecology: A wide variety of nesting areas are used by this species, but in the state under consideration vertical cliffsides and the sides or undersides of bridges are perhaps most commonly used. The nests are gourd-like structures made of small dried mud globules that are gathered by the birds and carried back in their beaks.

Breeding Biology: At least in the northern states, cliff swallows begin to pair immediately upon arrival at their nesting grounds. This activity takes place at or near the nest, and the pair bond appar-ently consists primarily of mutual tolerance at the nesting site. Male "primary squatters" persistently return to specific perching places, and their singing attracts secondary visitors to that location, some of which are unpaired females. Both sexes defend the nest site, and both bring mud to construct the nest, which requires nearly two weeks of effort. When the nest is nearly completed, copulation occurs in the nest cup, and copulatory behavior continues until the middle of the laying period. Many cliff swallows occupy old nests if they are still usable; otherwise, they construct entirely new ones. The clutch size is usually four or five eggs, and the incubation period lasts 12–14 days. Incubation might begin before the clutch is complete, and males regularly participate. There is a relatively long nestling period in this species, averaging about 24 days, and a relatively low proportion of females (27 percent in one study) attempt a second clutch. In at least some cases, females change mates for their second nesting, and a considerable amount of court-ship activity is evident between broods.

Suggested Viewing Opportunities: This species occurs throughout the state, especially where there are sheer rock cliffs, concrete bridges, or even large metal culverts, and is very probably the most abundant swallow in Nebraska. The sides of cement bridges are favored nest locations for cliff swallows in Nebraska, and colonies of 1,000 nests or more sometimes occur. These colonies were the basis for important long-term Nebraska studies of the benefits and costs of colonial nesting behavior (Brown and Brown 1996).

Selected References: BNA 149, rev. ed. (C. R. Brown, M. B. Brown, P. Pyle, and M. A. Patten 2017); Samuel 1971; Turner and Rose 1989; Brown and Brown 1996.

Family Paridae (Chickadees and Titmice)

BLACK-CAPPED CHICKADEE. *POECILE ATRICAPILLUS*

Status: A common resident in deciduous and coniferous forests throughout the state. National Breeding Bird Surveys from 1966 to 2015 indicate that this species' population underwent a survey-wide average annual increase of 0.61 percent (Sauer et al. 2017). However, between 2002 and about 2012 the national population dropped about 5.6 percent annually owing to an epidemic of West Nile disease, from which it has not yet recovered in Nebraska to approach the high numbers seen during the late 1980s. It is concentrated in the eastern third of the state, with fewer breedings in the western Sandhills, the Panhandle, southwestern Nebraska, and the region between the Platte and Republican Rivers. There were 226 possible to confirmed nesting records during the fieldwork for *The Second Nebraska Breeding Bird Atlas* (Mollhoff 2016).

Habitats and Ecology: Associated with a wide variety of wooded habitats of coniferous and hardwood types, chickadees breed wherever suitable nesting cavities exist. These typically consist of old woodpecker holes, but sometimes the birds excavate their own nest cavities in the rotted wood of dead stumps. Birdhouses built for wrens are also occasionally used.

Breeding Biology: Chickadees are largely nonmigratory, but winter flocking does occur. Pair bonds are weak or absent during this time, although there is enough contact to allow frequent re-pairing with past mates. Courtship is apparently simple, consisting mainly of the loud *phoe-be* song by males. Territories are not established until later, when the pair begins to excavate a nest site. Both sexes excavate, the female taking the lead, and both birds work intermittently during daylight. Eggs are laid daily and are covered with nesting material by the female, who also sleeps in the nest but does not begin incubating until the clutch of six to eight eggs is complete. The incubation period lasts 12–13 days. Only the female incubates, but the male feeds her at intervals. During the first week after hatching, the behavior of adults is similar to their behavior during incubation, but the male stops feeding the female, and both parents feed the young. After brooding is terminated, both sexes feed the young at about an equal rate, and the young birds leave the nest at 16–17 days of age. Fledglings are able to forage for themselves about ten days after leaving the nest, but they remain with their parents for three or four weeks. A small proportion of adults attempt a second brood.

Suggested Viewing Opportunities: This eastern-oriented chickadee is most common along wooded river valleys in the eastern quarter of Nebraska.

Selected References: BNA 39, rev. ed. (J. R. Foote, D. J. Mennill, M. Ratcliffe, and S. M. Smith 2010); Odum 1941, 1942; Bent 1946; Harrap and Quinn 1996.

Family Sittidae (Nuthatches)

RED-BREASTED NUTHATCH. *SITTA CANADENSIS*

Status: An uncommon resident in coniferous forests of the Pine Ridge (Sioux to Sheridan Counties), the Wildcat Hills (Scotts Bluff, Banner, and Morrill Counties), and the Niobrara Valley east from central Cherry County probably to western Holt County. There have been scattered breedings along other river drainages. There were 28 possible to confirmed nesting records during the fieldwork for *The Second Nebraska Breeding Bird Atlas* (Mollhoff 2016). National Breeding Bird Surveys from 1966 to 2015 indicate that this species' population underwent a survey-wide average annual increase of 0.72 percent (Sauer et al. 2017).

Habitats and Ecology: This nuthatch is limited largely to coniferous forests in Nebraska, primarily those of relatively tall pines. To a much more limited degree, deciduous riparian woodlands are sometimes also used.

Breeding Biology: During the winter, red-breasted nuthatches might remain paired if the food supply is good or if the birds are close to a bird-feeding site. At this season the birds maintain contact by uttering location calls, but by late winter unpaired males begin singing a series of plaintive *waa-aan* notes from tall trees that probably serve both territorial and courtship functions. A major behavior during pair formation is courtship-feeding of the female by her mate, which continues through the incubation period. Breeding occurs in nesting cavities of dead trees or the rotting portions of live trees, with the birds typically excavating their own nesting holes. Pairs seek out nesting sites together, the female making the final choice and doing the initial excavating. Courtship chases of the female are frequent during nest-building and might end in copulation. When the nest excavation is nearly finished, both pair members bring resin in their bills and spread it above and below the hole. This sticky material probably deters other animals from entering the hole. The clutch size is usually five to six eggs, and the incubation period lasts 12 days. The female does all the incubation, but both sexes feed the young, which fledge in periods that have been estimated to range from 14 to 21 days.

Suggested Viewing Opportunities: This nuthatch is most common in ponderosa pine forests of the Pine Ridge and Niobrara Valley.

Selected References: BNA 459 (C. K. Ghalambor and T. E. Martin 1999); Bent 1948; Kilham 1973; Harrap and Quinn 1996.

WHITE-BREASTED NUTHATCH. *SITTA CAROLINENSIS*

Status: An uncommon resident of deciduous forests and woodlands throughout western Nebraska, extending east in the Pine Ridge from Sioux to Sheridan County (*Sitta carolinensis nelsoni*). *S. c. cookei* extends west in the Niobrara drainage from the Missouri River confluence to Cherry County, in the Platte Valley to Deuel County, and in the Republican drainage to Dundy County. There were 242 possible to confirmed nesting records during the fieldwork for *The Second Nebraska Breeding Bird Atlas* (Mollhoff 2016). National Breeding Bird Surveys from 1966 to 2015 indicate that this species' population underwent a survey-wide average annual increase of 1.71 percent (Sauer et al. 2017).

Habitats and Ecology: Largely confined in eastern Nebraska (*Sitta carolinensis cookei*) to mature deciduous forests, but in the west this species (including *S. c. nelsoni*) is associated with lower-elevation coniferous forests, especially the ponderosa pine zone and the juniper woodland zone. Nesting occurs in old woodpecker holes or self-excavated holes in the rotted wood of dead or dying trees.

Breeding Biology: Apparently, white-breasted nuthatches maintain their pair bonds throughout most of the year and perhaps permanently, although during winter the paired birds roost in different areas and maintain little contact with each other. The male begins to sing in late winter, uttering early morning "rendezvous songs" from tall trees to attract the female. Males also sing and display directly to their mates when they arrive and might keep in touch with them during foraging by uttering a series of *wurp* notes. The female takes the initiative in choosing a nest site and does all the nest-building, but both sexes participate in "bill-sweeping" in and around the nest. This behavior is of uncertain significance but consists of lateral arc-like movements of the bill near the tree or cavity surface, sometimes while holding an insect or other object. It has been suggested that the odors thus spread might repel squirrels, which often eat bird eggs. The clutch size is usually five to nine eggs, and the incubation period lasts 12–14 days. The female does the incubating, but the male feeds her during egg-laying and incubation and later helps feed the young. The fledging period is approximately two weeks.

Suggested Viewing Opportunities: Almost any mature deciduous or coniferous woodland in the entire state is likely to support this nuthatch, but it is most common along the lower Niobrara and lower Platte Valleys.

Selected References: BNA 54, rev. ed. (T. C. Grubbs Jr. and V. V. Pravosudov 2008); Bent 1948; Kilham 1968, 1972; Harrap and Quinn 1996.

PYGMY NUTHATCH. *SITTA PYGMAEA*

Status: An uncommon resident in western areas, mainly in ponderosa pine woodlands of the Pine Ridge (Sioux, Dawes, and Sheridan Counties) and Wildcat Hills (Scotts Bluff, Banner, and Morrill Counties). There were 22 possible to confirmed nesting records during the fieldwork for *The Second Nebraska Breeding Bird Atlas* (Mollhoff 2016). National Breeding Bird Surveys from 1966 to 2015 indicate that this species' population underwent a survey-wide average annual decline of 0.43 percent (Sauer et al. 2017). This species is on the Nebraska Natural Legacy Project's Tier 2 list of threatened species in Nebraska (Schneider et al. 2018).

Habitats and Ecology: Primarily associated with ponderosa pines, pygmy nuthatches also occur locally in junipers. This tiny nuthatch generally forages fairly high in tall pines but nests closer to the ground in snags or stubs that have rotted trunks providing excavation opportunities. Mature trees with snags or rotting portions are preferred (Kingery 1998).

Breeding Biology: Pygmy nuthatches are more or less permanently territorial; males hold small territories and limit most defensive behavior to the nest site. This site might be an existing tree cavity, or the pair might excavate a new one. Sometimes three or more birds have been seen excavating a single site, and at least in some cases the extra birds are males, which may persist to help with nesting. Up to a month or more might be needed for excavation, and it requires one day to lay each egg. The clutch size is usually five to nine eggs, and the incubation period lasts 15–16 days. Only females incubate, but both sexes sleep in the cavity at night, and among the observed threesomes, all the birds roosted there. The male (or males) feed the female on or off the nest. The eggs typically hatch within a 24-hour span, and the young are fed by both adults and, when present, additional helpers, which are likely to be older siblings. The young fledge in 20–22 days but do not gain independence from the adults until they are approximately 45–50 days old.

Suggested Viewing Opportunities: This nuthatch can easily be found in mature ponderosa pine forests of the Pine Ridge and Wildcat Hills.

Selected References: BNA 567 (H. E. Kingery and C. K. Ghalambor 2001); Bent 1948; Norris 1958; Harrap and Quinn 1996.

Family Certhiidae (Treecreepers)

BROWN CREEPER. *CERTHIA AMERICANA*

Status: An uncommon resident in mature forested areas, probably throughout the state, but with only a few nesting records from the Pine Ridge (Dawes County) and the Niobrara (Brown County) and Missouri (Sarpy and Washington Counties) Valleys. There were eight possible to confirmed nesting records during the fieldwork for *The Second Nebraska Breeding Bird Atlas* (Mollhoff 2016). National Breeding Bird Surveys from 1966 to 2015 indicate that this species' population underwent a survey-wide average annual increase of 0.55 percent (Sauer et al. 2017). This species is on the Nebraska Natural Legacy Project's Tier 2 threatened species in Nebraska (Schneider et al. 2018).

Habitats and Ecology: Brown creepers are associated with forests throughout the year, including both deciduous and coniferous forests. Virtually all foraging is done on the trunks of fairly large trees, where the birds forage for insects in bark crevices and grooves. Dense stands of mature trees are favored for nesting, making the birds very sensitive to logging effects (Kingery 1998).

Breeding Biology: Although a few creepers might remain in the northern states through the winter, they are generally migratory, and in early spring small groups might be encountered foraging in loose flocks, maintaining contact by delicate *cree-cree-cree-ep* notes. It is known that during cold winter nights, the European species of creeper (*Certhia brachydactyla*) frequently roosts in cracks of tree trunks. Clinging woodpecker-like to the bark and supported by the tail, creepers can withstand subfreezing temperatures even when partly covered by snow. When spring returns the pair begins to work intermittently on the nest, which might require a month to finish. Both sexes bring materials, but only the female does the construction. The clutch size is usually six eggs, and the incubation period lasts 14–15 days. Observations in North America suggest that only the female incubates, whereas in England it has been reported that the male participates to some extent. Both sexes feed the young, which are ready to leave the nest in 13–14 days. Even though their short tail feathers do not provide them any rear support, young fledglings are able to cling to vertical branches and move about like adults.

Suggested Viewing Opportunities: This inconspicuous bird occurs almost statewide and can often be found clinging to large trees during winter in most towns and cemeteries.

Selected References: BNA 669, rev. ed. (J. Poulin, E. D'Astous, M. Villard, S. J. Hejl, K. R. Newton, M. E. McFadzen, and C. K. Ghalambor 2013); Bent 1948; Braaten 1975; Harrap and Quinn 1996.

Family Troglodytidae (Wrens)

ROCK WREN. *SALPINCTES OBSOLETUS*

Status: An uncommon spring and fall migrant and summer resident throughout the western half of the state in rocky areas, especially in brush-dominated rocky localities of the Pine Ridge (east to northwestern Sheridan County), the Wildcat Hills and North Platte Valley (east to Keith County), the southwestern corner of Nebraska (Kimball and Cheyenne Counties), and the cutbanks of the loess-covered hillsides of south-central Nebraska (north to Lincoln County). There were 56 possible to confirmed nesting records during the fieldwork for *The Second Nebraska Breeding Bird Atlas* (Mollhoff 2016). National Breeding Bird Surveys from 1966 to 2015 indicate that this species' population underwent a survey-wide average annual decline of 0.65 percent (Sauer et al. 2017).

Habitats and Ecology: Rock wrens are closely associated with eroded loess slopes, badlands, rocky outcrops, and similar rock-dominated habitats. Crannies in cutbanks are favorite nesting sites, but the birds also nest in crevices among rocks or boulders.

Breeding Biology: This species nests among slopes of loose rocks and boulders rather than on vertical cliff or canyon walls. Eggs are laid at the rate of one a day, typically totaling five or six, with incubation starting when the clutch is complete. The incubation period lasts 12–14 days. Only the female incubates, but the male usually feeds her, and both sexes feed the nestlings. When the young leave their nest at about 14 days of age the adults soon begin gathering nest material for their second brood, or they might begin a second clutch in the same nest.

Suggested Viewing Opportunities: This species occurs widely in the bluff and rimrock country of western Nebraska. In the vicinity of Kingsley Dam (Keith County), the species' loud territorial song can be heard during much of the summer.

Selected References: BNA 486 (P. E. Loether, D. L. Kroodsma, and G. H. Farley 2000); Bent 1948; Tramontano 1964; Brewer 2001.

HOUSE WREN. *TROGLODYTES AEDON*

Status: A common spring and fall migrant and summer resident in woodlands, backyards, suburbs, parks, and other habitats almost throughout the state, with few records in the western Sandhills and dry western uplands. There were 468 possible to confirmed nesting records during the fieldwork for *The Second Nebraska Breeding Bird Atlas* (Mollhoff 2016). National Breeding Bird Surveys from 1966 to 2015 indicate that this species' population underwent a survey-wide average annual increase of 0.26 percent (Sauer et al. 2017).

Habitats and Ecology: Generally most common in riparian deciduous forests, house wrens also favor areas of human habitations. Nesting occurs in natural tree cavities, old woodpecker or nuthatch holes, and artificial cavities such as birdhouses.

Breeding Biology: When house wrens arrive on their breeding grounds in the spring, adults tend to precede immature birds, and males arrive about nine days before females. An adult male that has nested previously normally returns to its old territory or establishes a new territory adjacent to it, and females also have a strong tendency to return to previous nesting areas. Males sing at least three kinds of songs, including a territory song, a mating song, and a nesting song, and both sexes utter a variety of call notes. Males typically have two or three possible nest sites within their territories and might have as many as seven, thus allowing the females considerable choice. When establishing nest sites, house wrens often destroy the nearby eggs, nests, or young of their own or other nearby nesting species, and there is a good deal of territorial shifting owing to nest-site competition and to the frequent changing of mates between broods. The clutch is usually of six to eight eggs, incubation lasts 13–25 days, and the nestling period takes approximately 15 days. In addition to mate changes between nesting efforts, a second female might mate with a male and nest within his territory. In one study it was found that about 6 percent of the matings were polygynous and that about 40 percent of second matings were with the same mate. There was likewise about a 40 percent incidence of mating with the same individual in the following year, when both birds returned to the same locality.

Suggested Viewing Opportunities: This ubiquitous species can be found almost anywhere, from backyards where wren houses have been erected to brushy woods and wooded riparian edges.

Selected References: BNA 380, rev. ed. (L. S. Johnson 2014); Kendeigh 1941; Bent 1948; Brewer 2001.

Family Polioptilidae (Gnatcatchers)
BLUE-GRAY GNATCATCHER. *POLIOPTILA CAERULEA*

Status: An uncommon and local spring and fall migrant and summer resident of western and central Nebraska (*Polioptila caerulea amoenissima*) in juniper woodlands of the Pine Ridge (Sioux and Dawes Counties) and the Wildcat Hills (Scotts Bluff, Banner, and Morrill Counties). Local in Garden, Keith, and Hitchcock Counties. Also breeds (*P. c. cerulea*) in mature riparian deciduous forests of eastern Nebraska from Harlan county east to the Missouri River and northeast along the Missouri and lower Platte Valleys to Knox County and the lower Niobrara River to Cherry County. There were 67 possible to confirmed nesting records during the fieldwork for *The Second Nebraska Breeding Bird Atlas* (Mollhoff 2016). National Breeding Bird Surveys from 1966 to 2015 indicate that this species' population underwent a survey-wide average annual increase of 0.38 percent (Sauer et al. 2017), and it is extending its range in western and central Nebraska.

Habitats and Ecology: Gnatcatchers breeding in the western Great Plains occur in junipers, mixed woodland, or sagebrush areas. There, arid, park-like areas with scattered thickets are preferred for foraging, and nests are usually placed in low junipers. In eastern Nebraska, the tall riparian deciduous forests of the Missouri, Platte, and Niobrara Valleys are used for breeding.

Breeding Biology: Shortly after they arrive on their breeding grounds, male gnatcatchers acquire territories; the time depends on the abundance of foliage-dwelling arthropods in the locality. All the breeding activities occur within the territory, which is defended by the male and sometimes also by the female. Pair bonds can be established almost immediately after territoriality begins, or it might develop later. When a female appears on the territory of an unmated male he accompanies her to various potential nest sites, frequently perching in an upright posture and singing an elaborate but whispered song sequence. Both members of the pair build the nest, and they frequently obtain materials by dismantling old nests. When only new materials are used, the birds need about two weeks to complete a nest, but when old materials are already available, they might finish one in three to six days. The clutch is usually of four or five eggs, and the incubation period lasts 15 days. Both sexes incubate about equally, and both sexes brood the young, which remain in the nest for 12–13 days. They are at least occasionally fed by their parents for as long as 19 days after leaving the nest. In regions with long breeding seasons, two or rarely even three broods might be raised during a single summer.

Suggested Viewing Opportunities: This species is slowly expanding its range in Nebraska and now can be expected almost anywhere except perhaps the Sandhills. Possibly the densest breeding concentration occurs in the tall floodplain forests of Richardson County, but the western shrub-adapted race, *Polioptila caerulea amoenissima*, is rapidly extending its range and abundance in western, northern, and southern Nebraska.

Selected References: BNA 23, rev. ed. (E. L. Kershner and W. G. Ellison 2012); Bent 1949; Fehon 1955; Root 1969.

Family Regulidae (Kinglets)
GOLDEN-CROWNED KINGLET. *REGULUS SATRAPA*
RUBY-CROWNED KINGLET. *REGULUS CALENDULA*

Status: The ruby-crowned kinglet is an uncommon seasonal migrant throughout the state but is more widespread and more numerous than the golden-crowned kinglet, which is an overwintering migrant. National Breeding Bird Surveys from 1966 to 2015 indicate that the golden-crowned population underwent a survey-wide average annual decline of 0.66 percent, whereas the ruby-crowned population underwent a survey-wide average annual increase of 0.47 percent (Sauer et al. 2017).

Habitats and Ecology: During winter, kinglets move from montane breeding habitats toward lower elevations, including prairie river-bottom woodlands, and sometimes into city parks and suburbs.

Suggested Viewing Opportunities: In Nebraska, these two species occur in all woodland habitats. The ruby-crowned kinglet's historic spring migration records are centered from April 13 to May 10, and the fall records extend from September 23 to October 28. The golden-crowned kinglet's historic migration records are concentrated from October 19 to April 10.

Selected References: Ruby-crowned kinglet: BNA 119, rev. ed. (D. L. Swanson, J. L. Ingold, and G. E. Wallace 2008). Golden-crowned kinglet: BNA 301, rev. ed. (D. L. Swanson, J. L. Ingold, and R. Galati 2012); Bent 1949.

Family Turdidae (Thrushes)

EASTERN BLUEBIRD. *SIALIA SIALIS*

Status: A common spring and fall migrant and summer resident across the eastern half of the state, with extensions west in the Niobrara Valley to Sioux County and the Pine Ridge, in the Platte Valley to Scotts Bluff County, and in the Republican Valley to Dundy County. There were 334 possible to confirmed nesting records during the fieldwork for *The Second Nebraska Breeding Bird Atlas* (Mollhoff 2016). Bluebirds favor riparian corridor forests, upland woodlands, backyards, suburbs, parks, and other partly wooded habitats. National Breeding Bird Surveys from 1966 to 2015 indicate that this species' population underwent a survey-wide average annual increase of 1.5 percent (Sauer et al. 2017).

Habitats and Ecology: The species frequents open deciduous woods, especially where they are interspersed with or adjacent to grasslands. Breeding birds all commonly use upland and floodplain forest edges, city parks and gardens, shelterbelts, and farmsteads.

Breeding Biology: Studies in Arkansas, where bluebirds are mostly permanent residents, indicate that wintering birds form pair bonds between November and the end of January, and courtship is closely associated with the visiting of nest boxes or other suitable nest cavities. After pair bonds are established they seem to last throughout the year. A territory is established around the nesting site and is retained until the last brood fledges. Courtship-feeding of the female by her mate is an important part of breeding activities; this starts before nest-building and continues into the nestling stage. Nests are typically in old woodpecker holes or natural cavities of dead trees, dead limbs, or utility poles. In many areas, birdhouses are used, especially where natural cavities are lacking. Bluebirds prefer nest boxes that are placed in open areas 8–12 feet high, with entrances no larger than 1.5 inches in diameter. There should be a suitable tree perch nearby with a view of the nest entrance, and, if possible, nests should face east or south to avoid exposure to the northwesterly spring rains common to the Great Plains. The cavity is filled with weed stalks and grasses to form a loose cup, sometimes with a lining of finer grasses or a few feathers. From three to six eggs are a typical clutch, and the incubation period is 12–15 days. Only females incubate, but males sometimes enter the nest box briefly when their mates are absent. The nestling period is usually 17–18 days, and in some southern states there are commonly three and rarely four nesting attempts, and up to three broods are reared, whereas single-brooding is usual in Canada.

Suggested Viewing Opportunities: Nesting-box programs have made this species once again common, and bluebirds are now likely to be seen along many eastern Nebraska country roads wherever nesting boxes have been erected. Historic spring arrival records are centered around March 23, with fall departures around November 5.

Selected References: BNA 381, rev. ed. (P. A. Gowaty and J. H. Plissner 2015); Thomas 1946; Bent 1949; Clement 2001.

MOUNTAIN BLUEBIRD. *SIALIA CURRUCOIDES*

Status: A local spring and fall migrant and summer resident of the Panhandle, especially in the Pine Ridge (Sheridan to Sioux Counties) and Wildcat Hills (Scotts Bluff, Banner, and Morrill Counties), and locally in Kimball County. There were 21 possible to confirmed nesting records during the fieldwork for *The Second Nebraska Breeding Bird Atlas* (Mollhoff 2016). National Breeding Bird Surveys from 1966 to 2015 indicate that this species' population underwent a survey-wide average annual decline of 0.54 percent (Sauer et al. 2017).

Habitats and Ecology: Breeding by mountain bluebirds in western states occurs in open woodlands and forest-edge habitats from mountain meadows downward into the juniper woodland zone. Typically, the birds favor nesting where either dead trees are available for nest cavities or where rock crevices or other suitable sites are present.

Breeding Biology: Bluebirds arrive relatively early on northern breeding grounds and immediately begin searching for nesting sites. Paired birds often displace unmated males, which defend their territories only weakly. When a nesting pair has established a territory they both defend it vigorously, the male defending the periphery and the female the actual nest site. One instance has been described of a male having two mates within his territory, nesting about 50 yards apart. Only the female builds the nest, which requires four to six days, and only the female incubates. The clutch is usually five or six eggs, and the incubation period lasts 12–15 days. Females brood their nestlings for about six days after hatching, and both parents actively feed the young. They fledge in 22–23 days, after which the female usually begins a second clutch, and the male remains with the fledglings for about ten days. Young of the first brood of mountain bluebirds have rarely been observed feeding the second brood.

Suggested Viewing Opportunities: The closer one approaches Wyoming, the greater the chances of seeing breeding mountain bluebirds, but during migration the birds might be seen anywhere in the Panhandle. Then they can often be seen in small mixed-sex flocks, slowly working their way from fence post to fence post, catching insects as they go. Historic spring arrival records are centered around March 11, with fall departures around October 16.

Selected References: BNA 222 (H. W. Power and M. P. Lombardo 1996); Bent 1949; Power 1966; Clement 2001.

TOWNSEND'S SOLITAIRE. *MYADESTES TOWNSENDI*

Status: A common and widespread wintering migrant in juniper woodlands, especially western and southwestern Nebraska. Breeding probably rarely occurs in the Pine Ridge (Sioux and Dawes Counties) (Mollhoff 2016). National Breeding Bird Surveys from 1966 to 2015 indicate that this species' population underwent a survey-wide average annual increase of 0.28 percent (Sauer et al. 2017). This species is on the Nebraska Natural Legacy Project's Tier 2 list of threatened species in Nebraska (Schneider et al. 2018).

Habitats and Ecology: In the winter these birds feed almost entirely on junipers or other kinds of berries, but while breeding the usual thrush diet of insects is the most important source of food.

Breeding Biology: Although much of a male's territorial singing is done from high trees, the nest location is usually low, in a tree stump cavity, a rock crevice, or a tangle of roots, much like the nest sites chosen by veerys but unlike the elevated tree nests of many thrushes. The nest is an open cup made of weeds, grasses, rootlets, and other vegetation and is built by the female over a period of six to ten days. The clutch is usually of four eggs, and the incubation period is probably 12 days and is probably done by the female. Both sexes tend the young, and the fledging period is 10–12 days.

Suggested Viewing Opportunities: During summer the loud songs of this species make it easy to find, but in winter juniper woodlands in central and western Nebraska are the places to search. Fall arrival records center around September 26, and spring departure records cluster around March 29.

Selected References: BNA 269 (R. V. Bowen 1997); Bent 1949; Lederer 1977; Clement 2001.

VEERY. *CATHARUS FUSCESCENS*

Status: An uncommon and widespread migrant in wooded areas of the state, declining westwardly. National Breeding Bird Surveys from 1966 to 2015 indicate that this species' population underwent a survey-wide average annual decline of 1.13 percent (Sauer et al. 2017).

Habitats and Ecology: In Nebraska the favored habitats of migrant veerys consist of wooded river valleys. Locations with heavy and thickety undergrowth that are difficult for humans to penetrate are this species' favorite habitats, and most of its foraging is done on the ground in shady conditions.

Suggested Viewing Opportunities: This species is most likely to be detected by its glorious song during its relatively brief spring migration presence, which is centered around May 15–18.

Selected References: BNA 142, rev. ed. (C. M. Heckscher, L. R. Bevier, A. F. Poole, W. Moskoff, P. Pyle, and M. A. Patten 2012); Bent 1949; Lederer 1977; Clement 2001.

SWAINSON'S THRUSH. *CATHARUS USTULATUS*

Status: A common and widespread migrant in wooded areas of the state, the most common of the migratory thrushes, and a rare local resident in the deep canyons of the Pine Ridge (Dawes County) (Mollhoff 2016). National Breeding Bird Surveys from 1966 to 2015 indicate that this species' population underwent a survey-wide average annual decline of 0.84 percent (Sauer et al. 2017).

Habitats and Ecology: On migration these birds are likely to be found in almost any fairly dense woodlands, but during the breeding season the birds are found at higher and cooler elevations. There they use shaded canyons where there are also fairly large areas of tangled brushy undergrowth, permitting safe ground-level foraging.

Breeding Biology: Nesting often occurs near streamside thickets, and the nest is typically built in a small tree and often placed less than 10 feet above the ground but rarely as high as 30 feet. It is a compact cup constructed of diverse dead vegetation materials and is built by the female over a period as brief as four days. The usual clutch is three or four eggs, which, like those of many thrushes, are mostly blue, with darker brownish and lilac markings. Incubation lasts 10–13 days and is performed by the female. Both sexes tend the young, and the fledging period is 10–12 days. Double-brooding is typical.

Suggested Viewing Opportunities: These birds might be found in streamside thickets and aspen groves with dense understories. This species is the most frequently seen of the forest thrushes; its spring migration is centered around May 6–27.

Selected References: BNA 540 (D. E. Mack and W. Yong 2000); Bent 1949; Dilger 1956; Clement 2001.

HERMIT THRUSH. *CATHARUS GUTTATUS*

Status: An uncommon migrant in wooded areas throughout the state. National Breeding Bird Surveys from 1966 to 2015 indicate that this species' population underwent a survey-wide average annual decline of 0.33 percent (Sauer et al. 2017).

Habitats and Ecology: Shady and leaf-littered forest floors are favored for foraging by migrants.

Suggested Viewing Opportunities: This is one of the more rarely seen of the forest thrushes in Nebraska; its spring migration is unusually early and is centered from April 20 to 26.

Selected References: BNA 261, rev. ed. (R. Delinger, P. B. Woods, P. W. Jones, and T. M. Donovan 2012); Bent 1949; Clement 2001.

AMERICAN ROBIN. *TURDUS MIGRATORIUS*

Status: An abundant and widespread resident throughout the entire state, especially in open woodland areas, becoming less common and variably migratory in winter. A total of 528 possible to confirmed breeding records was reported for every county during fieldwork for *The Second Nebraska Breeding Bird Atlas* (Mollhoff 2016). This is probably the most abundant of all North American songbirds, with an estimated population of 320 million birds during the late 1990s (Rich et al. 2004) and an estimated survey-wide annual increase of 0.12 percent during the period 1969–2015.

Habitats and Ecology: Open woodlands, whether natural or artificial, such as suburbs, city parks, and farmsteads, are typical robin habitats. The birds tend to occur almost anywhere there are scattered trees, soft ground suitable for probing for insects and worms, and places where mud can be gathered for the nest. Nesting on human-made structures seems to be preferred over natural nest sites such as trees, at least in protected areas.

Breeding Biology: Very early spring migrants, male robins tend to arrive on the breeding grounds slightly before females, and both sexes tend to return to the area where they were hatched. Males often establish essentially the same territory they held the previous year; the size of the territory seems to vary greatly with habitat and population density. The time nesting begins is closely associated with latitude; both sexes apparently jointly select the nest site. The nest is sometimes completed in as little as 24 hours, with the male carrying much of the material, and the female doing the nest-shaping.

However, many nests are built much more slowly, especially early ones, which often require five or six days. The three to five eggs are deposited at daily intervals, and the female almost exclusively does the incubation, which lasts 11–14 days. The fledging period is usually about 13 days but varies from 9 to 16 days. The young are cared for until they are about a month old. Even at the northern edge of their range, robins typically raise two broods, and pairs normally remain intact for the second brood. At times the same nest is used for the second clutch, but often a new one is constructed near the old one.

Suggested Viewing Opportunities: Robins occur commonly and ubiquitously throughout the entire state and are most common in well-watered grassy areas such as lawns.

Selected References: BNA 462, rev. ed. (N. Vanderhoff, P. Pyle, M. A. Patten, R. Sallabanks, and F. C. James 2016); Howell 1942; Bent 1949; Clement 2001.

Family Mimidae (Thrashers, Catbirds, and Mockingbirds)

GRAY CATBIRD. *DUMETELLA CAROLINENSIS*

Status: An uncommon spring and fall migrant and summer resident in wooded habitats across the eastern half of the state, with extensions west in the Niobrara Valley possibly to Cherry County, in the Platte Valley probably to Garden County, and in the Republican Valley probably to Dundy County. Isolated populations are apparently present in the Pine Ridge and Scotts Bluff County. There were 336 possible to confirmed nesting records during the fieldwork for *The Second Nebraska Breeding Bird Atlas* (Mollhoff 2016). National Breeding Bird Surveys from 1966 to 2015 indicate that this species' population underwent a survey-wide average annual decline of 0.01 percent (Sauer et al. 2017).

Habitats and Ecology: Catbirds favor dense thickets, ranging from riverine forests or prairie coulees, city parks and suburbs, orchards, and woodland edges to shrubby marsh borders. Overgrown habitats that provide a combination of dense vegetation and transitional "edges" are ideal habitats for this species. Coniferous forests are avoided, the birds favoring other habitats that offer rich sources of insects and berries. In Colorado, catbirds usually nest in dense shrubbery along streams at fairly low elevations (Kingery 1998).

Breeding Biology: Although catbirds are distinctly territorial, active defense seems to be largely limited to the vicinity of the nest site, and much of the male's territorial advertisement is achieved by simply singing from within the dense vegetation that the birds frequent. Males frequently indicate possible nest sites to females by sitting on branches with their wings spread and manipulating twigs or other objects, as if nest-building. In one Michigan study, most nests were within 2 feet of the side or top of shrub cover in sites providing good visibility for the sitting bird. Once a nest is begun, the female does most of the actual building, although the male might bring her materials. The first egg is usually laid two days after the nest is finished, and thereafter eggs are laid daily until the clutch has been completed. The clutch size is usually of four eggs, and the incubation period lasts 12–13 days. Incubation is by the female alone, and the male apparently feeds her very little during this time. The young remain in the nest for an average of 11 days, and their parents care for them for approximately two more weeks. In many cases, the pair raises a second brood, but rarely if ever is a third brood successfully reared in the central or northern plains.

Suggested Viewing Opportunities: This species is a common summer resident in Nebraska and can fairly easily be seen foraging in brushy edge habitats, where its cat-like calls often betray its location.

Selected References: BNA 167, rev. ed. (R. J. Smith, M. I. Hatch, D. A. Cimprich, and F. R. Moore 2011); Bent 1948; Nickell 1965; Brewer 2001.

BROWN THRASHER. *TOXOSTOMA RUFUM*

Status: A spring and fall migrant and summer resident over the entire state. A total of 507 possible to confirmed breeding records was reported for every county during fieldwork for *The Second Nebraska Breeding Bird Atlas*, and the species was judged to be the state's eleventh most common breeding bird (Mollhoff 2016).

National Breeding Bird Surveys from 1966 to 2015 indicate that this species' population underwent a survey-wide average annual decline of 1.04 percent (Sauer et al. 2017).

Habitats and Ecology: Thrashers are associated with open, brushy woodlands, scattered clumps of low woods in open environments, shelterbelts, woodlots, and shrubby residential areas. At the western edge of their range in Colorado, breeding birds are most often found in rural plantings of trees and shrubs (Kingery 1998). The birds are mostly confined to shrubby coulees in grassland areas or to riparian forests that provide sources of berries, and they use foraging locations in more open, grassy areas.

Breeding Biology: Males of this migrant usually arrive on their breeding areas a few days ahead of females and apparently establish nesting territories almost immediately, although territorial singing might not begin for ten days or more. Once a territory has been established, the males become very sedentary, and all the nests of the season are built within this territory. Brown thrashers and gray catbirds have very similar territorial requirements, and at times thrashers will evict catbirds from their territory. The clutch size is usually four to six eggs, and the incubation period lasts 11–14 days. Incubation is primarily by the female, and both birds help brood the young, although males seem to be less involved than females. The average nestling period is 11 days, but in some cases, the female leaves the brood soon after hatching and begins a second nest. In other cases, the two parents might each take part of the brood after they fledge but meet later to begin a second nesting effort. Studies of banded birds have indicated that birds sometimes change mates between broods, even when the original mate is still available.

Suggested Viewing Opportunities: The species is most common in brushy areas along rivers, and its song, which is a series of distinctive once-repeated phrases, is similar to that of the more vocally inventive northern mockingbird, although it is easily recognizable.

Selected References: BNA 557, rev. ed. (J. F. Cavitt and C. A. Hass 2014); Erwin 1935; Bent 1948; Brewer 2001.

Status: A highly local spring and fall migrant and summer resident in arid brush habitats (sagebrush, rabbitbrush, greasewood) of the northwestern Panhandle, with probable nestings in the Oglala National Grasslands and a single confirmed nesting in Kimball County. National Breeding Bird Surveys from 1966 to 2015 indicate that this species' population underwent a survey-wide average annual decline of 1.2 percent (Sauer et al. 2017).

Habitats and Ecology: This species is closely associated with sage-dominated grasslands and to a much lesser extent other shrublands dominated by shrubs of similar shrubby growth forms, such as rabbitbrush and greasewood. A greater diversity of shrubby habitats is used in other seasons. In Colorado, over 80 percent of breeding-season observations were in shrublands, with sagebrush accounting for more than half (Kingery 1998). A greater diversity of shrubby habitats is used in other seasons. Most foraging is done on the ground, but nests are placed in shrubs.

Breeding Biology: Relatively little has been written on the breeding biology of this arid-adapted thrasher. Some accounts describe a territorial flight song, or courtship flight, with the bird zigzagging low over the ground, uttering a warbling song, and landing with upraised and fluttering wings. Apparently, both sexes incubate, and incubation probably begins the day before the final egg is laid. The clutch size is usually four or five eggs, the incubation period lasts 15 days, and the nestling period is 11–13 days. When bringing food to the young, the adults are highly secretive, usually landing on a bush about 10 feet away, then approaching the nest while hidden from view. Pairs often remain mated during successive years, and the birds are sometimes rather long-lived, with one banded individual known to have reached 13 years of age.

Suggested Viewing Opportunities: This species is likely to be seen only in those rare semidesert habitats of the northwestern Panhandle where big sagebrush (*Artemisia tridentata*) is present, such as on the arid tablelands of the Oglala National Grassland.

Selected References: BNA 462 (T. D. Reynolds, T. D. Rich, and D. A. Stephens 1999); Bent 1948; Killpack 1970; Brewer 2001.

Family Bombycillidae (Waxwings)

BOHEMIAN WAXWING. *BOMBYCILLA GARRULUS*

Status: An uncommon to rare winter migrant over most of the state, more common northwardly.

Habitats and Ecology: Outside the breeding season, waxwings move about opportunistically, seeking out sources of berries and small fruits in trees and hedges such as mountain ash, crab apples, pyracantha, junipers, and other fruiting trees or shrubs. The fruits they eat are mostly watery, so, like hummingbirds, waxwings often consume more than their body weight daily. They are also quite mobile in their foraging, and the undigested seeds that they excrete become highly scattered over the landscape and aid in the plants' dispersal.

Suggested Viewing Opportunities: Bohemian waxwings are most likely to be seen among flocks of cedar waxwings, often during migration or winter, when the birds gather at fruit- or berry-bearing trees. They now winter only rather rarely in Nebraska, generally remaining farther north as winters have recently ameliorated (Johnsgard 2015c).

Selected References: BNA 714 (M. C. Witmer 2002); Bent 1950.

CEDAR WAXWING. *BOMBYCILLA CEDRORUM*

Status: An uncommon spring and fall migrant and summer resident that is widespread over the woodlands over most of the state, with extensions west in the Niobrara Valley to Sioux County and the Pine Ridge, in the Platte Valley to Scotts Bluff County, and in the Republican Valley to Dundy County. There were 255 possible to confirmed nesting records during the fieldwork for *The Second Nebraska Breeding Bird Atlas* (Mollhoff 2016). National Breeding Bird Surveys from 1966 to 2015 indicate that this species' population underwent a slight average annual increase of 0.07 percent (Sauer et al. 2017).

Habitats and Ecology: Somewhat open woodlands, primarily of broad-leaved species, are used for nesting, including riparian forests, farmsteads, parks, cedar groves, shelterbelts, and brushy edges of forests. Areas that have abundant growths of berry-bearing bushes, such as junipers in Nebraska, are especially favored, although insects, buds, and other food sources are also consumed and provide proteins and other nutrients that are lacking in fruits. In Colorado, waxwings nest sparingly in deciduous riparian forests below 7,500 feet (Kingery 1998).

Breeding Biology: This rather common and highly gregarious species remains in flocks for much of the year. Adult birds often can be seen passing berries, insects, leaves, or other items back and forth from beak to beak, but whether this represents ritualized courtship-feeding or pair-bonding or is part of a more general social integration activity is uncertain. It is known that females prefer to mate with older males having more red "wax" on their secondaries' wing tips. Crest-raising by the male, mutual breast-preening, and bill-clicking are other reputed courtship activities, and during the period of nest-building the female performs begging behavior and is fed by her mate. Territoriality is virtually absent in cedar waxwings. The nests are frequently situated somewhat colonially, perhaps because of locally rich food sources, and breeding seems to be timed around the period when berries and fruit ripen. Both sexes build the nest, which requires two to six days. The clutch size is usually three to five eggs, and the incubation period lasts 12–14 days. The female does the incubating, but her mate frequently feeds her, and she also intensively broods the young for several days after hatching. The young birds leave the nest when about 16 days old and might remain in the nest's vicinity for about a month.

Suggested Viewing Opportunities: Waxwings occur throughout the state, especially in riparian deciduous forests during summer. During their migrations they appear in small flocks, descend on berry- and fruit-bearing trees, and abruptly leave again immediately after harvesting them. Their fall appearance typically begins in

October, and they are often gone by December, only to reappear in late February as winter wanes.

Selected References: BNA 309, rev. ed. (M. C. Witmer, D. J. Mountjoy, and L. Elliot 2014); Bent 1950.

Family Motacillidae (Pipits)

AMERICAN PIPIT. *ANTHUS RUBESCENS*

Status: A common spring and fall migrant statewide.

Habitats and Ecology: During the nonbreeding seasons this species occurs on very open terrain that usually has only sparse vegetation and often a moist substrate. Migrants and wintering birds commonly use shorelines, flooded fields, river edges, and similar wetland habitats.

Suggested Viewing Opportunities: In Nebraska, this species is widespread around open muddy or sandy shorelines of wetlands during migration, which is historically centered in spring in late April and during fall in October.

Selected References: BNA 95, rev. ed. (P. Hendricks and N. A. Verbeck 2013); Bent 1950.

SPRAGUE'S PIPIT. *ANTHUS SPRAGUEII*

Status: An uncommon and inconspicuous spring and fall migrant. Nebraska is directly south of the species' breeding range, which extends south from the Canadian prairies to northern South Dakota.

Habitats and Ecology: This is one of many grassland-dependent species suffering long-term population declines. National Breeding Bird Surveys from 1966 to 2015 indicate that this species' population underwent a survey-wide average annual decline of 3.01 percent, reflecting a serious long-term population reduction (Sauer et al. 2017). This species is on the Nebraska Natural Legacy Project's Tier 1 list of threatened species in Nebraska (Schneider et al. 2018).

Suggested Viewing Opportunities: At Spring Creek Audubon Prairie near Lincoln this pipit is present only briefly during spring and especially fall migration and is most likely to be seen along the higher prairie hilltops between late September and the end of October. One must travel to the Dakotas to witness the species' spectacular territorial flight-song display.

Selected References: BNA 439, rev. ed. (S. K. Davis, M. B. Robbins, and B. C. Dale 2014); Bent 1950; Johnsgard 2001b.

Family Fringillidae (Finches)

GRAY-CROWNED ROSY-FINCH. *LEUCOSTICTE TEPHROCOTIS*

Status: A rare and local winter visitor in the western Panhandle.

Habitats and Ecology: During the breeding season, these birds inhabit montane cirques, talus slopes, alpine meadows with nearby cliffs, and adjacent snow and glacier surfaces, where foraging for frozen insects is common. During fall and winter, the birds move to lower elevations and to habitats that include meadows, grasslands, brushlands, and agricultural lands.

Suggested Viewing Opportunities: At Scotts Bluff National Monument as many as 200 of these small finches have been seen roosting in cliff swallow nests during cold weather. Rosy-finches also sometimes visit Panhandle bird feeders during winter, but they are erratic winter visitors even in far western Nebraska, especially around Harrison.

Selected References: BNA 559 (S. MacDougal-Shackleton, R. Johnson, and T. Hahn 2000); Austin 1968; Clement 1993.

HOUSE FINCH. *HAEMORHOUS MEXICANUS*

Status: An abundant resident across Nebraska. National Breeding Bird Surveys from 1966 to 2015 indicate that this species' population underwent a survey-wide average annual increase of 0.12 percent (Sauer et al. 2017). There were 215 possible to confirmed nesting records during the fieldwork for *The Second Nebraska Breeding Bird Atlas* (Mollhoff 2016).

Habitats and Ecology: House finches are generally associated with human habitations over most of their range, often nesting on building ledges. Otherwise, they nest in open woods, river bottom woodlands, scrubby desert or semidesert vegetation such as sagebrush, and tree plantings. Deciduous underbrush, preferably close to water, is favored over dense coniferous woods, and sources of seeds, berries, or fruits are also needed throughout the year.

Breeding Biology: This species has a courtship display similar to that of the purple finch, during which the male approaches the female with his tail spread and cocked and his wings lowered while uttering chirps and trills. Courtship-feeding of the female also occurs during pair formation and during incubation. Both sexes help in nest-building, which requires from two to eleven days; males help mainly in the early stages. The clutch size is usually four or five eggs, and the incubation period lasts 12–14 days. The female incubates and broods alone, but both sexes feed nestlings. Young fledged at an average of 15 days of age in one study and at nearly 18 days in another. It has been reported that females frequently begin to gather nesting materials for their second brood while being followed by begging young that are still partially covered with down. Fledging house finches tend to beg from any nearby songbird for food; northern cardinals seem especially willing to feed them. House finches accept cowbird eggs and feed the chicks, but the seeds they provide are unsuitable for nourishing cowbirds, and the cowbirds starve.

Suggested Viewing Opportunities: House finches occur statewide at lower elevations and probably can be found in any village or town, especially around bird feeders.

Selected References: BNA 46, rev. ed. (A. V. Badyaev, V. Belloni, and G. E. Hill 2012); Austin 1968; Clement 1993.

RED CROSSBILL. *LOXIA CURVIROSTRA*

Status: A local resident in pine forests of the Pine Ridge (Sioux to Sheridan Counties), the Wildcat Hills (Scotts Bluff, Banner, and Morrill Counties), and locally elsewhere, especially the Nebraska National Forest, Bessey Division (Thomas County). There were 28 possible to confirmed nesting records during the fieldwork for *The Second Nebraska Breeding Bird Atlas* (Mollhoff 2016). National Breeding Bird Surveys from 1966 to 2015 indicate that this species' population underwent a survey-wide average annual decline of 0.93 percent (Sauer et al. 2017).

Habitats and Ecology: Breeding is associated with coniferous forest habitats, especially those comprised of pines. In Wyoming, five call types (out of the nine known in North America) have been found; some of these different call type populations might act as biologically distinct species (Smith and Benkman 2007; Faulkner 2010). In Colorado, crossbills with different bill shapes and call types specialize on eating different conifer seeds (Douglas-fir, ponderosa pine, and lodgepole pine) during late winter and spring (Kingery 1998). Although usually found in conifers, crossbills also feed on ripe box elder seeds in late summer (Scott 1993).

Breeding Biology: Observations of this species in Colorado documented colonial nesting; about 24 pairs were found within a square mile of forest, but few were found elsewhere. The clutch is usually of three or four eggs. Hatched young and nests in progress have been found in Colorado as early as mid-January. Bailey, Niedrach, and Bailey (1953) reported that the females did the nest-building, which required about five days, and another four days elapsed before the first egg was laid. Four days were spent in egg-laying, 14 more in incubation, and fledging occurred 20 days after hatching. Fledging times of 16–25 days have been reported; such variations probably relate to variable local food supplies. The nesting cycles of crossbills are highly irregular. Some pairs raise two broods in rapid succession, or two widely spaced breedings might occur during a single year, and young birds might breed the same year in which they were hatched.

Suggested Viewing Opportunities: Although the Pine Ridge and Wildcat Hills regions no doubt attract the largest number of these birds, they can also often be seen at the Bessey Division of the Nebraska National Forest, Thomas County. During winter they occur statewide in coniferous plantings.

Selected References: BNA 256, rev. ed. (C. W. Benkman and M. A. Young 2019); Bailey, Niedrach, and Bailey 1953; Nethersole-

Thompson 1975; Benkman 1983; Clement 1993; Benkman, Parch-man, and Mezquida 2010.

Status: A common resident in pine forests of the Pine Ridge (Sioux to Sheridan Counties), the Wildcat Hills (Scotts Bluff, Banner, and Morrill Counties), and locally elsewhere throughout the state during population irruptions. There were 23 possible to confirmed nesting records during the fieldwork for *The Second Nebraska Breeding Bird Atlas* (Mollhoff 2016). National Breeding Bird Surveys from 1966 to 2015 indicate that this species' population underwent a survey-wide average annual decline of 3.67 percent (Sauer et al. 2017). This species is on the Nebraska Natural Legacy Project's Tier 2 list of threatened species in Nebraska (Schneider et al. 2018).

Habitats and Ecology: Nesting by siskins preferentially occurs in conifers of almost any type but has also been observed nesting in cottonwoods, willows, and even lilacs. Their foods are mainly conifer seeds but also include those of alders, birches, and various weeds, and during summer they feed on flower buds and insects.

Suggested Viewing Opportunities: Siskins are common at bird feeders in towns and cities statewide during the occasional winters in which irruptive populations are present.

Selected References: BNA 280, rev. ed. (W. R. Dawson 2014); Weaver and West 1943; Austin 1968; Clement 1993.

AMERICAN GOLDFINCH. *SPINIS TRISTIS*

Status: A common summer resident almost throughout the state. There were 476 possible to confirmed nesting records during the fieldwork for *The Second Nebraska Breeding Bird Atlas* (Mollhoff 2016). National Breeding Bird Surveys from 1966 to 2015 indicate that this species' population underwent a survey-wide average annual decline of 0.17 percent (Sauer et al. 2017).

Habitats and Ecology: Breeding by goldfinches often occurs in weedy, lightly grazed country and shrubby edge habitats, especially where thistles are abundant or where cattails are to be found. The seeds of thistles and other composites are used for feeding the young, and the "down" of thistles or cattails is used in nest construction. In Colorado, nesting surveys revealed a total of 26 different habitat types used by breeding goldfinches (Kingery 1998). Habitats used in Nebraska are similarly diverse, with habitat patch sizes of 2–10 acres the most commonly used (Mollhoff 2016). Riparian woodlands near weedy fields having abundant thistles provide an ideal nesting situation. During winter the birds range widely over weedy fields and farmlands and often visit urban bird feeders.

Breeding Biology: These gregarious birds remain in flocks well into late spring, and pair formation begins among flocked birds by May, if not earlier. It is marked by female flights accompanied with varying numbers of males, male-to-female courtship singing, a hovering flight song by the male, and an extended male song resembling that of a canary (*Serinus canaria*). Pair bonds are maintained by courtship-feeding of the female, which occurs from egg-laying through the nestling period. Nesting usually occurs when there is an abundant supply of small composite seeds such as thistles to feed the young. The clutch is usually of five eggs, and the incubation period lasts 12–14 days. The female alone performs nest-building and incubation, but both sexes feed the young by seed regurgitation. The nestlings fledge in 10–16 days, when the male takes over most of the feeding. This frees the female to begin a new nest, which sometimes happens as soon as three days after the brood's fledging.

Suggested Viewing Opportunities: During winter, goldfinches are likely to appear at any bird-feeding station if thistle seed (niger) is provided.

Selected References: BNA 80, rev. ed. (K. J. McGraw and A. L. Middleton 2017); Stokes 1950; Austin 1968; Clement 1993.

EVENING GROSBEAK. *COCCOTHRAUSTES VESPERTINUS*

Status: A rare wintering migrant, primarily in the Panhandle. National Breeding Bird Surveys from 1966 to 2015 indicate that this species' population underwent a survey-wide average annual decline of 6.36 percent, causing a sharp decline in national populations (Sauer et al. 2017), and one of the steepest rates of decline in all the boreal finches. Evening grosbeaks are too rare in Nebraska to estimate population changes.

Habitats and Ecology: During the breeding season, this species is primarily associated with mature coniferous forests, although nesting has also been observed in riparian willow thickets and even in city parks and orchards. During fall and winter, the birds often occur in flocks that feed on large and nutritious seeds such as maples, ashes, and sunflowers.

Suggested Viewing Opportunities: These conspicuous birds might appear anywhere in the state during winter, when they regularly visit bird feeders, but they are now too rare to expect, even in westernmost Nebraska.

Selected References: BNA 599 (S. W. Gillihan and B. Byers 2001); Austin 1968; Clement 1993.

Family Calcariidae (Longspurs and Snow Buntings)
CHESTNUT-COLLARED LONGSPUR. *CALCARIUS ORNATUS*

Status: An uncommon spring and fall migrant and summer resident in mixed-grass prairies of the Panhandle, mainly in the counties bordering the western edge of the state (from Sheridan probably to northern Scotts Bluff; southern Banner and Kimball Counties), locally east to Cherry, Keya Paha, and Boyd Counties. Somewhat farther east of the breeding range it is a local migrant. There were four possible to confirmed nesting records during the fieldwork for *The Second Nebraska Breeding Bird Atlas* (Mollhoff 2016). National Breeding Bird Surveys from 1966 to 2015 indicate that this species' population underwent a survey-wide average annual decline of 4.19 percent (Sauer et al. 2017). This species is on the Nebraska Natural Legacy Project's Tier 1 list of threatened species (Schneider et al. 2018).

Habitats and Ecology: Primary breeding habitats consist of grazed or hayed mixed-grass prairies, shortgrass plains, the meadow zones around alkaline ponds or lakes, mowed hayfields, heavily grazed pastures, and similar medium-height grasslands. In Colorado, this longspur breeds in taller and damper grasslands than does the

McCown's and on sites having less bare ground and more singing posts provided by tall forbs (Kingery 1998). Outside the breeding season the birds often are found in weedy cultivated fields.

Breeding Biology: Males establish territories shortly after their spring arrival; they prefer grassy plains that have rather sparse vegetation and at least one large rock or fence post to serve as a singing post. Such singing points often form a central part of the territory; the nest is usually located within 25 feet, and the total territory might be only about 100 feet in diameter. However, in some marginal areas, territories of up to 10 acres have been estimated. Although flight songs in this longspur are used in territorial advertisement, they are not as frequent or as stereotyped as in the McCown's longspur. Both species gradually gain altitude with rapid wingbeats, but the McCown's longspur then glides back downward fairly quickly with its wings upstretched, whereas the chestnut-collared longspur circles and undulates for a time while singing, then gradually descends. The female builds the nest alone and also does the incubation. The clutch size is usually three to five eggs, and the incubation period lasts 11–13 days. Both sexes feed the young, which leave the nest in 9–11 days or rarely as late as the fourteenth day. At two weeks the young can fly very well, and by about 26 days they are independent of their parents.

Suggested Viewing Opportunities: These birds favor mixed-grass prairies rather than the short-stature grasslands that are used by McCown's longspurs, and in spring they often can be found by looking and listening for their flight song while slowly driving country roads with one's car windows open and stopping periodically to listen and watch.

Selected References: BNA 288, rev. ed. (B. K. Blehn, K. Ellison, D. P. Hill, and L. K. Gould 2015); Moriarty 1965; Austin 1968; Byers, Curson, and Olsson 1995; Rising 1996; Johnsgard 2001b.

MCCOWN'S LONGSPUR. *RHYNCHOPHANES MCCOWNII*

Status: A common spring and fall migrant and summer resident on the shortgrass plains in the western Panhandle, including Sioux and Kimball Counties and possibly Banner County. There were seven possible to confirmed nesting records during the fieldwork for *The Second Nebraska Breeding Bird Atlas* (Mollhoff 2016). National Breeding Bird Surveys from 1966 to 2015 indicate that this species' population underwent a survey-wide average annual decline of 5.9 percent (Sauer et al. 2017). This species is on the Nebraska Natural Legacy Project's Tier 1 list of threatened species in Nebraska (Schneider et al. 2018).

Habitats and Ecology: During the breeding season, this species is mostly limited to shortgrass prairies and grazed mixed-grass prairies but also breeds to some degree on stubble fields or newly sprouting grain fields. In Colorado, the species breeds almost exclusively in very sparse grasses, with a large amount of exposed bare soil and a low diversity of other arid land plants, including prickly pear cactus, lupine, and locoweed (Kingery 1998). While they are on migration and during the winter period, the birds occupy open grasslands, low sage-steppe, meadows, and similar open habitats.

Breeding Biology: Male longspurs arrive on their breeding grounds of western Nebraska in late April and soon begin to select territories. These are marked by conspicuous flight-song displays, as well as by singing from shrubs or rocks. As competition increases, territories gradually decrease in size to an area about 250 feet in diameter. The courtship display is remarkable; the male moves around the female in a narrow circle, holding his nearer wing erect and thus exposing its white lining. The female gathers the nesting material and makes any nest excavation needed for building a ground nest. The clutch size is usually three or four eggs, and the incubation period lasts 12–13 days. The female also performs all the incubation and does most of the brooding, although the male occasionally relieves her during the later brooding stages. The young leave the nest at ten days of age and after two more days are able to fly for short distances.

Suggested Viewing Opportunities: The best way to search for this species is by driving country roads of the Panhandle during May while watching and listening for males performing their flight-song display, just as during a search for the chestnut-collared longspur.

Selected References: BNA 96, rev. ed. (K. A. With 2010); Mickey 1943; Austin 1968; Greer and Anderson 1989; Byers, Curson, and Olsson 1995; Rising 1996; Johnsgard 2001b.

Family Passerellidae (New World Sparrows and Towhees)

GRASSHOPPER SPARROW. *AMMODRAMUS SAVANNARUM*

Status: A common spring and fall migrant and summer resident over the entire state in grassland habitats. In Nebraska, it is probably second to lark sparrows in abundance and broad distribution. During *The Second Nebraska Breeding Bird Atlas* surveys, there were 465 possible to confirmed breeding records for all Nebraska counties (Mollhoff 2016). Grasshopper sparrows are most abundant in tall and midheight grasslands, but breeding extends to upland shortgrass prairies. National Breeding Bird Surveys from 1966 to 2015 indicate that this species' population underwent a survey-wide average annual decline of 2.52 percent (Sauer et al. 2017). Nebraska surveys indicate a similar 1.91 percent annual decline, based on a sample of 51 routes.

Habitats and Ecology: This species is primarily associated with mixed-grass prairies but is also found in shortgrass and tallgrass prairie, sage prairie, and disturbed grassland habitats such as retired cropland, hayfields, and stubble fields. Areas that have grown up to shrubs are avoided, but the presence of some scattered low trees is acceptable. Not only might grasshopper sparrow songs be confused with the stridulation sounds of grasshoppers, but the species also feeds to a large extent on grasshoppers. One study revealed that of 170 stomachs examined, grasshoppers formed 23 percent of the total food items consumed over eight months of the year, with a peak of 60 percent in June (Austin 1968).

Breeding Biology: The grasshopper sparrow exhibits notable year-to-year variations in breeding densities. Territories are immediately established after males return in spring and are advertised by the species' familiar insect-like "song." Besides their inconspicuous grasshopper-like vocalizations, males also utter a more sustained song that apparently serves to attract and maintain a mate. Although the birds are sometimes apparently semicolonial nesters, their territories at times are quite large, of 2–3 acres, and are strongly defended through the incubation period. The nesting period in Nebraska is extended, with eggs reported from about mid-May to mid-August (Mollhoff 2016). Nests are built on the ground in dense herbaceous vegetation and are often at the base of grass clumps that are well concealed from above. From three to six eggs are typical, averaging about five. The incubation period is probably 11–12 days. The female alone incubates and broods the young, and the male might continue to sing as late as the time of hatching. Both sexes feed and tend the young, which remain in the nest about nine days. Two broods are typical in the central Great Plains, with estimated nesting success rates (at least one chick fledged per nest) of 35–52 percent (Johnsgard 2001b). Nevertheless, 66 of 190 nests (35 percent) were reported to be parasitized by cowbirds during *The Second Nebraska Breeding Bird Atlas* field studies (Mollhoff 2016).

Suggested Viewing Opportunities: Grasshopper sparrows can breed on a relatively small remnant of native prairie but are most often found on patches of at least 100 acres. Males are more prone to sing from low vegetation than from tall fence posts, but they sometimes sing from low strands of fence wires.

Selected References: BNA 239 (P. D. Vickery 1996); Smith 1963; Austin 1968; Byers, Curson, and Olsson 1995; Rising 1996; Johnsgard 2001b; Wright 2019.

LARK SPARROW. *CHONDESTES GRAMMACUS*

Status: A common spring and fall migrant and summer resident over the entire state in grassland habitats. This is Nebraska's most common native sparrow; there were 475 possible to confirmed nesting records during the fieldwork for *The Second Nebraska Breeding Bird Atlas* (Mollhoff 2016). National Breeding Bird Surveys from 1966 to 2015 indicate that this species' population underwent a survey-wide average annual decline of 0.78 percent (Sauer et al. 2017).

Habitats and Ecology: This species favors grasslands that have scattered trees, shrubs, and large forbs or that adjoin such vegetation;

thus, weedy fencerows near grasslands, open brushland, Sandhills prairie with tall yuccas, scrubby woodlands, orchards, and similar habitats are all suitable. Generally, unobstructed views and a variety of plants, including scattered woody or shrubby vegetation, are preferred.

Breeding Biology: Lark sparrow nests are constructed on the ground by the female with a base of thin twigs, walls of thick grasses, and a lining of finer grasses, rootlets, and sometimes hair. The usual clutch is three to five eggs, and the incubation period lasts 11–13 days. It is believed that the female does most of the incubating, but males have been seen covering eggs, and they also help rear the young. On average, the young remain in the nest for nine days, but they are semidependent on their parents until they are 30–35 days old.

Suggested Viewing Opportunities: Lark sparrows are widespread regionally and are especially common in the Sandhills, where they are ubiquitous, most often breeding in habitat patches larger than 100 acres (Mollhoff 2016).

Selected References: BNA 488 (J. W. Martin and J. R. Parrish 2000); Austin 1968; Newman 1970; Byers, Curson, and Olsson 1995; Rising 1996; Johnsgard 2001b; Wright 2019.

LARK BUNTING. *CALAMOSPIZA MELANOCORYS*

Status: A common spring and fall migrant and summer resident throughout the western third of the state, mainly on plains and drier grasslands. Breeding extends east in the Niobrara Valley possibly to Knox County, in the Platte Valley possibly to Lincoln County, and in the Republican Valley possibly to Hitchcock County. There were 122 possible to confirmed nesting records during the fieldwork for *The Second Nebraska Breeding Bird Atlas* (Mollhoff 2016). National Breeding Bird Surveys from 1966 to 2015 indicate that this species' population underwent a survey-wide average annual decline of 2.9 percent (Sauer et al. 2017).

Habitats and Ecology: This species favors mixed-grass prairies for nesting but also can be found in shortgrass and tallgrass prairies, as well as in sage grasslands, retired croplands, alfalfa fields, and stubble fields. Areas with abundant shrubs are avoided, but fence posts or scattered trees are favored song posts. In Colorado, over 40 percent of breeding birds were found in shortgrass prairie, while taller prairies plus croplands comprised an additional 42 percent of the total (Kingery 1998).

Breeding Biology: Lark buntings arrive on their breeding areas in flocks, within which courtship begins, and thus flock dispersal occurs gradually. In spite of intensive and conspicuous flight-song activity, there seems to be relatively little territorial development; nests are often placed only 10–15 yards apart, and males sometimes sing from adjacent fence posts. Both sexes incubate, but females evidently perform the most. The usual clutch is four or five eggs, and the incubation period lasts an average of 12 days. At least through the incubation period, until about the middle of July, the males continue to sing and perform flight-song displays. The abundance and local distribution of nesting birds in western Nebraska seem to vary considerably from year to year; areas with dense populations one year might be virtually deserted the next. By late August the males have lost their distinctive nuptial plumage, and fall migration begins soon afterward.

Suggested Viewing Opportunities: In western Nebraska, breeding usually occurs on open pastures. I once counted over 100 males displaying in May along the approximate 15-mile stretch of gravel road between Harrison and Agate Fossil Beds National Monument.

Selected References: BNA 542 (T. G. Shane 2000); Austin 1968; Butterfield 1969; Creighton 1971; Taylor and Ashe 1976; Byers, Curson, and Olsson 1995; Rising 1996; Johnsgard 2001b; Wright 2019.

CHIPPING SPARROW. *SPIZELLA PASSERINA*

Status: A common spring and fall migrant and summer resident throughout the state in nearly all wooded areas, but with little or no breeding in the western Sandhills and dry uplands of the Panhandle. There were 403 possible to confirmed nesting records during the fieldwork for *The Second Nebraska Breeding Bird Atlas* (Mollhoff 2016). National Breeding Bird Surveys from 1966 to 2015 indicate that this species' population underwent a survey-wide average annual decline of 0.6 percent (Sauer et al. 2017).

Habitats and Ecology: Breeding in this widely ranging species is done in open deciduous or mixed forests, the margins of forest clearings, riparian woodlands, juniper woodlands, and other diverse habitats. Scattered trees, a sunny forest floor, and a sparse ground covering of herbaceous plants seem to define the kinds of habitat conditions that are important.

Breeding Biology: Territorial establishment begins almost immediately after males return to their breeding grounds in spring. Territories average about an acre in area but sometimes are as small as half an acre. The female gathers all the nesting materials and constructs the nest, which is typically well hidden among ground vegetation. Usually she also does all the incubating, but incubation by males has rarely been reported. The usual clutch size is four eggs, and the incubation period lasts 11–14 days. The female broods the young, but both sexes feed them, and they fledge in about ten days, with an observed range of 8–12 days. By the time they are 14 days old the young are able to fly several feet.

Suggested Viewing Opportunities: This species can be found throughout Nebraska in most variably wooded habitats. In Wyoming, these birds breed mostly in open coniferous forests but with some use of junipers, aspens, and other woodland habitats (Faulkner 2010).

Selected References: BNA 334 (A. L. Middleton 1998); Austin 1968; Walkinshaw 1944; Byers, Curson, and Olsson 1995; Rising 1996; Wright 2019.

BREWER'S SPARROW. *SPIZELLA BREWERI*

Status: An uncommon and local spring and fall migrant and summer resident in the western Panhandle from Sioux to Kimball County, but with densest populations along the Wyoming boundary in arid shrub-grassland (shrub-steppe) habitats. There were 11 possible to confirmed nesting records during the fieldwork for *The Second Nebraska Breeding Bird Atlas* (Mollhoff 2016). The shrub component might consist of big sagebrush (*Artemisia tridentata*), sand sage (*A.*

32. Grassland songbirds: Henslow's sparrow (top left), dickcissel (top right), Cassin's sparrow (middle left), Brewer's sparrow (middle right), grasshopper sparrow (bottom left), and clay-colored sparrow (bottom right).

filifolia), rabbitbrush (*Chrysothamnus*), greasewood (*Sarcobatus*), or other tall arid- and alkali-adapted shrubs. National Breeding Bird Surveys from 1966 to 2015 indicate that this species' population underwent a survey-wide average annual decline of 1.01 percent (Sauer et al. 2017). This species is on the Nebraska Natural Legacy Project's Tier 1 list of threatened species in Nebraska.

Habitats and Ecology: This species invariably breeds in shortgrass prairies with sagebrush or other semiarid shrubs present in varying densities. It is generally considered to be a sage-obligate species. Most breeding reported during *The Nebraska Breeding Bird Atlas* studies occurred on shrubby grasslands of 100–1,000 acres (Mollhoff 2016).

Breeding Biology: There is little information on the territorial behavior of Brewer's sparrows, although in eastern Washington, as many as 47 pairs have occupied 100 acres of favorable habitat. In a Montana study, spray-killing all the sagebrush on a study area reduced the Brewer's sparrow population by about half. The sparrows will nest in dead sagebrush, although it provides considerably less concealment than do live plants. The usual clutch is of three or four eggs, incubation lasts 11–13 days, and fledging occurs at eight or nine days. Double-brooding has been reported.

Suggested Viewing Opportunities: Big sagebrush is an ideal habitat for Brewer's sparrows, but it is rare in Nebraska, so other shrubs of a similar life-form, including brushy sage species such as sand sage, rabbitbrush, and greasewood, provide substitutes, and Brewer's sparrows might be found among any of them.

Selected References: BNA 390 (J. T. Rotenberry, M. A. Patten, and K. L. Preston 1999); Austin 1968; Best 1972; Byers, Curson, and Olsson 1995; Rising 1996; Johnsgard 2001b; Wright 2019.

AMERICAN TREE SPARROW. *SPIZELLOIDES ARBOREA*

Status: An common overwintering migrant throughout the state, more often seen in grassy habitats than among trees.

Habitats and Ecology: While it is in Nebraska this species occupies brushy prairie areas, roadside thickets, farmsteads, old orchards, overgrown and weedy pastures, and similar relatively open habitats having some shrubs. The birds often associate with juncos and other gregarious and winter-hardy sparrows, and they forage on the ground or snow surface, industriously searching out small seeds.

Suggested Viewing Opportunities: During fall and winter (mid-October to mid-April), this species is likely to appear in open to somewhat brushy areas throughout the entire state.

Selected References: BNA 37, rev. ed. (C. T. Naugler, P. Pyle, and M. A. Patten 2017); Austin 1968; Byers, Curson, and Olsson 1995; Rising 1996; Wright 2019.

DARK-EYED JUNCO. *JUNCO HYEMALIS*

Status: A common migrant and winter visitor statewide and a local summer resident in the Pine Ridge area of Sioux, Dawes, and Sheridan Counties. Besides the typical and locally common slate-colored race (*Junco hyemalis hyemalis*), several other often intergrading races such as the pink-sided phenotype (*J. h. mearnsi*) are fairly common in the state. One of the Oregon racial group subpopulations, the gray-headed phenotype (*J. h. caniceps*), is a regular winter visitor in western Nebraska. A paler "white-winged" race (*J. h. aikeni*), with wide white-tipped wing coverts, is a rare breeder in the Pine Ridge. There were eight possible to confirmed nesting records during the fieldwork for *The Second Nebraska Breeding Bird Atlas* (Mollhoff 2016). An additional variant and possible *oreganus/hyemalis* intergrade has a black head but a contrasting pale gray back and is often called the Cassiar junco (*J. h. cismontanus*). Illustrated in *The Sibley Guide to Birds* as the "Canadian Rocky Mountains" form, it has been only reported rarely in Nebraska, although probably few observers recognize and separately identify this variant. During the five Nebraska Christmas Bird Counts ending in 2017–18, nearly 8,600 dark-eyed juncos were racially identified, including 7,719 slate-colored, 632 Oregon, 218 pink-sided, 26 white-winged, and 4 Cassiar. The junco listed on the Nebraska Natural Legacy Project's Tier 2 list of threatened species in Nebraska (Schneider et al. 2018) no doubt refers to the Panhandle's endemic white-winged race, which also breeds in South Dakota's Black Hills and adjacent Wyoming. National Breeding Bird Surveys from 1966 to 2015 indicate that the dark-eyed junco population underwent a survey-wide average annual decline of 1.38 percent (Sauer et al. 2017).

Habitats and Ecology: Breeding habitats include open coniferous forests, especially ponderosa pine forests, mixed forests, forest clearings, and similar habitats that offer ground-foraging and ground-nesting opportunities.

Breeding Biology: Juncos are notable for their sociable winter flocking behavior, which persists until the birds return to their breeding areas. In the Black Hills, territorial singing sometimes can be heard in early March. When a female enters a male's territory he follows her with his tail lifted and fanned and with both wings drooping. Several days are spent in establishing and strengthening the pair bond, during which the birds remain close together and the male continues to display frequently. The female builds the nest over a period of several days, and she apparently does all the incubation and brooding. The usual clutch is three to five eggs, and the incubation period lasts 12–13 days. Both parents bring food to the young, which fledge in 10–13 days. The young continue to be semidependent on their parents for about three weeks after leaving the nest, and juveniles have been seen with their fathers as late as 46 days after fledging. By that time the females might be incubating a second clutch if not already feeding a second brood.

Suggested Viewing Opportunities: Juncos are probably the most common wintering songbird in Nebraska and the entire Great Plains, with a continental abundance that has been estimated at 260 million birds (Johnsgard 2011a; Rich et al. 2004). They are abundant at feeding stations statewide from October to mid-April. Collectively, up to five races of juncos might be seen in Nebraska, including the endemic white-winged race, which winters southwardly over the Panhandle region.

Selected References: BNA 718 (V. Nolan Jr., E. D. Ketterson, D. A. Cristol, C. M. Rogers, E. D. Clotfelter, R. C. Titus, S. J. Schoech, and E. Snajdr 2002); Hostetter 1961; Austin 1968; Byers, Curson, and Olsson 1995; Rising 1996; Wright 2019.

WHITE-CROWNED SPARROW. *ZONOTRICHIA LEUCOPHRYS*

Status: A common spring and fall migrant, especially in western Nebraska, with local or occasional overwintering southwardly. National Breeding Bird Surveys from 1966 to 2015 indicate that this species' population underwent a survey-wide average annual decline of 0.4 percent (Sauer et al. 2017).

Habitats and Ecology: On migration and during winter, white-crowned sparrows are common in a variety of central and western Nebraska habitats that offer a combination of brushy cover and open ground for foraging.

Suggested Viewing Opportunities: White-crowned sparrows can be readily attracted to ground-level bird-feeding platforms during migration, which during fall is centered in late September to late October, and during spring is concentrated from late April to early May.

Selected References: BNA 183 (G. Chilton, M. C. Baker, C. D. Barrentine, and M. A. Cunningham 1995); Austin 1968; Byers, Curson, and Olsson 1995; Rising 1996; Wright 2019.

WHITE-THROATED SPARROW. *ZONOTRICHIA ALBICOLLIS*

Status: A common spring and fall migrant throughout the state, with local or occasional overwintering southwardly. National Breeding Bird Surveys from 1966 to 2015 indicate that this species' population underwent a survey-wide average annual decline of 0.93 percent (Sauer et al. 2017).

Habitats and Ecology: On migration and during winter, white-throated sparrows are found in a variety of habitats statewide that offer a combination of brushy cover and open ground for foraging.

Suggested Viewing Opportunities: White-throated sparrows can be readily attracted to ground-level bird-feeding platforms during migration, which during fall is centered in late September and October, and spring migration occurs during April and May.

Selected References: BNA 128, rev. ed. (J. B. Falls and J. G. Kopachena 2010); Austin 1968; Byers, Curson, and Olsson 1995; Rising 1996; Wright 2019.

VESPER SPARROW. *POOECETES GRAMINEUS*

Status: A common spring and fall migrant and summer resident almost throughout the state in grassland areas, with the densest populations in eastern Nebraska and the Pine Ridge region. Reduced breeding occurs in the western Sandhills, and little or none occurs in southernmost Nebraska. There were 135 possible to confirmed nesting records during the fieldwork for *The Second Nebraska Breeding Bird Atlas* (Mollhoff 2016). National Breeding Bird Surveys from 1966 to 2015 indicate that this species' population underwent a survey-wide average annual decline of 0.85 percent (Sauer et al. 2017).

Habitats and Ecology: During the breeding season, this species is found in overgrown fields, prairie edges, weedy croplands, grasslands with scattered shrubs and small trees, brushy areas with scattered and stunted plants, and similar mostly grassy habitats. In Colorado, nesting occurs widely but is most common where sagebrush is interspersed with a good grass cover (Kingery 1998).

Breeding Biology: Vesper sparrows occupy a considerably larger home range than do many prairie-adapted sparrows and frequently defend territories of as large as 2 acres. Most singing is done from fairly high perches, but flight songs are also rarely performed. The usual clutch is three to five eggs, and the incubation period lasts 11–13 days. The female evidently does most of the incubating, but males have been seen covering eggs, and they sometimes also brood the young. On the average, the young remain in the nest for nine days, but they remain semidependent on their parents until they are 30–35 days old. In one Michigan study, a pair hatched a second brood 29 days after the hatching of the first, and among another group of 29 pairs, 15 pairs raised a single brood, 13 raised two broods, and 1 pair raised three broods in a single season.

Suggested Viewing Opportunities: In Nebraska, this species breeds almost statewide on open grasslands and to a lesser extent on no-till agricultural lands. Open grasslands, especially where some shrubs and bare ground are both present, are good places to look for this large sparrow.

Selected References: BNA 624 (S. L. Jones and J. E. Cornely 2002); Austin 1968; Byers, Curson, and Olsson 1995; Rising 1996; Johnsgard 2001b; Wright 2019.

HENSLOW'S SPARROW. *CENTRONYX HENSLOWII*

Status: The known Nebraska nesting range of this rare sparrow is limited to the state's southeastern corner. It currently (2019) includes only five documented local sites (Silcock and Jorgensen 2007) extending as far west as Mormon Island, east to northern Seward County and Spring Creek Audubon Prairie, Lancaster County, and south to Pawnee County (Mollhoff 2016). National Breeding Bird Surveys from 1966 to 2015 indicate that this species' population underwent a survey-wide average annual decline of 1.53 percent (Sauer et al. 2017). Breeding Bird Survey data for Nebraska are lacking. This species is on the Nebraska Natural Legacy Project's Tier 1 list of threatened species in Nebraska (Schneider et al. 2018).

Habitats and Ecology: The species is primarily associated with weedy prairies and meadows, neglected grassy fields, and pasturelands, especially those that are rather low-lying and damp. Scattered low bushes are often present. Unburned grasslands with high densities of standing vegetation and low coverage by woody vegetation are preferred nesting habitat (Zimmerman 1993). Like the grasshopper sparrow, this species tends to be a localized and semicolonial nester; as many as ten pairs have been reported breeding on a half-acre field in Iowa. Territories averaged about an acre each in one Michigan study and gradually increased in size through the summer. Territories are established within such nesting aggregations, and most territorial disputes are limited to rather formal "songfests" rather than to physical encounters. The intensity of singing varies greatly and is highest during territorial establishment and nest-building.

Breeding Biology: This species' nests are usually among rank and often moist areas of taller grass and taller weeds or shrubs. They typically have a base of dead grassy litter and are either hidden under a tuft of overhanging grass or attached to the stems of herbage up to 20 inches above the ground. Nest-building requires five or six days and is done mostly or entirely by the female. The female incubates her clutch of three to five brown-blotched eggs for 11 days, and both parents attend the nestlings during their nine-to-ten-day fledging period. Double-brooding (producing two broods per season) is typical. Brood parasitism by cowbirds is evidently rare; in a Missouri study, the parasitism rate was only 5.3 percent, compared to 8.8 percent for dickcissels and an overall rate of 8.1 percent for all passerine birds (Winter 1989), which perhaps reflects the well-concealed nests and furtive behavior of this species.

Suggested Viewing Opportunities: Only in a few Nebraska locations having stands of tallgrass prairie and well-developed litter accumulations, such as Spring Creek Audubon Prairie near Lincoln, can this elusive sparrow be sought after with any hope of success. Spring arrival records cluster around April, and fall departure records cluster around September 26. Most of the few nesting records for Kansas and Nebraska are for middle to late June.

Selected References: BNA 672, rev. ed. (J. R. Herkert, P. D. Vickery, and D. E. Kroodsma 2015); Hyde 1939; Austin 1968; Robins 1971; Byers, Curson, and Olsson 1995; Zimmerman 1993; Rising 1996; Johnsgard 2001b; Wright 2019.

SONG SPARROW. *MELOSPIZA MELODIA*

Status: A common resident in suitable habitats in the eastern half of the state, extending west in the Niobrara Valley to Cherry County, in the Platte Valley to Garden County, and in the Republican Valley to Red Willow County. Little or no breeding occurs in the western Sandhills, southwestern Nebraska, or the Panhandle. There were 286 possible to confirmed nesting records during the fieldwork for *The Second Nebraska Breeding Bird Atlas* (Mollhoff 2016). National Breeding Bird Surveys from 1966 to 2015 indicate that this species' population underwent a survey-wide average annual decline of 0.76 percent (Sauer et al. 2017).

Habitats and Ecology: Breeding habitats include such woodland edge types as the brushy margins of forest openings, the shrubby edges of ponds or lakes, shelterbelts, farmsteads, and prairie coulees. Foraging occurs mostly on the ground in both open areas and leaf-covered ones, where the birds scratch to expose foods.

Breeding Biology: The song sparrow is one of America's best-studied songbirds, thanks in part to classic banding and behavior studies by Margaret M. Nice (1943). Males are highly territorial and often maintain the same territories year after year. In one study, about half of the females returned to their old territories the following year but only infrequently (8 of 30 cases in one study) mated again with their previous partners. They often settle into adjacent territories if their prior mates have established new mates, or they might move as far as a mile from their place of hatching. Usually the female builds the nest alone, but there are cases of unmated males building nests and of helping mates in nest construction. About three or four days are needed for nest-building, and the female incubates alone. The usual clutch is three to five eggs, and the incubation period lasts 12–14 days. Both sexes feed the young, which remain in the nest for about ten days and finally become independent when 28–30 days old. Intervals between the fledging of two successful broods ranged from 30–41 days in one study.

Suggested Viewing Opportunities: Song sparrows are common to abundant summer residents in and around shrubby wetlands across most of eastern Nebraska.

Selected References: BNA 704 (P. Arcese, M. K. Sogge, A. B. Marr, and M. A. Patten 2002); Nice 1943; Austin 1968; Byers, Curson, and Olsson 1995; Rising 1996; Wright 2019.

SWAMP SPARROW. *MELOSPIZA GEORGIANA*

Status: An uncommon spring and fall migrant in eastern and central Nebraska wetlands, becoming rare westwardly. Breeding occurs in most eastern Sandhills counties, from Cherry and Lincoln Counties east to the sandy-loess soils of Boone and Knox Counties. Breeding also occurs locally in the North Platte, Platte, and Loup Valleys and in Garden and Sheridan Counties. There were 43 possible to confirmed nesting records during the fieldwork for *The Second Nebraska Breeding Bird Atlas* (Mollhoff 2016). This species is on the Nebraska Natural Legacy Project's Tier 2 list of threatened species in Nebraska (Schneider et al. 2018).

Habitats and Ecology: In Nebraska, the swamp sparrow is almost exclusively limited to the Sandhills wetlands, where it has a preference for nesting in shrubby wild indigo (*Baptisia*) habitat bordering marshes and streams. All but 6 of 48 reports of habitats used by breeding birds during fieldwork for *The Second Nebraska Breeding Bird Atlas* were in wetlands, and the others were in shrubby willow, dogwood, and wild indigo habitats (Mollhoff 2016). Elsewhere, it also breeds in wet meadows, along swampy shorelines of lakes or streams, and to a limited degree in saline meadows.

Breeding Biology: Probably because it is restricted to wet, inaccessible habitats, rather little is known of the breeding biology of this relatively insectivorous sparrow. A breeding density of two pairs in about 9 acres of Maryland bog has been estimated. Males have been reported to be sometimes polygynous. Swamp sparrow nests are rarely placed on the ground but instead are built about a foot above the substrate, often in water up to 24 inches deep. They are often constructed among the stalks of cattails or in flooded bushes and are rather bulky structures of grass with a finer grass lining. From three to six eggs are laid, usually four or five. The incubation period is 12–13 days. Apparently the female incubates alone, although the male has been observed feeding his brooding mate. The nestling period is about 11–13 days. The species is normally single-brooded, but in some regions two broods are raised.

Suggested Viewing Opportunities: Playing recordings of this species' territorial songs at Sandhills wetlands is the most effective way of finding this inconspicuous species.

Selected References: BNA 279 (T. B. Mobray 1997); Austin 1968; Byers, Curson, and Olsson 1995; Rising 1996; Wright 2019.

Status: A common spring and fall migrant and summer resident over the western three-fourths of the state, with seemingly "pure" *Pipilo maculatus* phenotypes occurring mostly west of a line from Cherry County to Deuel County. Most of the rest of the state is influenced by hybridization with the eastern towhee, so that a species-level distinction between these taxa is dubious (Sibley and West 1959). There were 192 possible to confirmed nesting records during the fieldwork for *The Second Nebraska Breeding Bird Atlas* (Mollhoff 2016). National Breeding Bird Surveys from 1966 to 2015 indicate that the spotted towhee's population underwent a survey-wide average annual decline of 0.03 percent (Sauer et al. 2017).

Habitats and Ecology: Breeding occurs in brushy fields, thickets, woodland openings or edges, second-growth forests, city parks, and well-planted suburbs. Habitats that have a good accumulation of litter and humus and a protective screen of shrubby foliage above the ground are highly favored by these birds. Some overwintering occurs in southeastern Nebraska.

Breeding Biology: Territories and pair bonds might be established very early, up to two months before nesting starts. Pair formation is achieved by males singing persistently from a variety of locations in their territories. As pair bonds form, the rate of singing drops off, and both birds forage within the male's territory. The female builds the nest with little or no help from the male, although he sometimes carries about small twigs. The usual clutch size is three or four eggs. The female incubates during the incubation period of 12–14 days, and both sexes feed the young. The fledglings leave the nest in 9–11 days.

Suggested Viewing Opportunities: Breeding populations are probably highest in the Pine Ridge and Wildcat Hills, but migrants (and hybrids) occur statewide, and wintering birds can often be found in woodland-edge habitats of southeastern Nebraska.

Selected References: BNA 263, rev. ed. (S. Bartos Smith and J. S. Greenlaw 2015); Sibley and West 1959; Austin 1968; Byers, Curson, and Olsson 1995; Rising 1996; Wright 2019.

Status: A common spring and fall migrant and summer resident over the eastern edge of the state, but extensive hybridization westward with *Pipilo maculatus* blurs all their distinctions (Sibley and West 1959). Some *erythropthalmus*-like phenotypes extend west in the Niobrara Valley to possibly Rock County, in the Platte Valley to possibly Keith County, and in the Republican Valley possibly to Hitchcock County. There were 136 possible to confirmed nesting records during the fieldwork for *The Second Nebraska Breeding Bird Atlas* (Mollhoff 2016). National Breeding Bird Surveys from 1966 to 2015 indicate that the eastern towhee's population underwent a survey-wide average annual decline of 1.34 percent (Sauer et al. 2017).

Habitats and Ecology: The species is associated with brushy fields, thickets, woodland edges or openings, second-growth forests, and city parks or suburbs having trees and tall shrubbery. Breeding habitats are essentially the same as those described above for the spotted towhee.

Breeding Biology: Pair formation in eastern towhees is much like that described for the spotted towhee. Nests are usually built on the ground but might also be in shrubs or vine tangles, rarely as high as 12 feet above the ground. The nest is a rather bulky structure built of leaves, weed stems, and bark strips and is lined with fine grasses and other soft materials. Ground nests are often placed under vegetation or brush piles, concealing them from above, and in tree nests the canopy is also usually very dense above the nest. Clutch sizes vary from two to six eggs, usually three or four. The incubation period is 12–14 days. The young leave the nest in 9–11 days. Double-brooding in a single season sometimes occurs.

Suggested Viewing Opportunities: City parks and backyards with shrubbery attract this common species. Unlike spotted towhees, eastern towhees do not overwinter in Nebraska.

Selected References: BNA 262, rev. ed. (J. S. Greenlaw 2015); Sibley and West 1959; Austin 1968; Byers, Curson, and Olsson 1995; Rising 1996; Wright 2019.

Family Icteriidae (Yellow-breasted Chat)

YELLOW-BREASTED CHAT. *ICTERIA VIRENS*

Status: This species has long been known to be the most aberrant member of the New World warbler family Parulidae (Ficken and Ficken 1962) and in 2017 was recognized as representing a unique single-species (monotypic) family, although its evolutionary relationships are still controversial. It is a local and declining summer resident in shrubby and woodland-edge habitats of western and central Nebraska, extending from the Pine Ridge east in the Niobrara Valley to probably Knox County, in the Platte Valley from the Wildcat Hills east probably to Keith County, and in the Republican Valley from Dundy County east possibly to Furnas County. It also breeds along wooded Sandhills streams in the Loup and Elkhorn drainages. There were 105 possible to confirmed nesting records during the fieldwork for *The Second Nebraska Breeding Bird Atlas* (Mollhoff 2016). National Breeding Bird Surveys from 1966 to 2015 indicate that this species' population underwent a survey-wide average annual decline of 0.62 percent (Sauer et al. 2017). However, in Nebraska the decline has been much greater (an estimated 5.64 percent annually) and has resulted in the disappearance of chats from nearly all of eastern Nebraska, a pattern also occurring elsewhere in the eastern Great Plains.

Habitats and Ecology: During the breeding season, this species occurs in the shrubby eroded or hillside areas of prairies, along brush-lined creeks and forest edges, and in shrubby overgrown pasturelands.

Breeding Biology: Male chats are distinctive in their remarkable diversity of vocalizations. Males often utter six to ten different song phrases, which are almost randomly produced and which vary greatly in timbre, loudness, pitch, and duration. They sing loudly during flight display, with their throats conspicuously expanded and their feet dangling. They also sing frequently at night, especially early in the mating season (Canterbury 2007). The usual chat clutch is three to five eggs, and the incubation period lasts 11–12 days. Only the female incubates, but both parents brood the young, which leave the nest in 8–11 days. Chats are reportedly double-brooded and also are often nonmonogamous. Recent DNA studies in Kentucky revealed that 5 of 29 chat nestlings were not fathered by the male of the pair, and 3 of 9 broods contained at least one nestling that had been fathered by a nonpair male.

Suggested Viewing Opportunities: Because chats have nearly disappeared from eastern Nebraska, the best chances of finding them are in the Panhandle, especially in brushy riparian habitats, such as along the North Platte River, and in Pine Ridge brushlands and woodland edges. Their loud and distinctive vocalizations often betray their locations.

Selected References: BNA 575 (K. P. Ekerle and C. F. Thompson 2001); Bent 1953; Griscom and Sprunt 1957; Ficken and Ficken 1962; Curson, Quinn, and Beadle 1994; Canterbury 2007.

Family Icteridae (Blackbirds, Orioles, and Meadowlarks)

YELLOW-HEADED BLACKBIRD.

XANTHOCEPHALUS XANTHOCEPHALUS

Status: A common spring and fall migrant and summer resident in wetland habitats, especially in the Sandhills and Rainwater Basin. There were 120 possible to confirmed nesting records during the fieldwork for *The Second Nebraska Breeding Bird Atlas* (Mollhoff 2016). National Breeding Bird Surveys from 1966 to 2015 indicate that this species' population underwent a survey-wide average annual decline of 0.06 percent (Sauer et al. 2017).

Habitats and Ecology: Restricted during the breeding seasons to relatively permanent marshes, the marsh zones of lakes, and the shallows of river impoundments where there are good stands of cattails, bulrushes, or phragmites. Although sometimes breeding in the same wetlands as red-winged blackbirds, yellow-headed blackbirds occupy the deeper areas adjacent to open water rather than among the dense stands of shoreline vegetation.

Breeding Biology: The displays of the yellow-headed blackbird are very similar to those of the red-winged blackbird, but the species differs ecologically in that it forages more heavily on emerging aquatic insects such as damselflies and thus is more dependent on larger and deeper open-water marshes than are redwings. In both species, the males' conspicuous and prolonged displays seem to be related to the importance of territorial size and quality in attracting the maximum number of females. The usual clutch size is four eggs. Like the red-winged blackbird, only the female incubates during the incubation period of 10–13 days, but males more often remain to help feed the young, especially those of their first mate. The young leave the nest at 9–12 days.

Suggested Viewing Opportunities: Cattail marshes in central Nebraska and the Sandhills that are deep enough to have some open water are likely to support this conspicuous and noisy blackbird.

Selected References: BNA 192 (D. J. Twedt and R. D. Crawford 1995); Bent 1958; Orians and Christman 1968; Jaramillo and Burke 1999.

BOBOLINK. *DOLICHONYX ORYZIVORUS*

Status: An uncommon spring and fall migrant and summer resident in prairie and meadow habitats statewide. Bobolinks are most common in wet meadows, a habitat type that is vanishing rapidly in Nebraska. During both surveys for *The Nebraska Breeding Bird Atlas,* there was a concentration of breeding records in the eastern Sandhills. Breeding is increasingly rare in the west and in the southwest. It extends locally west to Sioux and Garden Counties in the Panhandle but is apparently absent from the Republican River drainage. There were 215 possible to confirmed nesting records during the fieldwork for *The Second Nebraska Breeding Bird Atlas* (Mollhoff 2016). National Breeding Bird Surveys from 1966 to 2015 indicate that this species' population underwent a survey-wide average annual decline of 2.06 percent (Sauer et al. 2017).

Habitats and Ecology: Bobolinks breed in moist lowland meadows of mixed grasses and tall forbs, building their nests in shallow hollows that are effectively hidden by surrounding vegetation. A notable 47 percent of 192 bobolink nests were found to be parasitized by cowbirds during *The Second Nebraska Breeding Bird Atlas* studies (Mollhoff 2016).

Breeding Biology: Males arrive on their breeding areas about a week before females and quickly spread out, although specific territorial establishment and defense seem to be weak or lacking. Although the nests are well scattered, males tolerate other males surprisingly near the nest site. The usual clutch is five or six eggs, and the incubation period lasts 11–13 days. The female incubates alone; the male seldom visits her and apparently never feeds her but helps feed the young. Broods usually remain in the nest for about 10–14 days but have been reported to leave when only 7–9 days old. Males often acquire second mates after their first mate has begun nesting. These secondary mates tend to lay smaller clutches than the primary mates, perhaps because they often are young birds or are renesting. This smaller clutch size of secondary mates is adaptive, since males less frequently assist in feeding their second broods, and unassisted females are more likely to be able to effectively tend smaller broods.

Suggested Viewing Opportunities: Wet meadows in the Sandhills are prime habitats for bobolinks, and during spring their conspicuous flight songs over wet meadows are one of that region's prime natural attractions.

Selected References: BNA 176, rev. ed. (R. Renfrew, A. M. Strong, N. G. Peruit, S. G. Martin, and T. A. Gavin 2015); Bent 1958; Jaramillo and Burke 1999; Johnsgard 2001b.

EASTERN MEADOWLARK. *STURNELLA MAGNA*

Status: A common spring and fall migrant and summer resident in wetland habitats and a summer resident in suitable habitats in the eastern half of the state, extending west in the Niobrara Valley to Cherry County, in the Platte Valley to Garden County, and in the Republican Valley possibly to Franklin County. Breeding also commonly occurs in the western and central Sandhills along streams and in grassy pastures and wet meadows; apparent hybridization with the western meadowlark has occasionally occurred in such marginal populations (Rising 1983). There were 118 possible to confirmed nesting records during the fieldwork for *The Second Nebraska Breeding Bird Atlas* (Mollhoff 2016). National Breeding Bird Surveys from 1966 to 2015 indicate that this species' population underwent a survey-wide average annual decline of 0.31 percent (Sauer et al. 2017). This species is on the Nebraska Natural Legacy Project's Tier 2 list of threatened species in Nebraska (Schneider et al. 2018).

Habitats and Ecology: The species is associated with tallgrass prairies, meadows, and open croplands of small grain, as well as weedy orchards and other open, grass-dominated habitats. At the western edge of its range, where the western meadowlark also occurs, it is predominantly limited to lowland habitats, such as wet meadows and the edges of Sandhills marshes.

Breeding Biology: Nest-building begins early in eastern meadowlarks and evidently is performed by only the female over a period of three to eight days, the shorter periods being typical of later nests. Nests are built on the ground in scrapes or natural depressions, well hidden in grass clumps and with a canopy woven into the surrounding vegetation. There is a lateral entrance to the nest, which is constructed of coarse grasses and lined with finer grasses, and there is sometimes also a visible trail leading to the nest through the adjacent vegetation. Nest openings often face toward the east or north, away from the prevailing western winds and afternoon sun. Polygyny is fairly frequent in this species; about 50 percent of the males in one New York study had two mates, and one had three. The incubation period is 13–15 days, averaging 14. Only the female incubates, and she leaves the nest infrequently once it has begun. The young birds normally leave their nest 11–12 days after hatching but might leave sooner if disturbed. Within two or three days of nest departure, the female can start a second clutch, with the male remaining to tend the young of the first brood. The female does, however, periodically feed the first brood until she begins incubating again, when the young are about three weeks old.

Suggested Viewing Opportunities: Although during the 1960s the eastern and western meadowlarks were about equally common in the vicinity of Lincoln, the western meadowlark now substantially outnumbers the eastern, which has been declining at a greater rate. Only the relict tallgrass prairies of the easternmost counties of Nebraska currently support good populations of eastern meadowlarks.

Selected References: BNA 160, rev. ed. (L. A. Jaster, W. E. Jensen, and W. E. Lanyon 2012); Bent 1958; Jaramillo and Burke 1999; Johnsgard 2001b.

WESTERN MEADOWLARK. *STURNELLA NEGLECTA*

Status: A common spring and fall migrant and summer resident in prairie and other grassland habitats statewide. The state bird of Nebraska, western meadowlarks are still common in meadows and native prairies throughout the state. A total of 522 possible to confirmed breeding records was reported for every county during fieldwork for *The Second Nebraska Breeding Bird Atlas*, and the species was judged to be the state's ninth most common breeding bird (Mollhoff 2016). During those surveys 235 western meadowlark nests were reported, compared to 99 of the eastern meadowlark, and there were breeding records from every Nebraska county. National Breeding Bird Surveys from 1966 to 2015 indicate that this species' population underwent a survey-wide average annual decline of 1.29 percent (Sauer et al. 2017). Nebraska surveys indicate a similar 1.34 percent annual decline, based on a sample of 51 routes.

Habitats and Ecology: Mixed-grass prairies are prime habitat for western meadowlarks, but they extend west into shorter grasses and sage-steppe and east into wetter and taller prairies. The ecologies of the two meadowlark species are virtually identical, and the two provide an interesting problem in terms of their ecology and evolutionary relationships. Where they occur together they sometimes utter intermediate primary songs. However, this does not prove frequent hybridization, owing to the influences of early learning in song development; their different call notes are more diagnostic and indicative of ancestry. One area of apparent hybridization is the Platte Valley of Nebraska, where intermediate birds are several times more frequent than elsewhere in the Great Plains. In areas of overlap, there has been no evolutionary divergence in song types.

Breeding Biology: Meadowlark nests are typically well hidden by overhead grasses. The clutch is usually of five speckled white eggs, which the female incubates for 13–15 days. The nestling period lasts 11–12 days. Double-brooding is typical during the fairly long summer breeding period; western meadowlark egg dates in Nebraska extend from early April to late July (Mollhoff 2016). Among 23 major North American host species I reviewed (Johnsgard 1997a), this species was the most often parasitized by brown-headed cowbirds (47 percent of 294 nests). Of 235 western meadowlark nests, 100 (43 percent) had been parasitized by cowbirds during *The Second Nebraska Breeding Bird Atlas* field studies (Mollhoff 2016).

Suggested Viewing Opportunities: Prairie remnants of tallgrass or midgrass height are ideal habitat for this iconic prairie species, but it also uses overgrown grassy fields of any type. Its loud, melodious songs can be heard for hundreds of yards and are among the most beloved signs of spring for rural Nebraskans.

Selected References: BNA 104, rev. ed. (S. K. Davis and W. E. Lanyon 2008); Bent 1958; Jaramillo and Burke 1999; Johnsgard 2001b.

BALTIMORE ORIOLE. *ICTERUS GALBULA*

Status: A common spring and fall migrant and summer resident in wooded habitats statewide. There were 434 possible to confirmed nesting records during the fieldwork for *The Second Nebraska Breeding Bird Atlas* (Mollhoff 2016). National Breeding Bird Surveys from 1966 to 2015 indicate that this species' population underwent a survey-wide average annual decline of 1.49 percent, thus apparently declining slightly more rapidly than the Bullock's.

Habitats and Ecology: Favored habitats of Baltimore orioles include wooded river bottoms, upland forests, taller shelterbelts, and partially wooded residential areas and farmsteads. Trees with drooping outer branches hanging vertically downward are favored nest sites. Collectively, the Baltimore and Bullock's orioles are pandemic breeders across Nebraska. Typical Baltimore oriole phenotypes are generally found east of the 102nd meridian (along the western edges of Cherry and Keith Counties), and Bullock's oriole phenotypes predominate west of this line. However, there is a zone of frequent hybridization extending approximately 150 miles on either side of this division point. The Bullock's oriole is currently regarded as a species distinct from the Baltimore oriole, even though extensive hybridization in the Great Plains during the mid-1900s favored the view that they are biologically a single interbreeding species (Sibley and Short 1964). However, more recent evidence has suggested that hybrids are becoming less frequent in the overlap zone, supporting the current recognition of two species (Rising 1970).

Breeding Biology: The female builds the remarkable pendant nests of this species sometimes in as little as four or five days, although usually a week or so is needed. No true knots are tied in the process, but a loose tangle of fibrous materials is gradually pulled together and tightened, forming a seemingly woven structure. A new nest is made each year, but certain trees or territories from previous years seem to be favored, as the remains of old nests are often found near new ones. Nests are usually about 25 feet above the ground (range 9–70 feet) in rather large trees, especially elms and cotton-woods growing in open spaces, and are deep woven baskets of plant fibers, including grasses, although nests are not exclusively made of them, as in the orchard oriole. The nest is deeper than it is wide and sometimes has a lateral rather than an upper opening. There are some regional variations in nest placement and structure. The clutch size varies from two to six eggs, averaging about five. The incubation period is 12–14 days. Incubation is by the female, who is fed on the nest by her mate. The nestling period is approximately two weeks, and the young are dependent on the adults for another two weeks. This species is single-brooded.

Suggested Viewing Opportunities: Baltimore orioles are still very common summer residents of eastern Nebraska, and places such as riparian woodlands, city parks, and cemeteries offer excellent opportunities to find them.

Selected References: BNA 384 (J. D. Rising and N. J. Flood 1998); Bent 1958; Kroodsma 1970; Rising 1970, 1983; Clawson 1980; Jaramillo and Burke 1999.

BULLOCK'S ORIOLE. *ICTERUS BULLOCKII*

Status: A common spring and fall migrant and summer resident in wooded habitats of western Nebraska. The Bullock's oriole is a common summer resident in lightly wooded habitats of the Panhandle, with pure *bullockii* phenotypes mostly limited to the westernmost tier of Panhandle counties. Some *bullockii*-like birds occur east along the Niobrara River to Keya Paha County, along the North Platte Valley to Lincoln County, and in southwestern Nebraska to Red Willow and Frontier Counties. There were 72 possible to confirmed nesting records during the fieldwork for *The Second Nebraska Breeding Bird Atlas* (Mollhoff 2016). National Breeding Bird Surveys from 1966 to 2015 indicate that this species' population underwent a survey-wide average annual decline of 0.66 percent (Sauer et al. 2017). This species is on the Nebraska Natural Legacy Project's Tier 2 list of threatened species in Nebraska (Schneider et al. 2018).

Habitats and Ecology: During the breeding season, males of the Bullock's oriole especially favor river-bottom forests of willows and cottonwoods but also occur in city parks and on upland woodlands of aspen, poplars, and similar hardwood vegetation. In Colorado, mature native cottonwoods and nonnative landscaping trees provide a major breeding habitat for both orioles, but all the Colorado records of Baltimore orioles came from only three habitat types: cottonwoods plus some rural and urban habitats (Kingery 1998). During summer and fall, orioles are attracted to trees and bushes that provide ripe berries.

Breeding Biology: The early stages of breeding in this species are exactly like those of the Baltimore oriole. Incubation is likewise by the female, who is fed on the nest by her mate. The usual clutch is four or five eggs, and the incubation period lasts about 14 days. The nestling period and juvenile dependency periods of the two species are also comparable.

Suggested Viewing Opportunities: The Bullock's oriole is probably more common in the Pine Ridge than anywhere else in the state and the least affected by hybridization. However, even as far west as Scottsbluff there are occasional obvious Baltimore × Bullock's hybrids present in the general oriole population.

Selected References: BNA 416, rev. ed. (N. J. Flood, C. L. Schlueter, M. A. Reudink, P. Pyle, M. A. Patten, J. D. Rising, and P. L. Williams 2016); Bent 1958; Kroodsma 1970; Rising 1970, 1983; Jaramillo and Burke 1999.

RED-WINGED BLACKBIRD. *AGELAIUS PHOENICEUS*

Status: An abundant spring and fall migrant and summer resident in wetland habitats statewide. During state breeding bird surveys in the early 2000s, this species was judged to be the third most common breeding bird in Nebraska. A total of 537 possible to confirmed breeding records was reported for every county during fieldwork for *The Second Nebraska Breeding Bird Atlas* (Mollhoff 2016). This is one of the commonest of North American songbirds, with an estimated population of 210 million in the 1990s (Rich et al. 2004). National Breeding Bird Surveys from 1966 to 2015 indicate that this species' population underwent a survey-wide average annual decline of 0.93 percent (Sauer et al. 2017). Nebraska surveys indicate a similar 0.25 percent annual decline, based on a sample of 51 routes.

Habitats and Ecology: This is one of the most conspicuous and most thoroughly studied of all North American songbirds. Adult males arrive on their breeding marshes several days before females and begin to advertise their territories by flight songs and "song-spread" perch displays, both of which prominently exhibit the red upper wing coverts. Experiments with surgically muting males or painting these feathers black before they acquire mates resulted in the loss of territories by so-altered males. Pair bonds last only through the breeding season, and most territorial males manage to acquire at least two females. In one Wisconsin study, it was found that previously breeding males tend to return to their old territories in successive years and that first-year males are usually unable to hold territories long enough to breed. In that study, no more than three females were mated to a single male, but a few instances of double-brooding (producing two broods in a single nesting season) were found.

Breeding Biology: Although these birds will often nest in upland meadows, hayfields, and grasslands, their favorite breeding habitat consists of cattail-rich shallow marshes. Fragments of such marshes, such as wet and weed-choked roadside ditches, also provide an adequate substitute for breeding by a male and as many mates as he can attract. Females build sturdy nests around the stems of upright vegetation, such as cattails and rushes, using tightly woven leaves and sometimes also with some mud for added support. Three to five pale blue eggs with scrawled darker markings are laid and incubated by the female for 10–12 days, followed by another 10–11-day nestling period prior to fledging. Two breeding cycles per season are typical, and a vigorous male might be able to support two or three mates simultaneously. Nesting in Nebraska extends from mid-May to mid-July. This species is often parasitized by brown-headed cowbirds, with 47 percent of 45 nests found to be parasitized in *The Second Nebraska Breeding Bird Atlas* field studies (Mollhoff 2016). Among 31 studies cited by Ortega (1998), the median parasitism rate was 7.1 percent, the rates ranging from 0 to 76.5 percent.

Suggested Viewing Opportunities: One of the most abundant songbirds in North America, this species can be found breeding in marshlands, ditches, and weedy agricultural lands anywhere in the state. As many as 502,000 birds have been seen on Nebraska Christmas Bird Counts; some large marshes attract countless thousands of blackbirds (including grackles) during fall migration.

Selected References: BNA 184, rev. ed. (K. Yasukawa and W. A. Searcy 2019); Bent 1958; Willson 1964; Orians and Christman 1968; Kren 1996; Jaramillo and Burke 1999.

BROWN-HEADED COWBIRD. *MOLOTHRUS ATER*

Status: A common spring and fall migrant and summer resident in diverse habitats over the entire state; most often reported in woodlands and grasslands. National Breeding Bird Surveys from

1966 to 2015 indicate that this species' population underwent a survey-wide average annual decline of 0.93 percent (Sauer et al. 2017). Nebraska surveys indicate a 0.25 percent annual decline, based on a sample of 51 routes. During breeding bird surveys for *The Second Nebraska Breeding Bird Atlas*, the species was judged to be the second most common breeding bird species in Nebraska, and a total of 548 possible to confirmed breeding records was reported for every county during fieldwork for the atlas (Mollhoff 2016). The estimated North American cowbird population during the 1990s was 63 million (Rich et al. 2004), or enough to cause the loss of hundreds of millions of hosts' broods every year.

Habitats and Ecology: This is the only species of widespread North American bird that is an obligatory "brood parasite," one wholly dependent on one or more other species to incubate its eggs and rear its young. Before human colonization, the species was largely limited to the Great Plains, owing to its close foraging association with bison, foraging at their feet as these huge animals stirred up insects while moving about. More recently, the cowbird has expanded its geographic range and developed a similar commensal association with cattle and horses. It has thus come into contact with many new potential host bird species that have not yet had time to evolve defensive mechanisms. Ortega (1998) listed nearly 220 reported victim species, of which nearly 150 are known to have raised cowbird young. These cowbird-rearing victims include at least 44 warblers, 24 passerellid sparrows, 12 vireos, and 9 icterids. During *The Second Nebraska Breeding Bird Atlas* survey, 28 cowbird-parasitized species were found, the most important of which, in descending frequency of nests affected, were western meadowlark, bobolink, grasshopper sparrow, dickcissel, eastern meadowlark, and red-winged blackbird (Mollhoff 2016). Cowbirds rely on their host species to accept and incubate the alien eggs and to preferentially rear their young. Thereby they condemn the host's young to probable death by starvation, unless the hosts can recognize the intruding eggs and expel or otherwise render them harmless by ejecting them, piercing them, or burying them with nest materials.

There is no specific mimicry of host egg features in this parasitic species, but female cowbirds normally lay only one egg in each host's nest, usually during the host's egg-laying period. It is common for other cowbirds to add their own eggs to such nests; up to eight cowbird eggs have been found in a single host nest (Mollhoff 2016). The cowbird eggs usually hatch at about the same time as the host's eggs, and the cowbird's nestling period is roughly ten days, about the same as that for many of its hosts. However, the young grow much more rapidly than the host nestlings and thus consume more; therefore, the amount of food available for the host's young is correspondingly reduced, often causing their starvation.

Breeding Biology: This species is possibly the most destructive bird in North America, being responsible for the reproductive failures of millions of other songbirds annually as a result of its brood parasitism. Females lay their eggs almost indiscriminately in the nests of other passerine birds, especially other icterids, but all of Nebraska's common grassland passerines (Johnsgard 1997a). Cowbird eggs also have an incubation period of 11–12 days, which is the same as or slightly shorter than nearly all other North American icterids, and their slightly brown-spotted surface pattern also closely resembles those of many host vireos, warblers, and sparrows. Peer, Robinson, and Herkert (2000) found that western meadowlarks ejected 36 percent of artificial cowbird eggs inserted into their nests, and dickcissels ejected 11 percent. Ortega (1998) summarized data indicating that the burying of cowbird eggs by parasitized hosts is a defensive response most frequently performed by yellow warblers and occurred in 36 percent of 678 parasitized yellow warbler nests observed in ten studies. Nest desertion by the parasitized species is also a common, if extreme, defensive response, especially among smaller host species unable to pierce the alien eggs or too small to expel them. Ortega summarized nest desertion behavior in 19 frequently parasitized species, including dickcissels (17 percent of 18 nests), red-winged blackbirds (17 percent of 47 nests), grasshopper sparrows (22 percent of 9 nests), and eastern meadowlarks (46 percent of 28 nests). However, similar desertion rates of unparasitized clutches were found in 9 grasshopper sparrow and 12 eastern meadowlark nests.

Suggested Viewing Opportunities: The ubiquitous cowbird population is now centered where cattle grazing is common. Cowbirds can often be seen feeding on insects around the feet of various livestock.

Selected References: BNA 47 (P. E. Lowther 1993); Bent 1953; Newman 1970; Johnsgard 2001b; Kren 1996; Jaramillo and Burke 1999.

BREWER'S BLACKBIRD. *EUPHAGUS CYANOCEPHALUS*

Status: A spring and fall migrant and summer resident in the Panhandle, breeding in diverse habitats across the Pine Ridge east to northern Sheridan County and from western Sioux County south through the Wildcat Hills to Banner and Morrill Counties. There were 22 possible to confirmed nesting records during the fieldwork for *The Second Nebraska Breeding Bird Atlas* (Mollhoff 2016). National Breeding Bird Surveys from 1966 to 2015 indicate that this species' population underwent a survey-wide average annual decline of 2.25 percent (Sauer et al. 2017).

Habitats and Ecology: Low-stature grasslands are the primary breeding habitats of this species, including mowed or burned areas, farmsteads and residential areas, marsh edges (especially where scattered shrubs are present), the brushy banks of prairie creeks, and similar transitional habitats. Nesting occurs on the ground or in low shrubs such as sage, and fence posts also serve as singing posts where they are available. Outside the breeding season a wider array of open habitats is used, including grain fields, orchards, berry farms, and similar agricultural lands.

Breeding Biology: Brewer's blackbirds are often colonial nesters, but unlike red-winged and yellow-headed blackbirds, they are more frequently monogamous than polygynous. In contrast to these species, pair formation begins when the birds are still in winter flocks, and frequently mates of the previous season reestablish pair bonds. The female builds the nest, although the male might often accompany her as she gathers material. From 10 to 14 days are spent in building the nest and completing the clutch, typically five or six eggs. Males do not assist in incubation but occasionally visit the nest to feed the incubating female during the incubation period of 12–13 days. Both sexes feed the young, who leave the nest in about 13 days, and their parents might care for fledglings for at least three more weeks. At least in some areas, double-brooding is fairly frequent, and as many as three nesting attempts might be made in a single season by an unsuccessful pair.

Suggested Viewing Opportunities: This species is a common to abundant summer resident across westernmost Nebraska in ranchlands and agricultural lands. The Pine Ridge and Oglala National Grassland support good populations.

Selected References: BNA 626 (S. G. Martin 2002); Williams 1952; Bent 1953; Horn 1970; Jaramillo and Burke 1999.

COMMON GRACKLE. *QUISCALUS QUISCULA*

Status: A common spring and fall migrant and summer resident in diverse habitats over the entire state. National Breeding Bird Surveys from 1966 to 2015 indicate that this species' population underwent a survey-wide average annual decline of 1.75 percent (Sauer et al. 2017). During *The Second Nebraska Breeding Bird Atlas* field surveys there were confirmed breeding records from nearly all Nebraska counties and in 93 percent of all survey blocks, with a total of 276, the largest number for any of the icterids, followed by red-winged blackbirds and western meadowlarks (Mollhoff 2016).

Habitats and Ecology: Breeding habitats consist of woodland edges, areas partially planted to trees such as residential areas, farmsteads, shelterbelts, coniferous or deciduous woodlands of an open nature, woody shorelines around lakes, and riparian woodlands. Junipers, spruces, and other small and dense conifers are preferred for nesting, although hardwoods, shrubs, buildings, birdhouses, and even cattails are sometimes also used. In Colorado atlas studies, about 80 percent of breeding birds used various rural habitats, and the remainder used wooded riparian habitats comprised of diverse deciduous trees, including cottonwoods (Kingery 1998).

Breeding Biology: Males of this colonial-nesting species usually arrive on their breeding areas well before females and remain in flocks until the females arrive. There is a gradual breakup of migratory flocks as pairs are formed. A major component of pair formation is a flight involving a single female and up to five males, who follow her closely while holding their tails strongly keeled. After pairing, the female begins to select a nest site, and her mate defends only a small area of the nesting tree. The female gathers most of the material and does all the actual construction, which sometimes takes about a week but might occupy several weeks. The usual clutch size is four or five eggs, and the incubation period lasts 12–14 days. The female incubates alone and also does all the brooding. Both sexes feed their young, who leave the nest at 10–17 days.

Suggested Viewing Opportunities: Probably all Nebraska's cities, towns, and villages have resident flocks of grackles, who can be almost as much of a nuisance as starlings and tend to monopolize access to bird feeders.

Selected References: BNA 271 (B. D. Peer and E. K. Bollinger 1997); Bent 1953; Maxwell 1970; Maxwell and Putnam 1972; Jaramillo and Burke 1999.

Family Parulidae (New World Warblers)
ORANGE-CROWNED WARBLER. *OREOTHLYPIS CELATA*

Status: An uncommon spring and fall migrant statewide. National Breeding Bird Surveys from 1966 to 2015 indicate that this species' population underwent a survey-wide average annual decline of 0.61 percent (Sauer et al. 2017).

Habitats and Ecology: On migration these birds are found in a wide variety of brushy or wooded habitats but favor the brushy areas of river bottoms.

Suggested Viewing Opportunities: This is one of Nebraska's most common, if also most inconspicuous, warblers. Historic migration records indicate a clustering of spring occurrence records between April 30 and May 13 and a clustering of fall records between September 19 and October 15.

Selected References: BNA 101, rev. ed. (W. M. Gilbert, M. K. Sogge, and C. van Riper 2010); Bent 1953; Griscom and Sprunt 1957; Curson, Quinn, and Beadle 1994.

COMMON YELLOWTHROAT. *GEOTHLYPIS TRICHAS*

Status: A common spring and fall migrant and summer resident statewide. It breeds in wetlands, woodlands, and brushy areas throughout the state, second only to the yellow warbler as the state's most common breeding parulid. There were 430 possible to confirmed nesting records during the fieldwork for *The Second Nebraska Breeding Bird Atlas* (Mollhoff 2016). National Breeding Bird Surveys from 1966 to 2015 indicate that this species' population underwent a survey-wide average annual decline of 1.01 percent (Sauer et al. 2017).

Habitats and Ecology: Yellowthroats are generally associated with wetlands and riparian woodlands. Less often, they occupy dense deciduous woods or mixed woodland on upland slopes or mature river-bottom forests.

Breeding Biology: As soon as they arrive in spring, males establish territories that usually are less than 2 acres in area but in some instances might exceed 3 acres. Nests are usually built in the drier and more open parts of the territory, but water is always nearby. The female builds the nest over a period of two to five days and also performs all the incubation. The usual clutch size is four eggs. Feeding of the young is done by both sexes, and the chicks might leave their nest when only seven or eight days old. However, they don't fledge until they are 11–12 days old and don't begin feeding independently until about three weeks of age. One or both parents might tend them until they are four to five weeks old. Most or all females attempt to raise second broods, but apparently few are successful. Some females build at least three nests, and mate changing between broods evidently occurs occasionally.

Suggested Viewing Opportunities: This is a common species in brushy stream bottoms and wetlands throughout the state, and its easily recognized and syncopated *whichity* song makes it simple to localize, if not see, which is much harder to accomplish.

Selected References: BNA 448 (M. J. Guzi and G. Ritchison 1999); Bent 1953; Stewart 1953; Griscom and Sprunt 1957; Hofslund 1959; Morse 1989; Curson, Quinn, and Beadle 1994.

AMERICAN REDSTART. *SETOPHAGA RUTICILLA*

Status: A common spring and fall migrant and summer resident in deciduous woods over the eastern and northern parts of the state, extending along its entire Missouri River Valley, and west in the Niobrara Valley from the Missouri-Niobrara confluence to the Pine Ridge. There were 68 possible to confirmed nesting records during the fieldwork for *The Second Nebraska Breeding Bird Atlas* (Mollhoff 2016). National Breeding Bird Surveys from 1966 to 2015 indicate that this species' population underwent a survey-wide average annual decline of 0.28 percent (Sauer et al. 2017).

Habitats and Ecology: Breeding habitats of this species include moist bottomland woodlands, the margins or openings of mature forests, young or second-growth stands of various types of forests, and especially deciduous forests. Nearby water and brush seem to be important habitat components. In Colorado, redstarts typically nest in streamside habitats, with many undergrowth shrubs, but they might also nest in either deciduous or mixed deciduous-coniferous forests (Kingery 1998).

Breeding Biology: The American redstart is one of the few warblers whose social behavior has been extensively analyzed, including both its courtship and more general aggressive behavior. Males arrive on the breeding grounds before females and immediately become territorial. Females typically arrive at night and might have obtained a mate by the following morning, presumably being attracted to the males by nocturnal singing, a behavior also found in yellow-breasted chats. Up to 60 hours can elapse between the formation of the pair bond and the start of nest-building, during which time the female investigates the territory and begins to restrict her activities to the eventual nest site. The female selects the site and immediately begins to build a nest, which might require about three days. The usual clutch is of four eggs, and the incubation period lasts 12 days.

The female alone performs incubation and brooding, although males are typically highly attentive to the young during the nestling period, which usually lasts eight or nine days.

Suggested Viewing Opportunities: This is one of the more common of the state's breeding warblers. Historic migration records indicate a clustering of spring arrival records around May 12 and a clustering of fall departure records around September 10.

Selected References: BNA 277, rev. ed. (T. W. Sherry, R. T. Holmes, P. Pyle, and M. A. Patten 2016); Bent 1953; Griscom and Sprunt 1957; Ficken 1962; Hubbard 1969; Morse 1989; Curson, Quinn, and Beadle 1994.

YELLOW WARBLER. *SETOPHAGA PETECHIA*

Status: A common spring and fall migrant and summer resident throughout the state in woodland edge and shrubby habitats, probably breeding in every county, and the commonest Nebraska warbler. There were 442 possible to confirmed nesting records during the fieldwork for *The Second Nebraska Breeding Bird Atlas* (Mollhoff 2016). National Breeding Bird Surveys from 1966 to 2015 indicate that this species' population underwent a survey-wide average annual decline of 0.61 percent (Sauer et al. 2017).

Habitats and Ecology: Yellow warblers favor generally moist habitats, such as riparian woodlands and brush, as well as the brushy edges of marshes, swamps, or beaver ponds, but they also select drier areas, including roadside thickets, hedgerows, orchards, and forest edges. A combination of open areas and dense shrubbery seems to be important for breeding, although migrant birds are rather more widely distributed.

Breeding Biology: This widespread and abundant warbler seems to have rather generalized territorial needs, including a suitable nest site, tall singing posts, concealing cover, and foraging areas in shrubs and trees, all within an area of about two-fifths of an acre. Territorial behavior begins soon after the males arrive, and pairs might be formed in one to four days. Approximately four days are needed for nest construction, which is done mostly or entirely by the female. The female also does all the incubating, often beginning before the clutch has been completed. The usual clutch is of four eggs, and the incubation period lasts 11 days. Both sexes feed the young, and occasionally males have been seen brooding them, but this seems to be atypical. The young fledge in 9–12 days and remain in the general vicinity for another 7–10 days. Yellow warblers are commonly parasitized by brown-headed cowbirds but retaliate by burying their own and the cowbird's eggs, then starting a new clutch above the old one.

Suggested Viewing Opportunities: This is the most abundant and widespread of the state's breeding warblers. Historic migration records indicate a clustering of spring arrival records around May 12 and a clustering of fall departure records around September 10.

Selected References: BNA 454 (P. E. Lowther, C. Celada, N. K. Klein, C. C. Rimmer, and D. A. Spector 1999); Schrantz 1943; Bent 1953; Griscom and Sprunt 1957; Frydendall 1967; Hubbard 1969; Morse 1989; Curson, Quinn, and Beadle 1994.

YELLOW-RUMPED WARBLER. *SETOPHAGA CORONATA*

Status: A common spring and fall migrant and widespread summer resident in the Pine Ridge, with possible local breeding also in the Wildcat Hills. There were 12 possible to confirmed nesting records during the fieldwork for *The Second Nebraska Breeding Bird Atlas* (Mollhoff 2016). National Breeding Bird Surveys from 1966 to 2015 indicate that this species' population underwent a survey-wide average annual decline of 0.4 percent (Sauer et al. 2017).

Habitats and Ecology: This species breeds in a wide array of coniferous forests but in Nebraska occurs in extensive, park-like stands of ponderosa pine. During winter, the habitats used are more varied, and foraging might range from berry-eating to aerial fly-catching.

Breeding Biology: This warbler is among the first of the warblers to move northward in spring and often is fairly abundant as the first tree leaves appear in April. Nesting typically occurs in coniferous woodlands, but sometimes nests are placed in aspen groves or among scattered conifers. The female builds the nest on branches

4–50 feet above the ground. The usual clutch size is four or five eggs. Only the female incubates over the 12–13 days of incubation, but both sexes tend the young. The young fledge at about 12–14 days.

Suggested Viewing Opportunities: The western yellow-throated race *auduboni* of this species can be commonly observed in the pine forests of the Pine Ridge during summer, and both eastern and western races occur throughout the state during migration. There is a clustering of spring arrival records around April 23 and a clustering of fall departure records around October 22.

Selected References: BNA 376 (P. D. Hunt and J. Flaspohler 1998); Bent 1953; Griscom and Sprunt 1957; Ficken and Ficken 1966; Hubbard 1969; Morse 1989; Curson, Quinn, and Beadle 1994.

WILSON'S WARBLER. *CARDILLINA PUSILLA*

Status: A common spring and fall migrant in woodlands and brushlands of eastern Nebraska, becoming somewhat less common westwardly. National Breeding Bird Surveys from 1966 to 2015 indicate that this species' population underwent a survey-wide average annual decline of 1.8 percent (Sauer et al. 2017).

Habitats and Ecology: Migrants are associated with rank stands of weeds and low, shrubby vegetation, often near streams.

Suggested Viewing Opportunities: Wilson's warblers occur statewide during migration. Historic migration records indicate a clustering of spring occurrence records centering on May 12–19 and a clustering of fall records between September 1 and September 26.

Selected References: BNA 478 (E. M. Ammon and W. H. Gilbert 1999); Bent 1953; Griscom and Sprunt 1957; Morse 1989; Curson, Quinn, and Beadle 1994.

Family Cardinalidae (Cardinals, Tanagers, and Grosbeaks)

SCARLET TANAGER. *PIRANGA OLIVACEA*

Status: An uncommon spring and fall migrant in eastern Nebraska and a summer resident in the Missouri River's forested valley north to South Dakota, westward in the Niobrara Valley to possibly Brown County, and west in the lower Platte Valley to about

Saunders County. There were 48 possible to confirmed nesting records during the fieldwork for *The Second Nebraska Breeding Bird Atlas* (Mollhoff 2016). National Breeding Bird Surveys from 1966 to 2015 indicate that this species' population underwent a survey-wide average annual decline of 0.22 percent (Sauer et al. 2017).

Habitats and Ecology: In Nebraska this species is restricted primarily to mature hardwood forests in river valleys, hill slopes, and valleys; it less frequently breeds in coniferous forests, city parks, and orchards.

Breeding Biology: The relatively late spring arrival of this species, combined with its typical foraging characteristics of remaining high in the canopy of mature trees, keeps most of its behavior obscured from normal view. Nests are usually in tall trees, often oaks, from 8 to 75 feet above the ground, but usually are between 35 and 50 feet high and placed well out on horizontal limbs. The nest is rather small and loosely constructed of twigs and rootlets, lined with grasses and weed stems. From three to five eggs constitute a clutch, usually four, and the incubation period is 13–14 days. It is known that the female incubates alone, and in some cases, the male participates very little, even in feeding the young. The young leave the nest when about 15 days old and remain with their mother in the general vicinity of the nest for ten days or more. The species is single-brooded, attempting to produce only one brood per breeding season.

Suggested Viewing Opportunities: This species can be commonly observed in the mature Missouri Valley floodplain forests, such as at Fontenelle Forest and Indian Cave State Park. There is a clustering of spring arrival records around May 10 and a clustering of fall departure records around August 23.

Selected References: BNA 479 (T. B. Mowbray 1999); Bent 1958; Prescott 1965.

WESTERN TANAGER. *PIRANGA LUDOVICIANA*

Status: An uncommon local spring and fall migrant and summer resident in ponderosa pine forests of the Pine Ridge. National

Breeding Bird Surveys from 1966 to 2015 indicate that this species' population underwent a survey-wide average annual increase of 1.29 percent (Sauer et al. 2017).

Habitats and Ecology: Breeding in western North America occurs in various habitats, including riparian woodlands, aspen groves, ponderosa pine forests, and occasionally Douglas-fir forests and foothills woodlands. Its breeding range extends farther north, into Canada's Northwest Territories, than that of any other tanager. In Nebraska, it is usually found in areas having a predominance of pines, and usually in stands that are fairly open, but it occasionally occurs in fairly dense woods. There were 14 possible to confirmed nesting records during the fieldwork for *The Second Nebraska Breeding Bird Atlas* (Mollhoff 2016).

Breeding Biology: Little has been written on the breeding biology of this beautiful species, but it apparently closely resembles that of the better-studied scarlet tanager. In spite of the bright coloration of the males, breeding pairs are not conspicuous, since the olive-colored females remain high in the trees, and they tend to be very elusive during nesting. The nest is placed on the branch of a tree at from 8 to 60 feet above the ground. The female incubates the clutch of three to five eggs alone throughout the 13-day incubation period, and the male evidently rarely if ever approaches the nest during this time. After hatching, he helps feed the young, which leave the nest in 10–11 days and probably are fully fledged in about two weeks.

Suggested Viewing Opportunities: This beautiful species can be most easily observed in the pine forests of the Pine Ridge. There is a clustering of spring arrival records around May 19 and fall departure records clustering around September 15.

Selected References: BNA 431 (J. Hudon 1999); Bent 1958.

ROSE-BREASTED GROSBEAK. *PHEUCTICUS LUDOVICIANUS*

Status: A common spring and fall migrant and summer resident and variably common breeder over eastern Nebraska. Local breeding also occurs in the Sandhills along wooded streams. There were

194 possible to confirmed nesting records during the fieldwork for *The Second Nebraska Breeding Bird Atlas* (Mollhoff 2016). National Breeding Bird Surveys from 1966 to 2015 indicate that this species' population underwent a survey-wide average annual decline of 0.86 percent (Sauer et al. 2017).

Habitats and Ecology: This species is associated with open deciduous floodplain forests of eastern Nebraska, with relatively pure *ludovicianus* phenotypes extending west in the Niobrara Valley to Keya Paha County, in the Platte Valley possibly to Lincoln County, and in the Republican Valley possibly to Harlan County. In a South Dakota study, most grosbeak nests (rose-breasted and black-headed combined) were 10–19 feet above the ground, and box elders were favored nesting trees.

Breeding Biology: Immediately after the males return to their breeding areas in spring, they establish territories and begin to announce them with a warbled song and aggressive encounters with other males. Females arrive a few days later and are initially chased aggressively by males. Soon the male stops chasing the female, and she might attack him instead. Courtship-feeding of the female is apparently uncommon in this species. Generally, the forks and crotches of various deciduous trees are used for nest sites, and the nest is poorly constructed of twigs and grasses, lined with fine twigs and rootlets. The female builds the nest with the help of the male, and the male regularly participates in incubation. The clutch size varies from three to five eggs, averaging about four, and the incubation period is 12–14 days. Although usually single-brooded, double-brooding has been reported in semicaptive birds. Both sexes care for the young, and at least in two cases the males have been known to take over the care of young birds while the female began a second nesting. The young remain in the nest for 9–12 days.

Suggested Viewing Opportunities: State parks in southeastern Nebraska, such as Mahoney and Platte River State Parks, have excellent populations of this grosbeak. There is a clustering of spring arrival records around May 7 and fall departure records clustering around September 10.

Selected References: BNA 692, rev. ed. (V. E. Wyatt and C. M. Francis 2007); Bent 1958; West 1962; Dunham 1966; Kroodsma 1970.

BLACK-HEADED GROSBEAK. *PHEUCTICUS MELANOCEPHALUS*

Status: A common spring and fall migrant and variably common summer breeder over western Nebraska; the eastern limits are blurred by hybridization with the rose-breasted grosbeak (West 1962). Seemingly "pure" *melanocephalus* phenotypes extend east from the Pine Ridge in the Niobrara Valley to Keya Paha County, in the Wildcat Hills and Platte Valley east possibly to Buffalo County, and in the Republican Valley possibly to Harlan County. Local breeding also occurs in the Sandhills along wooded streams. There were 82 possible to confirmed nesting records during the fieldwork for *The Second Nebraska Breeding Bird Atlas* (Mollhoff 2016). National Breeding Bird Surveys from 1966 to 2015 indicate that this species' population underwent a survey-wide average annual decline of 0.72 percent (Sauer et al. 2017).

Habitats and Ecology: During the breeding season, this species is associated with open deciduous woodlands having fairly well-developed shrubby understories and usually on floodplains or upland areas. It extends into wooded coulees and riparian forests of cottonwoods, elms, and other hardwoods and sometimes also nests in orchards and on farmsteads.

Breeding Biology: So far as is known, the breeding biology of this species is essentially identical to that of the rose-breasted grosbeak, and hybrids are fairly common in Nebraska's central Platte Valley (West 1962). Studies of these two closely related forms in North Dakota indicate that the courtship behaviors of the two are very similar, and thus the color differences among the males are likely to be important in avoiding more widespread hybridization than occurs (Kroodsma 1970). Males apparently do not discriminate between the songs of their own and the counterpart species, but they do make visual discriminations when confronted with mounted males placed in their territories. Their nests are built by the females and are placed from about 6 to 12 feet above the ground at the fork of a small branch. The usual clutch is three to five eggs, incubated by the female. When the chicks hatch after about 13 days, they are tended by both sexes over the fledging period of 10–11 days.

Suggested Viewing Opportunities: Probably the only place to find only outwardly "pure" black-headed grosbeaks in Nebraska would be in the Pine Ridge, although vagrants have turned up as far east as Lancaster County. There is a clustering of spring arrival records around May 14 and fall departure records clustering around August 29.

Selected References: BNA 143, rev. ed. (C. Ortega and G. E. Hill 2010); Bent 1958; West 1962; Kroodsma 1970.

LAZULI BUNTING. *PASSERINA AMOENA*

Status: An uncommon spring and fall migrant and summer resident in brushy habitats of the Panhandle, mainly in the Pine Ridge and Wildcat Hills-North Platte Valley region, but also locally in southwestern Nebraska along the Republican River drainage of Dundy County. Hybrids with indigo buntings are frequent in regions of overlap. There were 35 possible to confirmed nesting records during the fieldwork for *The Second Nebraska Breeding Bird Atlas* (Mollhoff 2016). National Breeding Bird Surveys from 1966 to 2015 indicate that this species' population underwent a survey-wide average annual increase of 0.21 percent (Sauer et al. 2017).

Habitats and Ecology: On the Nebraska plains these birds are usually found in riparian woodlands supporting a mixture of shrubs, low trees, and herbaceous vegetation. Plant diversity and discontinuity of cover seem to be important habitat characteristics. In Colorado, the birds mostly breed in riparian habitats having an abundance of shrubs (Kingery 1998).

Breeding Biology: The breeding biology of this species can be considered nearly identical to that of the indigo bunting. In the western Great Plains, including western Nebraska, these two species overlap appreciably and sometimes hybridize (Emlen, Rising, and Thompson 1975; Baker and Boylan 1999). Playbacks of songs indicate that in some areas males respond only to the song of their own species and ignore that of the other, but in one area of Nebraska range overlap, males responded to both song types. Mixed matings in areas of breeding overlap are infrequent, and such pairs seem to exhibit delayed breeding characteristics compared with nonmixed pairings. The clutch size is usually of four eggs but varies from three to four. Incubation is performed by the female and lasts 12 days. Both sexes tend the young, which fledge in 10–15 days.

Suggested Viewing Opportunities: Among the best places to find phenotypically "pure" lazuli buntings in Nebraska are around Fort Robinson State Park and Scotts Bluff National Monument. There is a clustering of spring arrival records around May 16 and fall departure records clustering around August 25.

Selected References: BNA 232, rev. ed. (E. Greene, V. R. Muether, and W. Davison 2014); Bent 1958; Sibley and Short 1959; Kroodsma 1970; Emlen, Rising, and Thompson 1975; Baker and Boylan 1999.

INDIGO BUNTING. *PASSERINA CYANEA*

Status: A common spring and fall migrant and summer resident in the eastern half of Nebraska. Overlaps and hybridizes with lazuli buntings in western Nebraska. Seemingly "pure" *cyanea* phenotypes extend west in the Niobrara Valley to Cherry County, in the Platte Valley possibly to Keith County, and in the Republican Valley possibly to Dundy County. Local breeding also occurs in the Sandhills, mostly along wooded streams. There were 287 possible to confirmed nesting records during the fieldwork for *The Second Nebraska Breeding Bird Atlas* (Mollhoff 2016). National Breeding Bird Surveys from 1966 to 2015 indicate that this species' population underwent a survey-wide average annual decline of 2.44 percent (Sauer et al. 2017).

Habitats and Ecology: The indigo bunting is associated with relatively open hardwood forests on river floodplains or in uplands. Although the species occasionally breeds inside forests, it is most often associated with open, drier woods, favoring sites with reduced forest canopy and sapling density and increased shrub density. Thus, orchards, weedy fields, forest edges, second growth, and similar successional habitats are all widely used. The western limits of this species are confused by hybridization with the lazuli bunting (Emlen, Rising, and Thompson 1975). Hybrids were quite common around Cedar Point Biological Station, Keith County, during the 1980s and early 1990s (personal observation).

Breeding Biology: In this species, it is apparently the female that not only selects the specific nest site but also does all the actual construction. Sometimes nesting concentrations are very dense, such as 14 nests found in a 3-acre cotton patch in Mississippi. Nests are placed in the crotches of shrubs, in vine tangles, or in low trees at heights of 2–12 feet above the ground. They are built of grasses, twigs, bark strips, and weeds, lined with grasses and other soft materials. Clutches range from two to four eggs, averaging about three, and the incubation period is 12–13 days. Both sexes help feed the young, though only the female broods them, and sometimes the male will take charge of older nestlings, allowing the female to begin a second nesting. Usually a new nest is built for the second clutch, but sometimes the same one is used again.

Suggested Viewing Opportunities: Indigo buntings breed commonly in most if not all upland and floodplain deciduous forests of eastern Nebraska. There is a clustering of spring arrival records around May 10 and a clustering of fall departure records around August 28.

Selected References: BNA 4, rev. ed. (R. B. Payne 2006); Bent 1956; Sibley and Short 1959; Kroodsma 1970; Emlen, Rising, and Thompson 1975; Baker and Boylan 1999.

DICKCISSEL. *SPIZA AMERICANA*

Status: An uncommon spring and fall migrant and very common summer resident in the eastern two-thirds of Nebraska, becoming more local and less common in the Panhandle. Breeding also occurs throughout the Sandhills. During *The Second Nebraska Breeding Bird Atlas* surveys there were 459 possible to confirmed breeding records from every Nebraska county except for Sioux County (Mollhoff 2016). The estimated North American dickcissel population during the 1990s was 32 million (Rich et al. 2004). National Breeding Bird Surveys from 1966 to 2015 indicate that this species' population underwent a survey-wide average annual decline of 0.36 percent (Sauer et al. 2017). Nebraska surveys indicate a similar 1.04 percent annual decline, based on a sample of 51 routes.

Habitats and Ecology: This species is primarily associated with mixed-grass prairies but is also found in shortgrass and tallgrass prairies, sage prairies, and disturbed grassland habitats such as retired croplands, hayfields, and stubble fields. Areas that have grown up to shrubs are avoided, but scattered small trees offer acceptable habitat components and are often used as song perches.

Breeding Biology: Dickcissel nests might be on the ground among grassy vegetation or in a shrub or low tree as high as 6 feet off the ground; the latter site is probably much easier for cowbirds to find. The female alone builds the nest in about four days; male dickcissels are often polygynous and might be otherwise occupied. A clutch is typically of four eggs, which the female incubates for 11–12 days. Males usually obtain second mates during the laying or incubation phases of the first nesting cycle. Only the female feeds and tends the young, which might leave the nest in 7–10 days, although they do not fledge until they are 11–12 days old. This is one of the rather few species of North American passerine birds that regularly practices polygyny; a Kansas study indicated that 18 percent of the males had more than one mate, 40 percent were monogamous, and 42 percent were unmated. The variable success of males in attracting females seems to be related to the nest site characteristics that are available in their individual territories. Double-brooding is probably fairly common, which might help compensate for the high rate of cowbird-caused or other nest failures. Sixty-two of 92 (67 percent) dickcissel nests were parasitized during *The Second Nebraska Breeding Bird Atlas* field studies (Mollhoff 2016). Ortega (1998) reported cowbird parasitism rates in 12 studies to vary from 7 to 100 percent of the dickcissel nests found, with a median rate of 52.9 percent, one of the highest rates reported for any host species. Dickcissel egg dates in Nebraska extend from mid-May to late July. Luckily, by late June, when second nesting efforts by dickcissels might begin, cowbirds have typically stopped laying. After depositing some 40–

50 eggs, perhaps on a nearly daily basis, the cowbird females are probably exhausted (Johnsgard 2001b).

Suggested Viewing Opportunities: There are many places in eastern Nebraska where, if one drives slowly along country roads during late May, the songs of dickcissels are almost constantly in hearing range, and more than half of the songbirds perched on telephone wires or fence wires will probably be dickcissels. Densities of dickcissels in one tallgrass site of the Great Plains averaged about 210 birds per square mile (81 birds per square kilometer), a high density that is second only to the eastern meadowlark's among grassland birds (Johnsgard 2001b). There is a clustering of spring arrival records around May 16, and fall departure records cluster around August 22.

Selected References: BNA 703 (S. A. Temple 2002); Bent 1958; Zimmerman 1983, 1993; Johnsgard 2001b.

CHAPTER 4

Reptiles and Amphibians (Herpetiles)

AN INTRODUCTION TO NEBRASKA'S REPTILES AND AMPHIBIANS

Nebraska supports 14 native amphibians (2 salamanders, 11 frogs and toads) and 48 reptiles (9 turtles, 10 lizards, and 29 snakes), according to Ballinger, Lynch, and Smith (2010), who have produced Nebraska's first comprehensive book on the state's herpetiles. Besides providing species keys and biological narrative summaries for each species, the book also includes a history of Nebraska herpetology and a review of the relationships of Nebraska's geography and ecology to herpetile distributions in the state. Additionally, Fogell (2010) has authored a very useful field guide to all of Nebraska's reptiles and amphibians, illustrated with color photographs of each species, along with an identification key and county-based range maps for all the Nebraska species. Powell, Conant, and Collins (2016) have published a field guide with the most recent taxonomic treatment and excellent range maps of the eastern North American herpetiles (including all of Nebraska's species).

These welcome publications join a variety of regional books on the herpetiles of Kansas (Collins 1993), Colorado (Hammerson 1999), Wyoming (Baxter and Stone 2011), Missouri (Johnson 2000), South Dakota (Ballinger, Meeker, and Theis 2000; Kiesoe 2006), North Dakota (Wheeler and Wheeler 1966; Johnson 2015), and Oklahoma (Sievert and Sievert 2018). Compared with Nebraska's 62 herpetiles, Texas has more than 225 species (*The Herps of Texas*, https://www.herpsoftexas.org/), Oklahoma has 135 (Sievert and Sievert 2018), Kansas has 97 (Collins 1993), South Dakota has 44

(Ballinger, Meeker, and Theis 2000), and North Dakota has 28 (Johnson 2015). This marked species-diversity latitudinal gradient probably relates mostly to the strong influence of temperatures on distribution and abundance of these "cold-blooded" (ectothermic) vertebrates.

Within Nebraska, probable temperature-related trends can also be detected along its 210-mile north–south distance, judging from Fogell's (2010) county-record maps. Considering the northernmost tier of counties (Sioux to Dixon), there are county records for 10 amphibian and 25 reptile species, whereas in the southernmost tier (Dundy to Richardson) there are 11 amphibians and 41 reptiles. In Nebraska, the average annual precipitation more than doubles from west to east, from a minimum of about 15 inches in north-western Nebraska to about 36 inches in southeastern Nebraska, a distance of about 450 miles. A similar west–east near doubling of species diversity is suggested in Nebraska's herpetiles. Among the westernmost tier of counties (Sioux to Kimball), with more arid conditions and greater daily and seasonal temperature extremes, there are county records for 7 amphibian and 15 reptile species, whereas in the more mesic, more habitat-diverse, and more bio-logically complex easternmost tier (Dakota to Richardson), there are county records for 12 amphibians and 31 reptiles.

Lynch (1985) had earlier reviewed the taxonomy, species status, and geographic distributions of Nebraska's herpetiles. No Nebraska herpetiles have gone extinct, but one (sagebrush lizard) is known from a single county; the narrow-mouthed toad, slider, five-lined skink, and slender glass lizard have been documented from two counties; and the American toad, eastern glossy snake, and ven-omous copperhead from three (Fogell 2010). Ballinger, Lynch, and Smith (2010) indicated that the timber rattlesnake had been reported in four counties, but Fogell (2010) reported it from five. Other species with five county records include the western worm snake and coachwhip.

Based on county records, the most widespread Nebraska amphibian is the Woodhouse's toad (reported from all but four counties), and the most widespread Nebraska reptile is the Plains gartersnake (reported from all but ten counties). Lynch commented that ten Nebraska herpetile species occur virtually statewide, including the barred tiger salamander, Rocky Mountain (Woodhouse's) toad, striped treefrog (boreal chorus frog), snapping turtle, painted turtle, North American racer, milk snake, gophersnake, Plains gartersnake, and common gartersnake.

SELECTED SPECIES PROFILES

The following profile summaries include 24 of Nebraska's 62 total reptiles and amphibians. Nebraska supports 14 amphibians (2 salamanders, 6 toads, and 6 frogs) and 48 reptiles (8 turtles, 10 lizards, and 29 snakes), according to Ballinger, Lynch, and Smith (2010). The Latin nomenclature used here follows these authors. English vernacular names are primarily based on the 2012 checklist of the Society for the Study of Amphibians and Reptiles. However, some other English vernacular and Latin names that have often appeared in Nebraska's herpetile literature are shown parenthetically. Taxa here are arranged alphabetically within families, first by genus and second by species. Except for turtles, for which lengths refer to the length of the carapace (the dorsal bony shell), measurements given refer to body plus tail length or to snout to vent length, if so indicated.

AMPHIBIANS (SALAMANDERS, TOADS, AND FROGS)

Order Caudata (Salamanders)

Salamanders are unique among amphibians in that they retain their long tails into adulthood and develop four legs of equal length. As aquatic larvae they also have external gills, which in most species are lost by adulthood, after which lung-breathing occurs, and their moist skin might also allow for some oxygen exchange. In a few permanently aquatic species (and rarely also in tiger salamanders), the external gills are retained throughout the animal's lifetime.

Family Ambystomatidae (Mole Salamanders)

Mole salamanders remain below ground (thus, living like moles) for most of the year but move into pools and ponds for breeding. In contrast to frogs and toads, salamanders are mute.

BARRED TIGER SALAMANDER. *AMBYSTOMA MALVORTIUM*. UB, CO

Identification: This is the only salamander that occurs in nearly all of Nebraska, but the closely related smallmouth salamander (*Ambystoma texanum*) is additionally present in Otoe and Cass Counties. That small and very rare salamander lacks the contrasting black and yellow spots and stripes typical of the tiger salamander. Over most of Nebraska, the color pattern (figure 33) consists of a few large olive to yellowish spots on the back and olive to yellowish bands on the tail of a dark brown or black animal (Lynch 1985), but in parts of western Nebraska and the Sandhills, the tiger salamander's barred and spotted yellow and black color is reduced, and the resulting pattern is more one of dark blotches and spots over a yellow to greenish body (so-called blotched tiger salamanders). Adult males average 194 mm (7.6 in.) in length; females average 178 mm (7 in.) (Lynch 1985).

Status: Tiger salamanders are common and widespread in the state and have been reported from all but 19 scattered counties. The status of the very similar eastern tiger salamander (*Ambystoma tigrinum*) in the state is still undetermined, but it might occur among the counties bordering the Missouri River (Fogell 2010). Tiger salamanders are nocturnal, and when living on land they emerge from their burrows to forage only at night or during rains.

Habitats and Ecology: In western Nebraska, tiger salamanders sometimes live in the burrows of prairie dogs. Sandy areas or loose soils with nearby water are favored habitats; summer rains provide other water sources (Baxter and Stone 2011). During wet periods and breeding, these salamanders can be found in diverse wetlands. They often can be found in livestock water tanks during their spring breeding period. They prey on any animals small enough to swallow,

33. Tiger salamander, adult and larvae (top), and adults, eggs, and larvae of spadefoot (middle), northern leopard frog (bottom left), and Plains leopard frog (bottom right).

including insects, fish, frogs, toads, other salamanders, and mice. They are also prey for snakes and owls (Collins 1993).

Breeding Biology: Tiger salamanders begin breeding with the onset of early spring rains. At that time, small ponds and stock tanks might become full of salamanders, but they avoid ponds with large fish; predatory fish often eat salamanders. Breeding occurs in early spring, usually during March and April in Nebraska. Mating occurs in water by the male depositing a spermatophore (sperm packet), which the female picks up with her cloacal lips. The eggs are also laid in water either singly or in clusters of up to 1,000 eggs. The larvae hatch within a few weeks and soon develop into voracious predators, even eating other salamander larvae. The larvae are apparently distasteful to some predatory fish, allowing the larvae to survive in fish-rich waters. During the summer months, most of the larvae metamorphose into adults, gradually losing their gills and tail fins, and move onto land. Other individuals retain their larval structures and continue an aquatic existence, overwintering in that form. These larva-like (larviform or neotenous) individuals are commonly known in Nebraska as mudpuppies. Sexual maturity is reached the following spring, and mating begins.

Selected References: Baxter and Stone 2011; Collins 1993; Stebbens and Cohen 1995; Ballinger, Lynch, and Smith 2010.

Order Anura (Frogs and Toads)

Frogs and toads are easily recognized on the basis of their lack of a tail as adults (Anura means "lacking a tail"), moist, glandular skin, and four legs, the rear pair being much larger than the front pair. Except for spadefoot frogs, all the frogs and toads of Nebraska have horizontal pupils. Like other amphibians, frogs and toads produce gelatinous eggs that develop in water, resulting in tailed, legless larvae having external gills and mouthparts adapted for foraging on algae.

Family Pelabatidae (Spadefoots)

Spadefoot toads are unique among Nebraska's frogs and toads in having vertical pupils in addition to black nail-like tubercles (spades) on both of their hind feet; thus, they are not typical toads. Unlike typical toads, spadefoots also lack the obvious parotoid glands (glands that are typically visible behind the eyes in toads and exude distasteful or poisonous substances). However, some people have strong allergic reactions after handling them. Spadefoots are found in sandy soils and use the horny "spades" on their rear legs to dig into the sand and rapidly become hidden. Spadefoots can also inflate their bodies with air when threatened, making them more difficult to be swallowed by predators. Males produce trilling, rasping, or snoring calls in spring, usually while in temporary pools.

PLAINS SPADEFOOT. *SPEA BOMBIFRONS.* UB, CO

Identification: The Plains spadefoot is unique among Nebraska's amphibians in having vertical pupils and a horny "spade" on both of its hind feet that are used in digging. Males differ from females in having black throats and cornified fingers; two of the fingers have brown nuptial pads during the breeding season for use when clasping females. Snout–vent length from 38 to 50 mm (1.5 to 2 in.); females are slightly heavier than males.

Voice: The Plains spadefoot's call is a short, loud *squack* that lasts about one-half second.

Status: The Plains spadefoot is associated with sandy grasslands, open prairie, and floodplains with loose, sandy soil. It has been reported from all but 24 Nebraska counties.

Habitats and Ecology: In Kansas, spadefoots are active from late April to September. Adults feed on insects and spiders, especially nocturnal species, whereas small larvae forage on plankton and organic matter. Breeding occurs opportunistically between May and July, when temporary ponds, playas, and perhaps other water sources are fed by summer rains (Baxter and Stone 2011).

Breeding Biology: During breeding, males typically call in chorus to attract females. Mating begins after the first substantial warm spring rain (air temperature at least 52° F and at least 3.5 inches of rain by a Kansas estimate, and air temperatures above 50° F and rainfall greater than 0.7 inch by a Colorado estimate). Breeding continues through the summer by the males calling while standing around pools or on the open prairie. The male grasps the female's groin with his front legs during copulation. As the female releases her eggs in the water, the male fertilizes them by arching his body and depositing sperm on them. The eggs are deposited on submerged vegetation or other objects. They are laid as elliptical masses containing 250–2,000 eggs. After hatching, metamorphosis occurs after widely varied periods, depending on the water temperature, oxygen content, and food availability. Except when breeding, spadefoots are rarely seen, as they are nocturnal and spend daytime hours under sand or mud.

Selected References: Baxter and Stone 2011; Collins 1993; Stebbens and Cohen 1995; Ballinger, Lynch, and Smith 2010.

Family Bufonidae (Toads)

Toads, as distinguished from frogs, are semiterrestrial amphibians that have warty skin and a large parotoid gland on each side of the neck above the tympanum (ear drum) that secretes a highly poisonous mucus. The parotoid glands sometimes extend upward to the bony midpoint of the cranium (the cranial crest). Male toads vocalize in spring while inflating their throats, producing loud musical or metallic-sounding calls that serve to attract females.

WOODHOUSE'S TOAD. *ANAXYRUS (BUFO) WOODHOUSEI.* UB, CO

Identification: Easily identified as toads by their stocky shape and highly warty skin, the Woodhouse's and Great Plains toads are very similar. Both species occur widely, but Woodhouse's (figure 34) is more likely to be found in urban yards and gardens. There is a pale line down the middle of the toad's back and an unspotted abdomen, and the parotoid gland behind the eye extends farther forward, to the rear end of the bony cranial ridge along the top of the head. The tympanum is inconspicuous but is about as large as the eye. Breeding males are somewhat smaller than females and have dark throats and enlarged horny pads on the inner fingers of each hand. Snout–vent length 6.5–10 cm (2.5–4 in.).

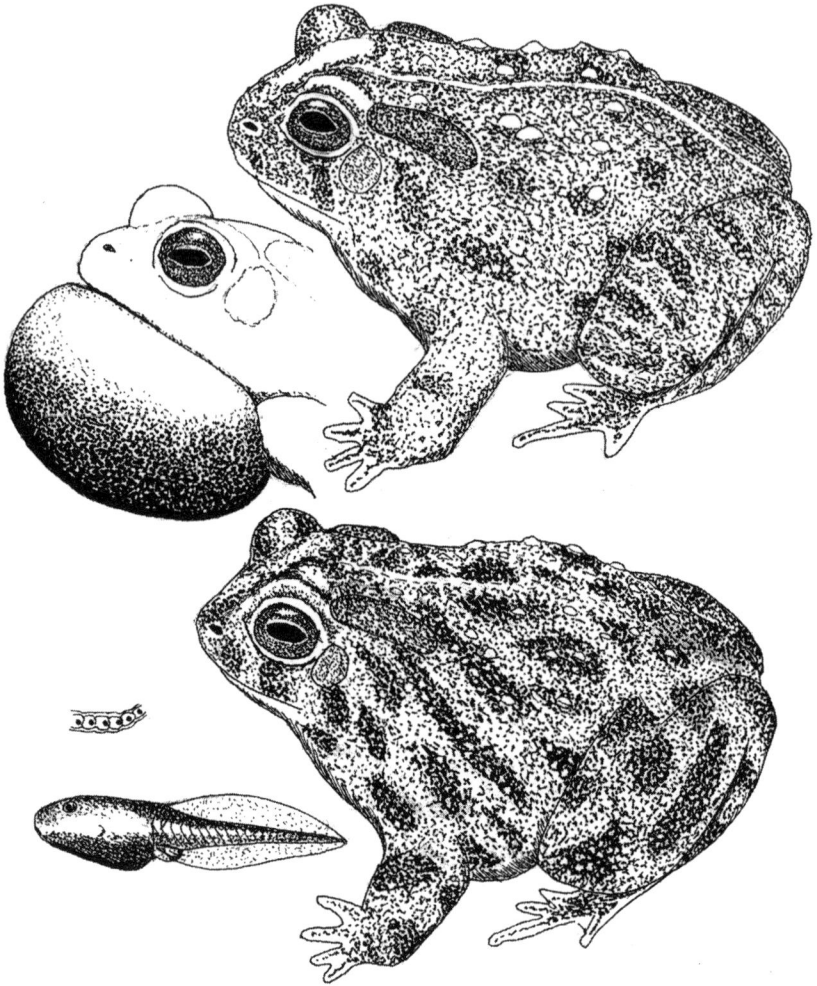

34. Woodhouse's (above) and Great Plains toads, showing throat inflation during male calling, eggs, and larvae.

Voice: This species of toad utters a nasal trill that is similar to the bleating of a calf or sheep and lasts one to three seconds, whereas the Great Plains toad produces a very loud, trilled scream that lasts up to 50 seconds and resembles riveting noises. When calling, the throat skin (the vocal sac) is strongly inflated, providing a resonating chamber.

Status: The Woodhouse's toad is widespread and has been reported from all but four Nebraska counties. The Great Plains toad is less widespread but has been reported from all but about 40 Nebraska counties.

Habitats and Ecology: This toad favors sandy areas but can be found anywhere that is fairly close to water. It is nocturnal, remaining in its burrow throughout the day. It emerges at night to feed on insects, especially beetles and ants, as well as snails, spiders, and other arthropods. The maximum published longevity record of the Great Plains toad is nearly 11 years (Collins 1993), although a female Woodhouse's toad at Lincoln's Pioneers Park Nature Center had survived for nearly 15 years as of 2019.

Breeding Biology: Breeding begins in late spring, with male choruses lasting until midsummer. Mating occurs by the male clasping the female between his forelegs. Eggs are laid in water in long strings of up to at least 25,000 eggs. They hatch in less than a week, and metamorphosis occurs 50–60 days after hatching. Under optimum conditions, two clutches might be produced in a single breeding season (Collins 1993). Krupa (1994) has described the breeding biology of the Great Plains toad in Oklahoma.

Selected References: Baxter and Stone 2011; Collins 1993; Krupa 1994; Stebbens and Cohen 1995; Ballinger, Lynch, and Smith 2010.

Family Ranidae (Typical Frogs)

Unlike tree frogs, typical frogs lack toe pads and have a large rounded tympanum (external ear drum) behind each eye. The North American species all have pointed toes that lack terminal pads, with extensive toe webbing on the muscular hind legs, and long legs. They are powerful jumpers. They are also excellent swimmers, and adults are carnivorous, eating anything they can capture and swallow. Males typically call in chorus during spring, and during that season they have swollen forearms and thumbs that enable them to grasp females firmly during mating. Eggs are laid in water, sometime in long strings or clusters that might contain up to nearly 50,000 eggs. From 6 to 24 months are needed for the hatched larvae to metamorphose and reach adulthood.

Identification: This is the largest frog in Nebraska, with adults at least 13 cm (5 in.) long. Adults vary from lime green to olive, with warty backs, and often have reddish-brown to blackish dorsal markings. The tympanum is large (especially in males) and is unmarked and conspicuous. Snout–vent length averages about 15 cm (6 in.). Rarely, adults might reach a total length of 20 cm (8 in.) and attain weights in excess of 453 gm (1 lb.).

Voice: The breeding call of the male has been described as a low snore that sounds like *jug-o-rum.* It can be heard from May into July and serves as both a territorial and a sexual advertisement signal.

Status: This frog has a statewide distribution, having been recorded from all but about 25 counties. It is probably most abundant in eastern Nebraska, where there is more available surface water.

Habitats and Ecology: Bullfrogs usually can be found along the banks of almost any wetland, from ponds and marshes to slow rivers, where they patiently wait for prey to come within reach of their huge mouths. They try to capture anything they are able to swallow, from insects to other amphibians, turtles, small snakes, mammals, and birds. Mammals such as opossums, raccoons, and skunks prey upon adult bullfrogs. Captive specimens are known to have survived for more than seven years.

Breeding Biology: Breeding begins in late spring, when males advertise and defend their territories against other males by a variety of fighting techniques, such as bumping, kicking, and biting. The male mates by clasping the female with his forelegs and releasing sperm into the water. After mating, females deposit clusters of 25,000–48,000 eggs as surface films on water. Following hatching, the larvae disperse and spend two years growing and completing their metamorphosis into adults and becoming torpid over the winters.

Selected References: Baxter and Stone 2011; Collins 1993; Stebbens and Cohen 1995; Ballinger, Lynch, and Smith 2010.

PLAINS LEOPARD FROG. *LITHOBATES (RANA) BLAIRI.* WI, CO

NORTHERN LEOPARD FROG. *LITHOBATES (RANA) PIPIENS.* WI, CO

Identification: Recognizing leopard frogs is relatively easy; they are the commonest typical frogs in Nebraska. However, separating the Plains leopard frog from the northern leopard frog is another matter. Visual distinction between the two is difficult; both are smaller than bullfrogs, ranging in length from 8.9 to 10.2 cm (2.5 to 4 in.), and their overall color is green to brown, with dark brown spots and blotches that are often edged with white. Two pale dorsolateral folds that extend from the back of the head are discontinuous toward the rear in the Plains leopard frog (upper right in figure 33), but these folds are continuous in the northern species. The Plains leopard frog also usually has a white spot in the middle of the tympanum, a pattern lacking in the northern leopard frog, and the latter species also has a generally more slender body conformation. Length Plains leopard frog 51–95 mm (2–3.75 in.), rarely to 100 mm (3.9 in.); northern leopard frog 51–90 mm (2–3.5 in.), rarely to 107 mm (4.2 in.).

Voice: Males of the two leopard frog species differ in their mating calls. In the Plains species, the calls are a long (35–40 seconds) series of *chuck* notes that end in a longer *cu-u-u-uck*, the series sounding like a finger being rubbed over a balloon. In the northern species, the call is a long snore that is followed by two or three *chuck* notes, each series lasting only two or three seconds. Females of *Lithobates pipiens* showed differential responses to the calls of *L. blairi*, a presumed hybrid, and conspecifics (Kruse 1981). Littlejohn and Oldham (1968) related the mating calls of the *Rana pipiens* complex to their taxonomy.

Status: Plains leopard frogs are common in the eastern two-thirds of Nebraska but are absent in the Panhandle and over most of the western counties north of the Platte, where the northern species is common. Plains leopard frogs mostly occur south of a line from Perkins County to Boyd County. Northern leopard frogs occur mostly north of a line from Perkins County to Otoe County.

Habitats and Ecology: As adults, leopard frogs eat a variety of insects and other invertebrates, as well as vertebrates up to the maximum size that their mouths can accommodate. During summer the adults often wander about on land, and prior to winter they bury themselves in mud and leaves at the bottom of a pond.

Breeding Biology: Plains leopard frogs begin their mating activities very early, usually by the end of March. Males call while floating at the water surface, beginning after sunset. The newly fertilized female deposits a globular cluster of 4,000–6,000 eggs, which hatch in a few days. Metamorphosis into the adult frog occurs in 50–60 days.

Selected References: Littlejohn and Oldham 1968; Kruse 1978, 1981; Lynch 1978; Baxter and Stone 2011; Collins 1993; Stebbens and Cohen 1995; Ballinger, Lynch, and Smith 2010.

Family Hylidae (Chorus Frogs)

Tree frogs are small frogs that are often found in trees and are adapted for climbing by the presence of adhesive pads on their toes. One group of frogs (*Acris*) in this family doesn't climb trees; these species are variously known as chorus frogs, cricket frogs, and spring peepers. These species have reduced toe webbing and small toepads. Males of this family have round vocal sacs that expand greatly during calling, the different species producing buzzing (*Hyla*), cricket-like clicking (*Acris*), or sounds that resemble a comb's teeth being stroked (*Pseudacris*). Some tree frogs are able to closely match their background by adjusting their skin colors from gray, tan, or brown to bright green, depending on the background color.

COPE'S GRAY TREEFROG. *HYLA CHRYSOCELIS.* SD (EA)

Identification: This tiny tree-dwelling frog can change from bright green to brown in a few hours but often appears as a mottled gray or black dorsally. The inside of their thighs is bright yellow orange. Length 32–58 mm (1.25–2.25 in.).

Voice: Males utter a high-pitched buzzy trill in spring that lasts two to four seconds. A closely related and visually identical form,

Hyla versicolor, has a slower-pulsed or lower-pitched voice and has been found in Iowa. It has not yet been definitively reported from Nebraska (Ballinger, Lynch, and Smith 2010) but probably occurs here (Fogel 2010).

Status: This frog ranges over 14 counties of southeastern Nebraska, north to Dodge County and west to Lancaster and Jefferson Counties, with scattered occurrences west to Hall and Lincoln Counties and north to Knox County.

Habitats and Ecology: These are arboreal frogs and might call from treetop perches on warm summer days until late June or July. They are territorial, interacting with other males by uttering encounter calls and fighting. These frogs are active in Nebraska from about April through August, with breeding occurring from mid-April through June.

Breeding Biology: Tree frogs breed in permanent to temporary ponds near woods and ponds having weedy vegetation. Males call from the water's edge or surrounding vegetation and might perch up to 10 feet above water. Following mating, females deposit up to 3,800 eggs in deep water either singly or in clusters. The eggs hatch in four or five days, and metamorphosis into the adult body form occurs within two months. By August the froglets have left the water and are absorbing what is left of their tails. As adults, they eat both terrestrial and flying insects and have been reported to survive for as long as nearly eight years.

Selected References: Baxter and Stone 2011; Collins 1993; Stebbens and Cohen 1995; Ballinger, Lynch, and Smith 2010.

BOREAL CHORUS FROG. *PSEUDACRIS MACULATA.* UB, CO

Identification: This tiny frog barely exceeds an inch in length and is the only Nebraska frog having a white streak along the upper lip, three dark streaks along the back, and paired white stripes that extend from the nose through the eyes and back along the flanks. Length 19–39 mm (0.75–1.5 in.).

Voice: Males begin advertisement calling as soon as the snow melts, typically during early March in southern Nebraska. Their calls are uttered while clinging to vegetation and with the vocal sac greatly enlarged. Calling continues throughout both night and day. The call is a mechanical, high-pitched trill that is similar to the sound produced by running one's fingers over the teeth of a comb. This is distinctly different from the voice of the Blanchard's cricket frog, which resembles the sound made when shaking a few pebbles in one's hand. Singing in Nebraska might extend into August in reservoir-chilled water but normally ends by late April.

Status: This frog is widespread throughout Nebraska, having been reported from all but 15 counties that are mostly along or near the Kansas border.

Habitats and Ecology: Chorus frogs use water bodies of almost any size, including vernal ponds, flooded fields, sewage lagoons, and lakes. During summer these frogs also might be found on grasslands and woods far from water. They eat grubs, beetles, spiders, and ants and probably also aquatic and semiaquatic insects. In a Colorado study, some frogs were found to survive for as long as six years.

Breeding Biology: After early spring rains, males soon begin to congregate at breeding sites such as roadside ditches, ponds, marshes, lakes, and slow-moving streams. Chorusing might occur at temperatures as low as 35° F. Immediately after mating, females release their eggs, which are adhesive and cling to vegetation. Individual clusters might contain nearly 300 eggs, and the total seasonal ovarian production might be 500–800 eggs. Metamorphosis occurs after about six weeks (by June in Nebraska), and thereafter both juvenile and adult frogs feed on insects and other small arthropods.

Selected References: Baxter and Stone 2011; Collins 1993; Stebbens and Cohen 1995; Ballinger, Lynch, and Smith 2010.

REPTILES (TURTLES, LIZARDS, AND SNAKES)

Order Chelonia (Turtles)

All turtles are easily recognized by the presence of a bony dorsal "shell" (the carapace) and a corresponding but variably smaller ventral supporting bony structure (the plastron). In most turtles these protective structures are covered by thick, horny scutes. Turtles lack true teeth but have sharp rims along their upper jaws that serve to cut and tear their food. All turtles are egg layers (oviparous), typically depositing them in sandy soil and then abandoning them, requiring the hatchlings to independently dig their way out. All the North American turtles can retract their fairly long neck back under the carapace, some more fully than others.

Family Chelydridae (Snapping Turtles)

Snapping turtles are notable for their hard, bony, and rough dorsal carapace, which is raised into three keel-like enlargements that extend from front to rear at its center, but the ventral plastron is relatively small and cross-shaped. The tail is at least as long as the carapace, and the upper carapace scutes have saw-toothed keels. Snapping turtles have large heads and powerful jaws and are the largest (and most dangerous) of Nebraska's turtles. Like all turtles, snapping turtles are oviparous; females might lay clutches of up to more than 100 eggs. Egg mortality is high among turtles, as the eggs are not defended and represent valuable food sources for many animals, as well as humans.

EASTERN SNAPPING TURTLE. *CHELYDRA SERPENTINA*. WI, CO

Identification: Snapping turtles are unmistakable; the large head, long tail, and powerful front and hind legs set them apart from other Nebraska turtles. The triple-keeled rather than smoothly curved carapace is unique, and its posterior portion has uniquely serrated edges. Snapping turtles are the largest turtles in the state, with some individuals weighing up to about 33 kilograms (75 pounds). The species' maximum published weight is 39 kilograms (86 pounds) (Collins 1993), although a specimen in Nebraska's Schramm State Park aquarium was over 70 years old as of 2019 and was recently reported to weigh 89 pounds. Adult carapace length 203–360 mm (8–14 in.); females are slightly larger than males.

Status: Snapping turtles are widespread in Nebraska, and although they have not been reported from every county, they are likely to be found in almost any river or larger wetland in the state.

Habitats and Ecology: Snapping turtles are most often found in shallow bodies of water, especially where aquatic vegetation or debris is present. They must be treated with great caution when closely approaching them or when lifting them by the carapace; their long necks can extend surprisingly far forward or backward and could easily cut off a finger. I once found an incapacitated western grebe with one missing leg that had obviously been cleanly sheared off by a snapping turtle. Snapping turtles are sometimes hunted for food and are often killed simply because of their presumed threat potential.

Breeding Biology: In Nebraska, nesting begins in May and mostly occurs over a three-week period. Clutches of up to 109 eggs have been found but average about 50 eggs. Snapping turtles move to land to deposit their eggs, the females digging out holes in areas of sandy soil and open vegetation. The sandy substrate evidently provides the best incubation conditions, and the short vegetation makes it easier for the hatchlings to disperse from the nest site. The female does not protect them, and raccoons, skunks, and minks are all major egg predators. Hatching occurs from as early as August until as late as October. The sex of the hatchlings depends on incubation temperature; males are produced at intermediate temperatures and females at both high and low temperatures. From 10 to 12 years are required to reach sexual maturity; captive individuals are known to have survived for more than 70 years.

Selected References: Baxter and Stone 2011; Collins 1993; Ernst, Lovich, and Barbour 1994; Ballinger, Lynch, and Smith 2010.

Family Embydidae (Pond and Box Turtles)

All of Nebraska's turtles other than the snapping turtle and softshell turtles belong to the pond (or basking) and box turtle family. They have bony carapaces that are covered with smooth, horny scutes and ventral plastrons that are almost as large as their carapaces and that are often distinctively colored or patterned. Their tails are shorter than their carapaces. In three species (the ornate box turtle, Blanding's turtle, and yellow mud turtle), the plastron is hinged (doubly hinged in the mud turtle) to allow the plastron's front section to be elevated and permit the animal's head to be fully retracted when threatened. Most species are semiaquatic, and the aquatic species in this family tend to rest on floating logs or at the water's edge (thus the common name "basking" turtles). The terrestrial ornate box turtle rarely if ever enters water and is often found miles from the nearest water.

WESTERN PAINTED TURTLE. *CHRYSEMYS PICTA*. UB, AB

Identification: The painted turtle is well named. Its plastron is reddish, with an irregularly shaped black to brown central figure, and some red color is often present along marginal scutes of the carapace. In the similar Blanding's and false map turtles, the plastron is yellow, with varying amounts of black markings. The typical adult carapace length is 90–180 mm (3.5–7 in.); the maximum carapace length is 9.5 inches (Collins 1993).

Status: This is the commonest and one of the most widespread turtles in Nebraska, with records for all but 32 counties. It tends to favor large ponds and lakes rather than small ponds and shallow marshes.

Habitats and Ecology: Painted turtles can be found in larger, permanent ponds, in lakes, and in streams and larger slow-flowing rivers. Basking sites are an important part of the species' habitat, and individuals sometimes fight over preferred locations. The turtles are basically diurnal, with most foraging occurring in later morning and late afternoon, but nighttime activities have also been reported. They perform limited movements between ponds and have been found able to return home after displacements of up to 100 meters (400 feet). Longer translocations might result in orientations using a sun-compass guide and an internal clock. During winter they might remain active under the ice or burrow into mud at the bottom of the wetland.

Breeding Biology: Mating occurs during early spring and fall, and females lay clutches of about 13–14 eggs, with as many as three clutches produced per year in the Nebraska Sandhills. Iverson and

Smith (1993) found that clutch size increases with female size and that female survivorship exceeded 90 percent annually, with some individuals likely to survive at least 30 years. The eggs usually hatch at about 2–2.5 months. Warmer nests produce young that are all or mostly females, colder nests result in all males, and intermediate-temperature nests produce a mixture of the sexes. Apparently the temperature present during the middle part of the incubation period determines sex. In western Nebraska, the hatchlings spend their first winter in the nest and can survive freezing for at least 48 hours. Hatchling painted turtles are the only known reptiles that can survive a cold winter while freezing up to 54 percent of their bodily fluids (Collins 1993). Adult painted turtles eat a variety of large aquatic invertebrates, such as crayfish, insects, and insect larvae, as well as plant materials. They also consume small vertebrates, including salamanders, frogs, and fish. Once often sold as pets in general and sporting goods stores, captives are known to have survived for as long as 20 years. However, they make very poor pets, as they can transmit salmonella-based diseases to humans, and painted turtles released in Europe have become invasive and threatened native turtle populations.

Selected References: Baxter and Stone 2011; Collins 1993; Iverson and Smith 1993; Ernst, Lovich, and Barbour 1994; Ballinger, Lynch, and Smith 2010.

ORNATE BOX TURTLE. *TERRAPENE ORNATA.* WI, UN

Identification: Box turtles are easily identified. They are Nebraska's only entirely terrestrial turtle and have a hinge on the plastron that allows their front end to be lifted and their head withdrawn protectively behind the high-domed carapace. The Blanding's turtle also has a hinge and movable plastron similar to the box turtle's, and the yellow mud turtle's plastron is doubly hinged. However, the box turtle has a mixture of yellow lines and spots radiating outward from the center of the carapace, and the plastron is dark brown with yellow spots (figure 35), whereas the Blanding's similarly dark carapace is sprinkled with small yellow spots, and the plastron is

35. Greater short-horned lizard (top), tiger salamander (middle), and ornate box turtle (bottom).

yellow with black smudges (figure 36). The iris color is red in adult males and varies from green to yellow brown or maroon in females. Adult box turtles usually have carapace lengths of 100–125 mm (3–5 in.), with females averaging slightly larger than males.

Status: The box turtle has been reported from nearly all of the Panhandle counties but from relatively few counties in the eastern third of Nebraska. It has been reported from almost 60 counties but is most common in the Sandhills. Box turtles eat carrion, insects and their larvae, crayfish, fruit, and various plants, including prickly pear cacti.

Habitats and Ecology: Still common in the Nebraska Sandhills, box turtles tend to wander during spring and summer, probably in search of females, and have been proven able to find the way back to their home range after being displaced over distances of up to 9 kilometers (5.5 miles). During their travels the turtles often cross roads and frequently are accidentally (or purposefully) run over by motorists. Their home ranges in Kansas average about 5 acres, with no apparent territorial behavior. Legler (1960) has monographed this species' natural history in Kansas.

Breeding Biology: In Nebraska, box turtles become active as early as mid-April, when males begin to seek out females. Mating extends from late April to early June. Clutch sizes are small in box turtles, averaging only about four or five eggs, but females probably produce more than one clutch per year in some populations. The eggs hatch in a little more than two months. Like that of many other turtles, sex determination apparently depends on temperatures during the incubation period. Hatchlings overwinter in burrows below their nests in western Nebraska and perhaps also Wyoming, and during hot weather adults tend to be inactive and remain in burrows (Ballinger, Lynch, and Smith 2010). Sexual maturity occurs at seven to eight years of age. Although often captured as prospective pets, few captives live long, because their owners wrongly assume the animals are vegetarians and fail to provide dietary proteins. Coyotes and skunks are among the few predators of box turtles, and box turtles have been found to survive for up to at least 29 years in the Sandhills. Probably being run over by motor vehicle, accidentally or by design, is a significant mortality factor in these slow-moving turtles.

Selected References: Legler 1960; Baxter and Stone 2011; Collins 1993; Ernst, Lovich, and Barbour 1994; Ballinger, Lynch, and Smith 2010.

Family Trionychidae (Softshell Turtles)

Softshell turtles are easily defined by the fact that the carapace is covered by leathery skin rather than scutes, which overlie the bony supporting skeleton. Softshell turtles also have relatively long tails and large, webbed hind legs that make them powerful swimmers. Their webbed toes have only three claws. They have flattened, pancake-shaped carapaces, long necks, and heads with long, tapering noses, allowing them to breathe by lifting only the tip of the nose above water. They have fleshy lips and eat crayfish, snails, insects, and other animals.

MIDLAND SMOOTH SOFTSHELL TURTLE. *APALONE MUTICA.* SD (EA), CO

EASTERN SPINY SOFTSHELL TURTLE. *APALONE SPINIFERA.* WI, CO

Identification: The softshell turtles are unique in having a flattened, oval body shape, a very long neck, and a narrow head, with a nose that tapers to a point and in the spiny softshell extends as two projections. In the spiny softshell, the carapace is sandpaper-like, and its anterior margin has many spiny tubercles. The smooth softshell lacks these tubercles and has a smooth carapace surface rather than a rough one. The tail in male softshells is very long and thick, whereas females have very short tails. Smooth softshell carapace length is 115–306 mm (4.5–12 in.); spiny softshell carapace length is 125–432 mm (5–17 in.). The dorsal carapace in adult female spiny softshells might rarely be up to 400 mm (18 in.) in length and in males to 200 mm (9 in.); adult females are as much as twice as heavy as adult males. Females of the smooth softshell might reach up to 285 mm (11.25 in.) in carapace length and also are much larger than males.

Status: Both species occur in ponds, lakes, and large streams. The smooth softshell is mostly confined to the Missouri, Elkhorn, eastern Platte, and Big Blue Rivers, whereas the spiny softshell occurs in every major drainage basin of Nebraska.

Habitats and Ecology: Softshells are highly aquatic turtles that are rarely found on land except when basking on sand bars or while laying eggs. They are powerful swimmers; their forelimbs are adapted to swim in an effective rowing manner that increases the turtles' swimming efficiency. Female spiny softshells have large home ranges and might move as far as 7 kilometers (4.3 miles) to find a nesting site. They evidently can use the sun as a compass and have an associated internal clock. Softshells eat a variety of foods, including fish, frogs, tadpoles, crayfish, and aquatic insects. They are sometimes killed and eaten by larger turtles such as snapping turtles and, like horned lizards, have been reported to be able to squirt blood from their eyes, presumably defensively (Ballinger, Lynch, and Smith 2010).

Breeding Biology: In Nebraska, breeding occurs from early to mid-June (first clutch) and from late June to early July (second clutch). Clutch sizes range from 16 to 42 eggs, averaging about 30. Eggs incubated under moister conditions are larger than those in drier situations. Egg predation often affects hatching success rates (Ballinger, Lynch, and Smith 2010).

Selected References: Lynch 1985; Baxter and Stone 2011; Collins 1993; Ernst, Lovich, and Barbour 1994; Ballinger, Lynch, and Smith 2010.

Order Lacertilia (Lizards)

Family Scincidae (Skinks)

Skinks comprise a distinctive family of lizards that have small, short legs, smooth scales, and long, cylindrical bodies and tails. Many species have a brightly colored tail with a fracture plate that allows the tail to be broken off when grasped, thereby permitting the animal to escape with minimal damage and to gradually regenerate its tail. All skinks are egg layers, and females of the locally occurring species tend their clutches of up to about a dozen eggs during an incubation period of three to four weeks. Skinks are diurnal insectivores and are strongly patterned with multiple dark and white body stripes that extend into the tail. Nebraska's several skinks are mostly small, up to about 90 mm (3.5 in.) in snout–vent length, but the Great Plains skink is notably larger, reaching 140 mm (5.5 in.) in snout–vent length and up to 230 mm (9 in.) in total length. It is also unique in its dorsal color pattern, consisting of eight to ten alternating light stripes and narrow dark lines (figure 36). Its range barely reaches the southernmost counties of Nebraska.

MANY-LINED SKINK. *EUMECES (PLESTIODON) MULTIVIRGATUS.* DI (WE)

Identification: Skinks are notable among lizards for their very long and often colorful tails, which are easily detached and left behind when grabbed by a predator, letting the skink escape and later regrow a new tail. As this species' common name indicates, there are eight to twelve dark dorsal stripes and seven to nine light alternating stripes that extend from head to tail. A patternless gray to tan variant body pattern is also present in western Nebraska. Breeding males have red lips, presumably a breeding-status social signal. The legs are short, but the tail is very long (1.5 times the snout–vent length) and is bright blue in young individuals. Otherwise, the body is mostly dark brown to greenish and is longer in adult females than in males. Length 125–84 mm (5–7.6 in.).

Status: This skink occurs in western and central Nebraska, east to Pierce, Stanton, and Colfax Counties in the Elkhorn and lower Platte Valleys, and to Greeley County in the Loup Valley.

Habitats and Ecology: This lizard is closely associated with sandy soils, such as upland sand sage grasslands and sandy riverside habitats. It is often found in Nebraska under cow dung and in Colorado in rocky habitats. In Wyoming it occurs in eastern plains grasslands and in rocky woodlands and can often be found under loose boards and scrap of junkyards. Insects and insect larvae are typical foods, and spiders are also eaten.

Breeding Biology: Breeding occurs in spring and summer; in

Wyoming, a female with internal eggs was found in May, and in Nebraska, a female with enlarged follicles was found in late June (Ballinger, Lynch, and Smith 2010). In Colorado, the breeding season extends from late March or April to September or October. Little is known of the species' courtship, but in the closely related five-lined skink, the males seek out females by sight and smell and pursue them. When a male catches a female, he grasps her by her shoulder skin and orients his hindquarters so that their two cloacae meet. Females brood their eggs, and clutch sizes vary from three to seven eggs. The eggs hatch in one to two months.

Selected References: Fitch 1970, 1985; Lynch 1985; Baxter and Stone 2011; Ballinger, Lynch, and Smith 2010.

NORTHERN PRAIRIE SKINK. *PLESTIODON SEPTENTRIONALIS.* DI (EA)

Identification: This is the commonest (but rarely seen) lizard in eastern Nebraska, extending west to Rock and Phelps Counties. Its Nebraska range corresponds closely with that of the historic tall-grass prairie. In this skink, the light middorsal line is represented by a broad brown stripe, whereas in the other Nebraska skink, the five-lined skink, five white lines extend down the back. In both species, the chin and neck of males become bright reddish orange during the breeding season, and the tail is bright blue in both sexes of juveniles. One function of the blue tail color in juvenile skinks might be to inhibit adult males from aggressively attacking them as potential rivals (Clark and Hall 1970). Length 124–78 mm (5–7 in.), maximum 224 mm (8.75 in.) (Collins 1993).

Status: This skink occurs in eastern and central Nebraska, west to Rock County in the Niobrara Valley, and to Phelps County in the Platte Valley. It is often associated with open grasslands, where flat rocks and loose soil allow it to dig an extensive tunnel system. It also occurs along sandbanks and on gravelly glacial outwashes.

Habitats and Ecology: Fitch (1970, 1985) detailed the life history of the northern prairie skink in Kansas. This skink overwinters below the frost line under stones and debris, sometimes gathering in large groups as far as 6 feet below the surface. Prairie skinks consume

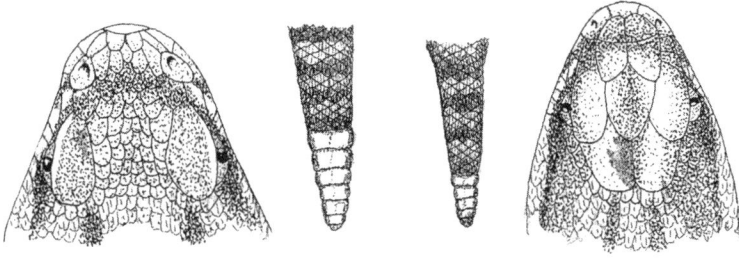

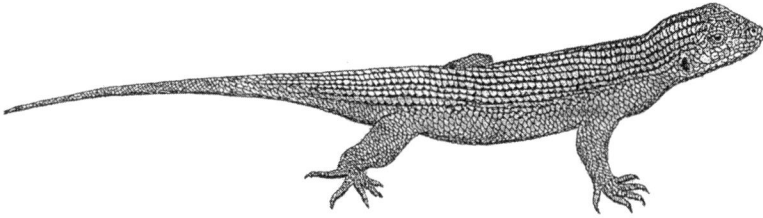

36. Prairie rattlesnake (top left), massasauga (top right), Blanding's turtle (upper middle), prairie skink (lower middle), and Great Plains skink (bottom).

insects, spiders, and the larvae of various arthropods. Grasshoppers and other orthopterans are an important part of their diet.

Breeding Biology: Skinks emerge from mid-April to early May, with mating occurring in late May. Almost nothing is known of this species' courtship, but in the Great Plains skink, the male approaches the female and flicks his tongue on her back. After a chase he grasps her shoulder skin and loops his hindquarters below hers to achieve mating, which might last several minutes (Collins 1993). During June, females lay clutches of 5–18 eggs, averaging 11, and brood them until hatching has occurred one or two months later. Females actively tend their nests to maintain proper environmental conditions for them. They also remove infertile eggs and eggs with dead embryos.

Selected References: Breckenridge 1943; Fitch 1970, 1985; Lynch 1985; Somma 1990; Collins 1993; Ballinger, Lynch, and Smith 2010.

Family Phrynosomatidae (Spiny, Earless, Tree, and Horned Lizards)

This large family of lizards contains over 130 species in North and Central America. They all have small teeth, and most of the U.S. species have keeled scales. Most of them are diurnal and active insectivores that might reach up to about 12 inches in length, although none of the Nebraska species exceeds 7 inches in length. There are three species in Nebraska, two of which are in the widespread and species-rich genus *Sceloporus*.

GREATER (MOUNTAIN) SHORT-HORNED LIZARD.
PHRYNOSOMA HERNANDESI. SD (WE)

Identification: This distinctive lizard (often popularly called a horned toad) is unique in having a flattened body covered by granular to spine-like scales and is fringed along its widest margins with soft spines. The head has a crown of blunt spines that point posteriorly. This species is gray to brown overall, with large, dark, and irregular spots on each side of the midline and a series of fainter and smaller spots more laterally. The underside is white, except for a spotted or mottled throat. Snout–vent length 57–77 mm (2.2–3 in.); females average slightly larger than males.

Status: This lizard occurs in the northern Panhandle from Sioux to Sheridan Counties and in the southern Panhandle south of the Wildcat Hills. It is especially associated with sagebrush and sage grasslands on sandy soils. It can quickly burrow and hide under sand, and its grayish mottled coloration closely resembles sandy or pebbly soil, making it hard to see even when it is fully exposed.

Habitats and Ecology: Short-horned lizards primarily consume ants but also eat grasshoppers, beetles, other insects, and spiders. Unlike the other Nebraska lizards in this family, they are slow-moving and rely on camouflage to catch their prey. This is the only Nebraska lizard that bears its young alive (vivipary) rather than laying a clutch of eggs that hatches before leaving the female's body (ovovipary) or that undergoes an incubation period outside her body (ovipary). Horned lizards are also the only lizards that squirt blood out of their eyes when threatened by a potential threat such as a coyote, which reportedly might serve as a deterrent to predators (Sherbrooke and Middendorf 2001). However, Collins (1993) found that overheated Texas horned lizards (*Phrynosoma cornutum*) often squirt blood when handled, and he saw a coyote lick the expelled blood off a lizard it had bitten and then consume the lizard. Horned lizards primarily eat ants, the acidic content of which (formic acid) reportedly makes the lizards distasteful to some mammalian predators. Although these lizards are often kept in captivity as pets, survival under captive conditions is usually very short because of the lizard's specialized diet. Nevertheless, a greater horned lizard, of unknown age when captured, survived in captivity for more than four years (Allison Johnson, personal communication).

Breeding Biology: Very few studies have been done on this inconspicuous species, and little is known of its mating behavior. It is active in Nebraska from May to September, and mating occurs in early spring, with a long hibernation period between. The young are born during late June or early July in southern Canada. Litter sizes vary geographically, with reports of 13–24 young in Colorado versus 5–6 in North Dakota, the number of young trending fewer northwardly (Ballinger, Lynch, and Smith 2010).

Selected References: Lynch 1985; Baxter and Stone 2011; Ballinger, Lynch, and Smith 2010.

PRAIRIE LIZARD. *SCELOPORUS (UNDULATUS) CONSOBRINUS.* DI (WE)

COMMON SAGEBRUSH LIZARD. *SCELOPORUS GRACIOSUS.* HL (WE)

Identification: The prairie lizard and sagebrush lizard can be separated by the scales on their posterior thighs, which are small and granular-like in the generally paler and widespread sagebrush lizard and are larger, keeled, and overlapping in the somewhat darker and more contrastingly patterned prairie lizard. Adult males of both the sagebrush and prairie lizards have bright blue or bluish-green throat and belly patches, but the prairie lizard differs in having mottled gray rather than blue throat markings and instead has bright blue patches on both sides of its belly, bordered medially with black. Additionally, the prairie lizard has a broad gray or brown dorsal stripe from its head to its hind legs, whereas in the sagebrush, the dorsal pattern is blotched with brown and gray. It should also be noted that in earlier literature (e.g., Baxter and Stone 1980; Collins 1993), the prairie lizard was earlier taxonomically classified as part of the widely ranging eastern fence lizard (*Sceloporus undulatus*) but is classified by recent authorities as being a distinct species (*S. consobrinus*) (Powell, Conant, and Collins 2016). However, the classification of the several populations once considered to be part of a broadly constituted *undulatus* group (which might consist of as many as four biological species) seems to be still unsettled (Ballinger, Lynch, and Smith 2010). Snout–vent length of sagebrush lizard (*graciosus*) about 55 mm (2.2 in.); of prairie lizard about 55–70 mm (2.2–2.8 in.). Females are larger than males in both species.

Status: The sagebrush lizard has so far been reported only from Morrill County, but the prairie lizard extends from the northern Panhandle (Sioux and Scotts Bluff Counties), east through the Sandhills to Pierce and Stanton Counties, and south of the Platte from the eastern Colorado border to Frontier and Furnas Counties. An isolated population also exists in Sherman, Buffalo, and Hall Counties. This species is on the Tier 1 list of the Nebraska Natural Legacy Project's list of threatened species in Nebraska (Schneider et al. 2018). The prairie lizard is widespread over western and central Nebraska, east to Stanton County.

Habitats and Ecology: The sagebrush lizard is closely associated with sagebrush, although in Nebraska it typically is found in rocky terrain and outcrops (Ballinger, Lynch, and Smith 2010). In Nebraska, the prairie lizard is often found in relatively open areas with sandy soils, although it too is sometimes found among rocky outcrops and steep terrain (Fogel 2010). The daily activities of prairie lizards typically begin in early morning and end at sunset, with reduced activity during the middle of hot days. Their home ranges vary from 700 to 800 square meters (7,500 to 8,600 square feet) in open areas and average about 400 square meters (4,500 square feet) in dense vegetation. They might include a favorite perch, as well as burrows into which the animals can retreat at night. Their foods are nearly all insects (94 percent by volume in one study) (Ballinger, Lynch, and Smith 2010).

Breeding Biology: All of the *Sceloperus* species are egg layers, often generating two or three laying cycles in a breeding season. The prairie lizard is the better studied of the two Nebraska species. In Nebraska, the breeding season is extended; the animals emerge from hibernation in late March or early April, and males quickly establish territories, which they maintain until the onset of winter. Mating occurs as early as late April, and a succession of clutches is produced until mid-July. Warmer than normal springs allow for an earlier start to egg-laying and usually results in three clutches being produced in a single season. Females born during the previous summer lay up to three clutches of four to six eggs; older females have clutch sizes that vary with their body size. Eggs hatch after about a month following their deposition, and hatching continues until mid-September. Mortality is highest during the overwintering period, and maximum longevity in the wild is about four years (Ballinger, Lynch, and Smith 2010).

Selected References: Jones and Droge 1980; Lynch 1985; Baxter and Stone 2011; Collins 1993; Ballinger, Lynch, and Smith 2010.

Order Serpentes (Snakes)

Snakes are easily distinguished from other reptiles; their scaled, legless bodies instantly separate them from all other reptiles except for a few legless lizards (glass lizards, family Anguidae), which, unlike snakes, have movable eyelids and external ear openings. All snakes are mute, but many can hiss or make mechanical noises through physical activity such as tail shaking. Snakes lack any external auditory structures, so they can perceive sounds only via vibrations received through their body. Both jaws possess teeth, which are backward slanted, and in some species, longer specialized teeth (fangs) exist that are adapted for injecting venom into prey, which is swallowed whole. Their skull bones and jaws are connected loosely, allowing for the swallowing of very large prey. Most snakes have ventral plate-like scales that can be tilted slightly to increase friction against a solid substrate; these scales are used to facilitate locomotion by means of muscular action and sinuous body movements. Most and probably all snakes can also swim, using similar sinuous body movements.

Family Colubridae (Harmless Egg-laying Snakes)

The snake family Colubridae is the largest of all snake families, numbering nearly 800 species worldwide, although the group is variously defined taxonomically, so the number of included species varies. Many genera of this family occur in North America and are all nonvenomous. They lack heat-sensitive (infrared) sensory pits, and most have round pupils. Most colubrids are egg layers (oviparous), producing clutches of up to about 25 eggs. None of the snakes in this family poses a serious threat to humans, although all of them prey on live animals, ranging in size and diversity from insects to large rodents and other snakes.

GOPHERSNAKE (BULLSNAKE). *PITUOPHIS CATENIFER.* WI, AB

Identification: This is Nebraska's most common large snake. Together with the western ratsnake and timber rattlesnake, the gophersnake is one of the state's three largest snakes. Its head is large and somewhat triangular, with alternating dark and yellow bands extending vertically downward from the eye to the jaw and a brown band extending diagonally from behind the eye to the angle of the jaw. More than 40 black to reddish-brown blotches are scattered along the back, sides, and tail, and the yellow belly is also spotted with black. The dorsal scales are strongly keeled. Average length of 15 Wyoming adults, 131 cm (51.7 in.); maximum length, 174 cm (68.5 in.) (Baxter and Stone 2011). Nebraska female gophersnakes often reach 143 cm (56 in.) in length; males attain at least 178.5 cm (70 in.) and rarely exceed 200 cm (79 in.).

Voice: Like other snakes, gophersnakes are voiceless, but when they are threatened they can make a hissing sound by exhaling and passing air over a unique laryngeal structure. They also often shake their tail when disturbed, which might resemble the rattle of a rattlesnake if the tail brushes through vegetation. A threatened bullsnake can even change the shape of its head to make it become more triangular, somewhat resembling a rattlesnake's. Such features suggest that gophersnakes might be mimicking rattlesnakes as a protective visual adaptation (Kardong 1980).

Status: Vying with the common garter snake for being Nebraska's most widely distributed snake, gophersnakes have been recorded statewide, except for 22 of Nebraska's east-central counties.

Habitats and Ecology: This is primarily a grassland snake, favoring taller prairies, especially those with rocky outcrops or rimrock, but it also often enters farmlands and urbanized areas and rarely occupies woodlands. Gophersnakes are largely diurnal and are active from about April to October. At night they return to their burrows, which are often gopher holes (thus the name gophersnake) or prairie dog burrows. Gophersnakes eat a variety of mammals, especially rodents, lizards, and ground-nesting and cavity-nesting birds, as well as their eggs and nestlings. They are remarkably adept at climbing trees and can even climb vertical concrete walls to reach the nests of culvert-nesting cliff swallows. They are also able to penetrate into soft soil by using head-digging movements.

Breeding Biology: Gophersnakes breed in spring, shortly after emerging from their winter burrows. Females lay eggs until as late as July. In Nebraska, clutches average 12–13 eggs but range from 8 to 17. There is a notably long incubation period of about 70 days, presumably the result of delayed fertilization.

Selected References: Baxter and Stone 2011; Kardong 1980; Collins 1993; Ballinger, Lynch, and Smith 2010.

Family Natricidae (Live-bearing Snakes)

Like the closely related colubrid snakes (and sometimes included as part of that family), members of this assemblage have round pupils and lack facial sensory pits. The included species are semiaquatic, they prey mostly on cold-blooded animals, and at least the North American species are live bearers rather than egg layers, producing litters of rarely up to about 80 offspring.

WANDERING (WESTERN TERRESTRIAL) GARTERSNAKE. *THAMNOPHIS ELEGANS.* HL (WE)

Identification: The wandering gartersnake can be easily separated from the other two species by the absence of red or bright yellow on its body. Instead, it has a brown or sometimes greenish body with three longitudinal stripes, a pale yellowish dorsal one and two white lateral ones, between which are two staggered rows of darker brown spots. The head is also brown with pale yellow spots, and the underparts are gray to pale blue. The total adult length is about 606–53 mm (24–27.5 in.).

Status: The wandering gartersnake has been reported only from Sioux County. All three *Thamnophis* species tend to favor locations close to water, but the common gartersnake seems to also have a preference for grassland-woodland transitional habitats where rocks and other escape cover are present. Wandering gartersnakes are most common near water, where they feed on frogs and small fish but also prey on small mammals and some invertebrates.

Habitats and Ecology: Gartersnakes are diurnal and seasonally active in Nebraska, emerging in March from underground hibernacula that are often shared by large numbers of others, sometimes even including other species of snakes. They return to their winter dens in late October but might briefly emerge during warm winter days.

Breeding Biology: Mating begins with spring emergence (Joy and Crews 1985). Males emerge from their hibernacula well before the females and then patrol the area while waiting for the females to appear. From trails left by odorous cloacal secretions produced by females (pheromones), the males can recognize those individuals originating from their own den and possibly prefer to court them. Males might also use visual clues in their choices of females to be courted. Writhing "knots" of several simultaneously mating snakes sometimes are often generated by competing animals. The average estimated litter size in Nebraska (25 litters) is 20.6 young, which is very similar to averages estimated for both the Plains and common gartersnakes (Ballinger, Lynch, and Smith 2010).

Selected References: Baxter and Stone 2011; Joy and Crews 1985; Rossman, Ford, and Seigel 1996; Ballinger, Lynch, and Smith 2010.

PLAINS GARTERSNAKE. *THAMNOPHIS RADIX.* UB, AB

COMMON GARTERSNAKE. *THAMNOPHIS SIRTALIS.* UB, CO

Identification: The common gartersnake closely resembles the equally common and even more widespread Plains gartersnake. Both species are distinctively patterned dorsally and laterally with dark and light stripes and complex lateral patterning, as well as a conspicuous yellow stripe down the middle of the back. The common gartersnake alternates brick-red and black vertical bars and wedges along its entire body. The Plains gartersnake has an orange dorsal stripe and a lateral series of black spots. However, a variant morph of the Plains gartersnake in eastern Nebraska also alternates red and black lateral markings, much like the common gartersnake's. The length of Plains gartersnake adults normally ranges from 380 to 710 mm (15 to 28 in.), with a maximum of 109 cm (43 in.). Common gartersnake adults normally range in length from 410 to 710 mm (16 to 28 in.), with a maximum of 149 cm (58.6 in.) (Collins 1993).

Status: The common gartersnake has been recorded in all but 22 Nebraska counties and the Plains gartersnake in all but 10. Both species tend to favor locations close to water, but the common gartersnake also often occurs in woodlands. Plains gartersnakes appear to attain the highest densities of all Nebraska's gartersnake species (Ballinger, Lynch, and Smith 2010).

Habitats and Ecology: During summer, common gartersnakes are surprisingly mobile, with males having home ranges of up to 14 hectares (35 acres) and females up to nearly 9.3 hectares (23 acres). Displaced common gartersnakes are able to orient themselves in the correct homeward direction, apparently using the sun as a compass. Adult common gartersnakes prey on a variety of animals, especially leopard frogs, but they also consume small rodents, earthworms, and others, evidently using chemical clues to identify potential prey. Plains gartersnakes likewise mostly feed on amphibians, fish, earthworms, and slugs. Various studies indicate that annual survival rates of gartersnakes range from 34 to 50 percent, meaning that few individuals live beyond three to four years in the wild. However, captive Plains gartersnakes have survived for more than eight years.

Breeding Biology: Gartersnake hibernacula are present near the visitor center areas at both the Crane Trust and Rowe Audubon Sanctuary; by mid-March common gartersnakes can often be seen moving in and out of the small entrances to their hibernacula. Movements of about 0.25 mile between winter hibernacula and summer ranges are typical of common gartersnakes in Kansas. There, Plains gartersnakes mate during April and May and sometimes also in the fall. Spring mating is typical in Nebraska. A male will crawl alongside a female while performing writhing, jerking movements and attempt to curl his tail underneath her until their cloacal openings meet and mating occurs. Sometimes more than one male mates with a female. Plains gartersnake litters in Kansas average about 17–18 young but range from 5 to 60 (Collins 1993), and common gartersnake litters in Nebraska similarly average about 20, ranging from 8 to 51 (Ballinger, Lynch, and Smith 2010). The young of all three Nebraska *Thamnophis* species hatch from July to September, and most adults and young probably move into hibernacula by October. Freezing and flooding during the overwintering period are probably important causes of mortality in northern populations (Ballinger, Lynch, and Smith 2010).

Selected References: Fitch 1980; Baxter and Stone 2011; Joy and Crews 1985; Collins 1993; Rossman, Ford, and Seigel 1996; Ballinger, Lynch, and Smith 2010.

Family Dipsadidae (Rear-fanged Snakes)

Rear-fanged snakes are mildly poisonous, having grooved teeth near the rear of the jaw that transmit toxic saliva to their prey while the snake is holding or swallowing it. Some species of this group, including hog-nosed snakes, are included in the family Colubridae, and in some sources (e.g., Collins 1993), the rear-fanged snakes are classified as the family Xenodontidae.

PLAINS (WESTERN) HOG-NOSED SNAKE. *HETERODON NASICUS.* WI, UN

EASTERN HOG-NOSED SNAKE. *HETERODON PLATYRHINOS.* DI, UN

Identification: Hog-nosed snakes are unique among American snakes in that they have an upturned and pointed snout (an enlarged and strongly keeled scale). Nebraska's two species can be distinguished by the fact that the pointed snout of the eastern species is only slightly upturned, whereas that of the western species is strongly upturned (figure 37). This shape provides a digging tool that helps in burrowing and in uncovering toads, their primary prey. Hog-nosed snakes also have enlarged teeth at the back of the upper jaw; these teeth puncture and deflate toads and frogs that have inflated their bodies by swallowing air to try to avoid being eaten. The two species also differ in underpart coloration; the western hog-nosed has black ventral markings, which the eastern species lacks. Eastern hog-nosed snakes are usually 520–760 mm (20–30 in.) long, with a maximum of 116 cm (45.5 in.), whereas Plains hog-nosed snakes average slightly shorter, typically 380–635 mm (12–25 in.) long, with a maximum length of 100 cm (39.5 in.) (Collins 1993).

Status: The Plains (western) hog-nosed snake occurs in the grass-

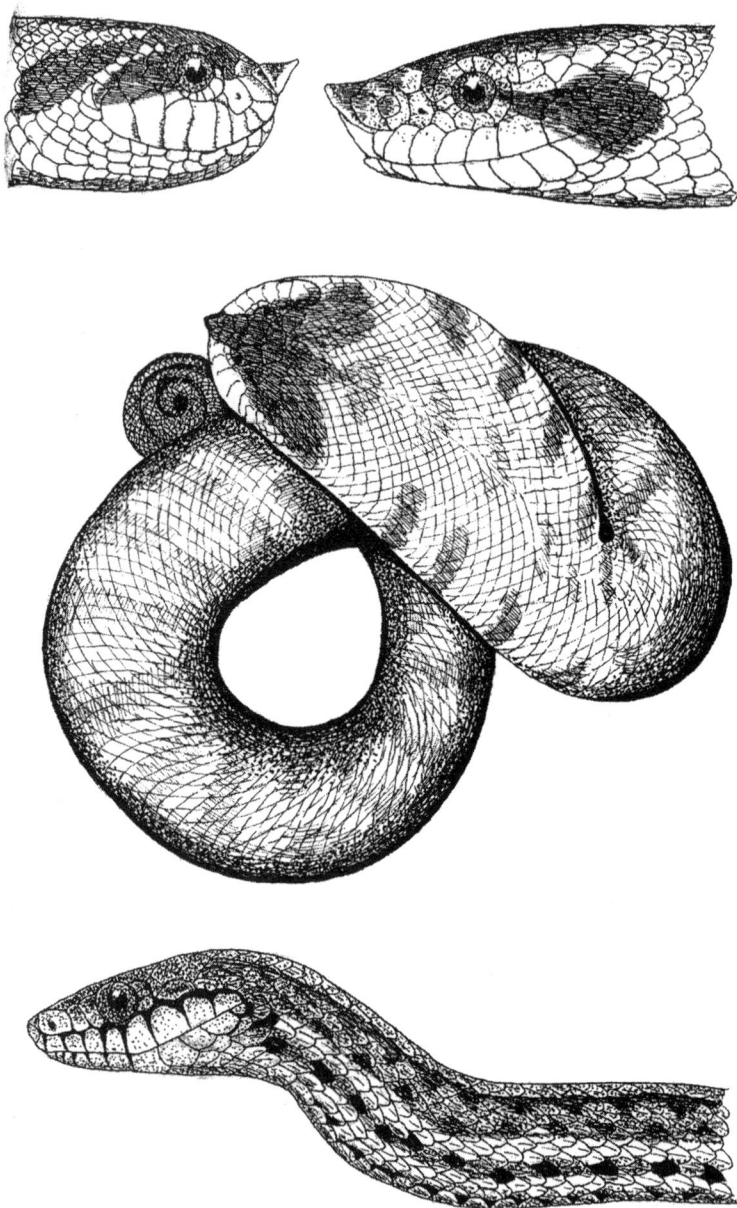

37. Plains (top left) and eastern (top right) hog-nosed snakes, defensive
posture of the eastern hog-nosed snake (middle), and Plains garter
snake (bottom).

lands of western and central Nebraska, throughout the Sandhills, and east to Hamilton County in the Platte Valley. The eastern hog-nosed snake occurs in sandy grasslands from the Missouri Valley west to Cherry County in the Niobrara Valley, to Lincoln County in the Platte Valley, and to Chase County in southwestern Nebraska (Fogell 2010).

Habitats and Ecology: This species is found where their favorite prey, toads, are abundant; thus, they often occur in areas of sandy soil. They also eat frogs, salamanders, lizards, turtles, turtle eggs, birds, and mammals. Hog-nosed snakes produce hissing sounds when threatened and might even strike defensively, often with the mouth closed. At least the eastern hog-nosed sometimes also assumes a rather cobra-like defensive posture by spreading its upper neck vertebrae (figure 37). With further provocation, the snake will writhe convulsively, turn over on its back, gasp, and become apparently lifeless. If it is turned upright, it will often turn over and again act lifeless. When left alone, the snake will soon turn over and crawl away, or if it is placed in water, it will swim away.

Breeding Biology: Hog-nosed snakes are oviparous. The western species lays clutches of 4–24 eggs, averaging about 9 in Kansas, whereas the eastern species produces clutches of 4–61 eggs, averaging about 22 (Collins 1993). Females frequently will skip reproducing for a year and develop a biennial reproductive cycle (Ballinger, Lynch, and Smith 2010). Hog-nosed snakes are active for about six months of the year, from early May to late October in Kansas (Ballinger, Lynch, and Smith 2010). In captivity, the western species has survived for as long as nearly 20 years, whereas the longevity record for the eastern species is less than 8 years.

Selected References: Baxter and Stone 2011; Collins 1993; Ballinger, Lynch, and Smith 2010.

Family Viperidae (Pitvipers)

All the members of this uniformly venomous family have large erectile and hollow upper teeth (fangs) that can inject venom, which is used to subdue their prey by destroying their red blood cells and is also used defensively when threatened. Pitvipers have large, triangular heads, eyes with vertical pupils (in association with their nocturnal activity), and paired infrared (heat-sensitive) pits located just behind their nostrils. All of these snakes specialize in eating warm-blooded vertebrates, although they also consume other snakes, lizards, frogs, and a variety of invertebrates. The four Nebraska pitvipers all bear their young alive. All of the most dangerous North American snakes belong to the pitviper family, of which the timber rattlesnake is the largest in Nebraska. Nebraska's timber rattlesnakes average 112–127 cm (44–50 in.), with a maximum of 152 cm (60 in.) (Fogell 2010). That species' maximum recorded length is 189 cm (74.5 in.) (Ballinger, Lynch, and Smith 2010). The average amount of venom injected by adult prairie rattlesnakes is only about one-eighth the lethal dose for an adult human (Baxter and Stone 2011), and even less is produced by the smaller and much rarer massasauga. The only other venomous Nebraska species, the copperhead, is limited to a few locations in Gage and Richardson Counties and is a reclusive, state-protected species that has very little contact with humans. There is an average of about five human deaths in the United States per year caused by rattlesnake bites, nearly all by diamondback rattlesnakes, but during the past century at least eight have been caused by timber rattlesnakes, and at least four by copperheads. Prairie rattlesnake bites have caused the deaths of at least three children in the United States since 1900.

PRAIRIE (WESTERN) RATTLESNAKE. *CROTALUS VIRIDIS.* WI, UN

Identification: The rattle of a rattlesnake is enough to identify it, although rattlesnakes also have the distinctive family traits of large, triangular heads, eyes with vertical pupils, and sensory pits behind their nostrils. Most Nebraska specimens are grayish to pale brown, with about 40 darker brown dorsal spots, the spots morphing into vertical bands posteriorly and finally becoming rings on the tail. Two light stripes extend from behind and below the eye back to the angle of the jaw. Together with the timber rattlesnake, gophersnake, and western ratsnake, prairie rattlesnakes are among Nebraska's longer snakes. They sometimes reach 120 cm (47 in.) and elsewhere have been reported as reaching 145 cm (57 in.) (Collins 1993). In a sample of 117 Wyoming adults, males averaged 905 mm (35.6 in.), and females averaged 787 mm (31 in.) (Baxter and Stone 2011).

Sound Production: The defensive rattle of a rattlesnake is easily recognized. It is sometimes mimicked by other species, such as gophersnakes, which shake their tails in a similar manner, and similar sounds are uttered by a few cavity-dwelling birds (such as burrowing owls) when they are threatened.

Status: This species is widespread in western Nebraska, extending east to Cedar County in the Niobrara Valley, to Buffalo and Kearney Counties in the Platte Valley, and to Furnas County in the Republican Valley. It is most common in rocky woodlands and foothills where rimrock and outcrops of limestone occur.

Habitats and Ecology: The primary food of rattlesnakes consists of rodents, making them a desirable addition to the state's fauna (from a farmer's standpoint), and they also eat rabbits, lizards, birds, and other snakes. They are mainly diurnal in activity and hibernate in large aggregations, often in deep, rocky limestone crevices or outcrops. They are often found among prairie dog colonies, using their burrows for shelter and the prairie dogs as a food source (Baxter and Stone 2011). Females often begin to gather at wintering sites as early as August, followed by newly born hatchlings.

Breeding Biology: Although courtship in Nebraska has been seen as late as autumn (Ballinger, Lynch, and Smith 2010), mating is much more likely to occur in the spring, soon after the snakes emerge from their winter dormancy. Females normally reach sexual maturity in their third year, and mean litter size in a Kansas study averaged 10 young, ranging from 5 to 18. Newborns average 23–25 cm (9–10 in.) long and are venomous from birth. The females give birth to their young from spring to September, depending on their time of mating, but some females produce litters only during alternate years. Yearling snakes average about 61 cm (24 in.) in length by late summer, and two-year-olds average about 76 cm (30 in.). Captives have been found to survive as long as 28 years (Collins 1993).

Selected References: Klauber 1972; Baxter and Stone 2011; Collins 1993; Holycross 1995; Ballinger, Lynch, and Smith 2010.

MASSASAUGA RATTLESNAKE. *SISTRURUS CATENATUS*. HL (EA), RA

Identification: All of Nebraska's venomous snakes can be recognized as such by their vertically oriented pupils and by the presence of paired sensory pits located immediately behind their nostrils. They also all have wide, rather triangular-shaped heads, although this shape is not so obvious in the massasauga. The massasauga is unique in having only nine unusually large scales on the top of its head; only three scales separate the eyes (figure 36). A broad dark brown streak through the eye is bordered below by a broad white line that loops down and encircles a second brown band. The snake's basic color is light brown to gray, with many dark brown blotches on the back and sides that transform into vertical bands toward the tail. Adults usually range from 46 to 76 cm (18 to 30 in.) in length. Males have longer tails than females; the tail rattles are relatively small in this species. The species' maximum reported length is 100 cm (39.5 in.) (Collins 1993). Because of their small size, these snakes are inconspicuous, and their rattling is hard to hear. Being bitten by one is serious but not fatal, as relatively small amounts of venom are injected. Apparently there are no records of human fatalities from being bitten by a massasauga.

Status: The massasauga has now been reduced to surviving apparently only in Pawnee County, although it is possible that relict populations still exist in other remnant tallgrass prairies of southeastern Nebraska. George Hudson (1942) last reported it from Lancaster County (Nine Mile Prairie). This species is on the Nebraska Natural Legacy Project's Tier 1 list of threatened species in Nebraska (Schneider et al. 2018).

Habitats and Ecology: Massasaugas are active in Nebraska from April to October in habitats varying from wet prairies to woodlands. The species' vernacular name is of Ojibwa origin and is derived from that culture's name for an Ontario river, the Mississagi. In Kansas, these snakes are active from April to October. There they are most abundant in grassy wetlands, and in Iowa, they have been reported to hibernate in marshes. Elsewhere, they have been found to winter in rock crevices or rodent burrows. They feed on frogs, lizards, other snakes, and rodents, as well as birds' eggs and invertebrates. Their maximum longevity in captivity has been reported as 20 years.

Breeding Biology: During courtship, competing males evidently engage in "combat dances," which also occur in other venomous snakes. Mating occurs during both spring and fall, with females producing young in alternate years in some regions. Mating behavior is like that of other snakes, with the male insinuating his tail beneath the female until their cloacal apertures come into contact. Like other pitvipers, the massasauga is viviparous, giving birth to living young. Litter sizes in a Colorado study ranged from 5 to 7 young and in Kansas varied from 3 to 13, averaging 6. The young are 180–240 mm (7–9.5 in.) long at birth (Ballinger, Lynch, and Smith 2010). In Kansas, they are born during July and August and are venomous at birth (Collins 1993). Massasaugas are rare in Nebraska but are not currently classified as threatened.

Selected References: Wright and Wright 1957; Klauber 1972; Collins 1993; Ballinger, Lynch, and Smith 2010.

CHAPTER 5

Species Checklist and
Status/Habitat Codes

DATA SOURCES AND ABUNDANCE, HABITAT, AND CONSERVATION CATEGORIES

Mammal distribution/abundance estimates are mostly based on Genoways et al. (2008), but some estimates are from other available national and state distribution maps (e.g., Benedict, Genoways, and Freeman 2000; Reid 2006). Bird abundance estimates are based on Johnsgard (2018a). Habitat categories (following distribution/abundance categories and separated by semicolons) were derived from those used by Armstrong, Fitzpatrick, and Meaney (2011) and Johnsgard (2019). The Nebraska Natural Legacy Project, 2018 (Schneider et al. 2018), identified those species of highest state conservation concern as Tier 1 species (T1-NNLP-18) and those of secondary concern as Tier 2 species (T2-NNLP-18).

MAMMALS (89 NATIVE SPP.)

Family sequence, binomial nomenclature, and vernacular names follow Genoways et al. (2008), but genera and species are sequenced alphabetically within families. Species with narrative text profiles are bold. Species extirpated during the nineteenth or early twentieth century are shown in brackets. The status code sequence (see the list of abbreviations in the front matter) following each species' name is Range (Nebraska geography); Abundance; Habitats (conservation status).

Order Marsupialia (Marsupials)

FAMILY DIDELPHIDAE (OPOSSUMS)

Virginia Opossum, *Didelphis virginiana.* Di, Co; RiDe, RiSS; WeSW, HuFR, HuTC, HuCL

Order **Cingulata (Armadillos)**

FAMILY DASYPODIDAE (ARMADILLOS)

Nine-banded Armadillo, *Dasypus novemcinctus.* HL (So, SC); VR; FoUD, RiDe, HuPG

Order **Lagomorpha (Rabbits and Hares)**

FAMILY **LEPORIDAE (RABBITS AND HARES)**

Black-tailed Jackrabbit, *Lepus californicus.* Ub; Un; GrAll, UpSh

White-tailed Jackrabbit, *Lepus townsendii.* Wi (No); Ra; GrAll, UpSh (T2-NNLP-18)

Desert Cottontail, *Sylvilagus audubonii.* SD (We); Un; GrSG, UpSh

Eastern Cottontail, *Sylvilagus floridanus.* Ub; Co; GrAll

Order Soricomorpha (Shrews and Moles)

FAMILY **SORICIDAE (SHREWS)**

Northern Short-tailed Shrew, *Blarina brevicauda.* Wi; Un; GrAll

Elliott's Short-tailed Shrew, *Blarina hylophaga.* SD (So); Un; FoUD, RiDe

North American Least Shrew, *Cryptotis parva.* SD; Un; GrAll, SRUP

Cinereus (Masked) Shrew, *Sorex cinereus.* Ub; Un; WeSW, GrAll

Merriam's Shrew, *Sorex merriami.* Lo (Sioux, Dawes, and Sheridan Counties); SaSG, SRUP (T2-NNLP-18)

Dwarf Shrew, *Sorex nanus.* HL (Sioux County); SaSG, RoBL (T2-NNLP-18)

FAMILY TALPIDAE (MOLES)

Eastern Mole, *Scalopus aquaticus.* Wi, Un; GrAll

Order **Chiroptera (Bats)**

FAMILY VESPERTILIONIDAE (VESPER BATS)

Townsend's Big-eared Bat, *Corynorhinus townsendii.* HL (Sioux County); Ra GrAll, UpSh, WoJu, FoWC (CC, T2-NNLP-18)

Big Brown Bat, *Eptesicus fuscus.* Wi; Co; UpSh, WoJu

Silver-haired Bat, *Lasionycteris noctivagans.* Wi; Un (migratory); UpSh (T1-NNLP-18)

Eastern Red Bat, *Lasiurus borealis.* Ub; Un (migratory); *RiWe* (T1-NNLP-18)

Hoary Bat, *Lasiurus cinereus.* Wi; Un (migratory); UpSh, WoJu (T1-NNLP-18)

Western Small-footed Myotis, *Myotis ciliolabrum.* Lo (We, Cen), Un; SaSG, RiWe, *RiDe*

Little Brown Myotis, *Myotis lucifugus.* SD (NW, Ea); Un; *RiDe* (T1-NNLP-18)

Northern (Long-eared) Myotis, *Myotis septentrionalis.* SD (No, SE); Ra (federally threatened); WoUp (T1-NNLP-18)

Fringed-tailed Myotis, *Myotis thysanodes.* SD (NW); Ra; WoJu (CC)

Long-legged Myotis, *Myotis volans.* SD (We); Ra; WoJu (CC, T2-NNLP-18)

Evening Bat, *Nycticeius humeralis.* SD (SE); Un; WoUp

American Pipistrelle (Tri-colored Bat), *Perimyotis subflavus.* Wi; Un; *RiWe, RiDe* (T1-NNLP-18)

FAMILY MOLOSSIDAE (FREE-TAILED BATS)

Brazilian Free-tailed Bat, *Tadarida brasiliensis.* Vagrant

Order Carnivora (Carnivores)

FAMILY FELIDAE (CATS)

Canada Lynx, *Lynx canadensis.* Vagrant

Bobcat, *Lynx rufus*. Ub; Un; RoBL, SRUP, WoUp

Cougar (Puma, Mountain Lion), *Puma concolor*. Di (NW); Ra; FoWC, RoBL, WoUp (CC, T2-NNLP-18)

FAMILY CANIDAE (DOGS)

Coyote, *Canis latrans*. Ub, Un; SaSG, SRUP, GrAll

[Gray Wolf, *Canis lupus*. Extirpated since late 1800s; vagrants from Wyoming or Minnesota probably occur rarely]

Gray Fox, *Urocyon cinereoargenteus*. Di (Ea); Oc

Swift Fox, *Vulpes velox*. Lo (We); GrAll (CC, T1-NNLP-18)

Red Fox, *Vulpes vulpes*. Wi (Ea), Un; FoUD, WoUp

FAMILY URSIDAE (BEARS)

Black Bear, *Ursus americanus*. Vagrant

[Grizzly Bear, *Ursus arctos horribilis*. Extirpated since the 1800s]

FAMILY **MUSTELIDAE (WEASELS AND RELATIVES)**

Wolverine, *Gulo gulo*. Vagrant

Northern River Otter, *Lontra canadensis*. Lo (reintro-duced); RiLo, RiWe (CC)

Least Weasel, *Mustela nivalis*. Wi (Ea, No), Un; RiDe SaSG, FoUD, GrAll

Long-tailed Weasel, *Mustela frenata*. Ub, Un; RiDe, SaSG, FoUD, GrAll (T2-NNLP-18)

[**Black-footed Ferret**, *Mustela nigripes*. Extirpated; SaSG]

American Mink, *Neovison vison*. Ub, Un; RiDe, RiHe, RiSS, RiWe

American Badger, *Taxidea taxus*. Ub, Un; SaSG, GrMG, GrTG

FAMILY MEPHITIDAE (SKUNKS)

Striped Skunk, *Mephitis mephitis*. Ub, Co; GrAll, WoUp, HuFR, HuCL

Eastern Spotted Skunk, *Spilogale putorius*. Ub, Oc; SRUP, GrAll, WoUp (T1-NNLP-18)

FAMILY PROCYONIDAE (RACCOON)

Raccoon, *Procyon lotor*. Ub, Co; WeDW, WeSW, WoUp, RiWe, RiDe, HuPG, HuTC

Order Artiodactyla (Even-toed Ungulates)

FAMILY **CERVIDAE (DEER)**

Moose, *Alces alces*. Vagrant

Elk, *Cervus canadensis*. Nearly extirpated by 1900; locally reintroduced and self-reestablished; Lo (Niobrara River, PR, and Lincoln County); Oc; FoWC, GrAll

Mule Deer, *Odocoileus hemionus*. Wi (We, Cen); FoUD, HuFR, RiDe, SRUP, RoBL

White-tailed Deer, *Odocoileus virginianus*. Ub, Co; FoUD, HuFR, HuCL, RiDe

FAMILY **ANTILOCAPRIDAE (PRONGHORN)**

Pronghorn, *Antilocapra americana*. Nearly extirpated by 1920; locally reintroduced and self-reestablished; Wi (We); GrSG, GrSH

FAMILY **BOVIDAE (CATTLE, SHEEP, AND GOATS)**

[**American Bison**, *Bison bison*. Extirpated by 1880s; many local preserves and commercially developed herds currently exist]

Bighorn (Mountain) Sheep, *Ovis canadensis*. Extirpated by early 1900s; locally reintroduced since the 1980s; Ra (PR, WH); GrAll, FoWC (T1-NNLP-18)

Order Rodentia (Rodents)

FAMILY **SCIURIDAE (SQUIRRELS, CHIPMUNKS, AND PRAIRIE DOGS)**

Black-tailed Prairie Dog, *Cynomys ludovicianus*. Di (We); Un; SaSG (T2-NNLP-18)

Southern Flying Squirrel, *Glaucomys volans*. Lo (SE); Oc; FoUD, RiDe (CC, T1-NNLP-18)

Thirteen-lined Ground Squirrel, *Ictidomys tridecemlineatus.*
Ub, Co; GrAll

Woodchuck, *Marmota monax.* Wi (Ea); Un; GrAll

Eastern Fox Squirrel, *Sciurus niger.* Wi, Co; FoUD, HuPG, HuTC

Eastern Gray Squirrel, *Sciurus pennsylvanicus.* Lo (SE) Un;
FoUD, HuPG, HuTC (T2-NNLP-18)

Wyoming Ground Squirrel, *Spermophilus elegans.* HL (SW, two
records); SaSG

Franklin's Ground Squirrel, *Spermophilus franklini.* Di (Ea);
Oc; GrAll

Spotted Ground Squirrel, *Spermophilus spilosoma.* Di (We);
Un; GrSG, GrSH

Least Chipmunk, *Tamias minimus.* SD (NW Pan), Un; FoWC,
RoBL, RoUc (T2-NNLP-18)

Eastern Chipmunk, *Tamias striatus.* **Lo (SE) Un**; FoUD, RiDe
(T2-NNLP-18)

FAMILY **CASTORIDAE (BEAVERS)**

Beaver, *Castor canadensis.* Ub, Co; RiLo, RiDe, WeDW

FAMILY **HETEROMYIDAE (POCKET MICE)**

Hispid Pocket Mouse, *Chaetodipus hispidus.* Wi (We, SE);
GrAll, SRUP

Ord's Kangaroo Rat, *Dipodomys ordii.* Wi (We, SH); GrSH,
SaSG

Olive-backed Pocket Mouse, *Perognathus fasciatus.* SD (Pan),
Co; GrAll (T2-NNLP-18)

Plains Pocket Mouse, *Perognathus flavescens.* Wi (Ea, Cen);
GrAll, SRUP, WoUp (*P. f. perniger*, T1-NNLP-18)

Silky Pocket Mouse, *Perognathus flavus.* Wi (We); GrAll,
SRUP, WoUp (T2-NNLP-18)

FAMILY **GEOMYIDAE (POCKET GOPHERS)**

Plains Pocket Gopher, *Geomys bursarius.* Wi (Ea); Co; GrAll
(T1-NNLP-18)

Hall's Pocket Gopher, *Geomys jugossicularis.* SD (SW); Co; GrAll

Sand Hills Pocket Gopher, *Geomys luetescens.* SD (SH); Co; GrMG, GrSH

Northern Pocket Gopher, *Thomomys talpoides.* SD (SW); Co; GrAll (includes *T. t. pierricolus* and *T. t. cheyennensis,* T1-NNLP-18)

FAMILY **DIPODIDAE (JUMPING MICE)**

Meadow Jumping Mouse, *Zapus hudsonius.* Wi (Ea, Cen); Co; GrMG RiHe

FAMILY **CRICETIDAE (NEW WORLD RATS AND MICE)**

Prairie Vole, *Microtus ochrogaster.* Ub, Co; GrAll, GrSH

Meadow Vole, *Microtus pennsylvanicus.* Ub, Co; GrAll, WeSW

Woodland Vole, *Microtus pinetorum.* Lo (SE); Co; GrAll (T2-NNLP-18)

Bushy-tailed Woodrat, *Neotoma cinerea.* SD (Pan); Co; RoBL, RoUc, RoCl (T2-NNLP-18)

Eastern Woodrat, *Neotoma floridana.* Lo (No, SW, SC, SE); Co; FoUD, RoBL, RoUc (includes *N. f. baileyi,* T1-NNLP-18)

Muskrat, *Ondatra zibethicus.* Ub, Co; WeSW, RiWe

Northern Grasshopper Mouse, *Onychomys leucogaster.* Ub, Co; GrAll

White-footed Deer Mouse, *Peromyscus leucopus.* Ub, Co; TeAll, GrAll

North American Deer Mouse, *Peromyscus maniculatus.* Ub, Co; TeAll, GrAll

Western Harvest Mouse, *Reithrodontomys megalotis.* Ub, Co; TeAll

Plains Harvest Mouse, *Reithrodontomys montanus.* Ub, Co; GrAll

Hispid Cotton Rat, *Sigmodon hispidus.* SD (So) Un; GrTG, HuCL, HuFR

Southern Bog Lemming, *Synaptomys cooperi.* Wi (Ea, Cen, SW); FoWC, WeSW

FAMILY MURIDAE (OLD WORLD RATS AND MICE)

House Mouse, *Mus musculus.* Introduced accidentally; Ub, Co; HuFR, HuTC, HuPG

Norway Rat, *Rattus norvegicus.* Introduced accidentally; Ub, Co; HuFR, HuTC, HuPG

FAMILY ERETHIZONTIDAE (PORCUPINES)

Porcupine, ***Erethizon dorsatum.*** Wi (We, Cen); Un; FoWC, FoDe

The English names, family and species sequence, and Latin nomenclature used here follow the American Ornithologists' Union's *Checklist of North American Birds*, with supplements through 2019. Species with associated narrative profiles are *bold-italic*. The sequence of status codes listed after each species (see the list of abbreviations in the front matter) is Abundance/Season/Status; Habitats. Habitat status codes shown for migrants relate to periods when the species is present in Nebraska; most species utilize more habitats than the ones indicated. Mollhoff (2016) provided a numerical breakdown of relative habitat use of 225 species documented as proven or probable breeders during the fieldwork for *The Second Nebraska Breeding Bird Atlas*. Hypothetical species lacking adequate occurrence documentation are parenthetically identified.

Additional information on Nebraska's birds can be found at *Birds of Nebraska—Online* (https://birds.outdoornebraska.gov/). This Nebraska Game and Parks Commission website has species accounts for over 500 species, including all 462 species of Nebraska birds currently accepted by the Nebraska Ornithologists' Union (NOU), plus many other reported but insufficiently documented species. It also provides regularly updated information on the abundance and distribution of both common and rare species. Likewise, the website of the NOU (https://www.noubirds.org/) provides a wealth of information on Nebraska birds, including its official state list of species and links with the Nebraska Birding Trails and with the online NEBirds Discussion Group. Individual county checklists of birds can also be generated at the NOU website.

Order Anseriformes (Waterfowl and Relatives)
FAMILY ANATIDAE (SWANS, GEESE, AND DUCKS)
Fulvous Whistling-Duck, *Dendrocygna bicolor*. Accidental
Black-bellied Whistling-Duck, *Dendrocygna autumnalis*.
Accidental
Emperor Goose, *Anser canagica*. Accidental, NT

Snow Goose, *Anser caerulescens.* AbMig; WeSW, HuCL

Ross's Goose, *Anser rossii.* CoMig; WeSW, HuCL

Greater White-fronted Goose, *Anser albifrons.* CoMig;
   WeSW, HuCL

Taiga Bean-Goose, *Anser fabalis.* Accidental

(Pink-footed Goose, *Anser brachyrhynchus.*) Hypothetical

Brant, *Branta bernicla.* Accidental

Cackling Goose, *Branta hutchinsii.* AbMig; WeSW, HuCL

Canada Goose, *Branta canadensis.* AbMig; WeSW, HuCL

Barnacle Goose, *Branta leucopsis.* RaMig; WeSW

Mute Swan, *Cygnus olor.* Accidental or extremely rare (self-
   introduced and feral)

Trumpeter Swan, *Cygnus buccinator.* RaRes; WeSW (CC, T2-
   NNLP-18)

Tundra Swan, *Cygnus columbianus.* OcMig; WeSW

Wood Duck, *Aix sponsa.* UnSuRes (Ea); RiDe

Garganey, *Spatula querquedula.* Accidental

Blue-winged Teal, *Spatula discors.* CoSuRes; WeSW

Cinnamon Teal, *Spatula cyanoptera.* UnSuRes (We); WeSW
   (T2-NNLP-18)

Northern Shoveler, *Spatula clypeata.* CoSuRes; WeSW

Gadwall, *Mareca strepera.* CoSuRes; WeSW

Eurasian Wigeon, *Mareca penelope.* OcMig; WeSW

American Wigeon, *Mareca americana.* UnSuRes; WeSW (T2-
   NNLP-18)

Mallard, *Anas platyrhynchos.* CoSuRes; HuCL, HuPG

American Black Duck, *Anas rubripes.* RaMig; WeSW

Mottled Duck, *Anas fulvigula.* Accidental

Northern Pintail, *Anas acuta.* CoSuRes; WeSW, HuCL (T2-
   NNLP-18)

Green-winged Teal, *Anas crecca.* OcSuRes; WeSW

Canvasback, *Aythya valisineria.* LoSuRes; WeSW (T2-
   NNLP-18)

Redhead, *Aythya americana.* LoSuRes; WeSW

Ring-necked Duck, *Aythya collaris*. UnMig; WeSW

Tufted Duck, *Aythya fuligula*. Accidental

Greater Scaup, *Aythya marila*. OcMig; WeSW, WeDW

Lesser Scaup, *Aythya affinis*. CoMig; WeSW, WeDW (T2-NNLP-18)

King Eider, *Somateria spectabilis*. Accidental

Common Eider, *Somateria mollissima*. Accidental

Harlequin Duck, *Histrionicus histrionicus*. Accidental

Surf Scoter, *Melanitta perspicillata*. RaMig; WeDW

White-winged Scoter, *Melanitta fusca*. OcMig; WeDW

Black Scoter, *Melanitta deglandi*. VRMig; WeDW

Long-tailed Duck, *Clangula hyemalis*. RaMig; Vul; WeDW

Bufflehead, *Bucephala albeola*. CoMig; WeDW

Common Goldeneye, *Bucephala clangula*. CoWiMig; WeDW

Barrow's Goldeneye, *Bucephala islandica*. RaWiMig; WeDW

Hooded Merganser, *Lophodytes cucullatus*. OcMig; WeDW, RiDe

Common Merganser, *Mergus merganser*. CoWiMig; WeDW, RiDe

Red-breasted Merganser, *Mergus serrator*. CoMig; WeDW

Ruddy Duck, *Oxyura jamaicensis*. LoSuRes; WeSW

Order Galliformes (Gallinaceous Birds)

FAMILY ODONTOPHORIDAE (NEW WORLD QUAIL)

Northern Bobwhite, *Colinus virginianus*. CoRes

FAMILY PHASIANIDAE (PARTRIDGES, GROUSE, AND TURKEYS)

Gray Partridge, *Perdix perdix*. LoRes; HuFR, GrAll (introduced in the early 1900s)

Ring-necked Pheasant, *Phasianus colchicus*. CoRes; GrAll, HuFR (introduced in the early 1900s)

Ruffed Grouse, *Bonasa umbellus*. Extirpated in the 1800s

Greater Sage-Grouse, *Centrocercus urophasianus*. Extirpated or extremely rare winter visitor (CC1), NT

Sharp-tailed Grouse, *Tympanuchus phasianellus*. CoRes; GrMG, GrSH

Greater Prairie-Chicken, *Tympanuchus cupido*. UnRes, GrTG (CC, CC1, NT, T2-NNLP-18, Vul)

Lesser Prairie-Chicken, *Tympanuchus pallidicinctus*. Extirpated in the early 1900s (CC1, Vul)

Wild Turkey, *Meleagris gallopavo*. CoRes; FoUD, RiDe

Order Podicipediformes (Grebes)

FAMILY PODICIPEDIDAE (GREBES)

Pied-billed Grebe, *Podilymbus podiceps*. LoSuRes; WeSW (T2-NNLP-18)

Horned Grebe, *Podiceps auritus*. UnMig; WeSW, WeDW

Red-necked Grebe, *Podiceps grisegena*. RaMig; WeSW, WeDW

Eared Grebe, *Podiceps nigricollis*. LoSuRes (We); WeSW, WeDW

Western Grebe, *Aechmophorus occidentalis*. LoSuRes (We); WeSW, WeDW (T2-NNLP-18)

Clark's Grebe, *Aechmophorus clarkii*. LoSuRes (We); WeSW, WeDW (T2-NNLP-18)

Order Columbiformes (Pigeons and Doves)

FAMILY COLUMBIDAE (PIGEONS AND DOVES)

Rock Pigeon, *Columba livia*. AbRes; HuTC, HuFR (introduced in the 1800s or earlier)

Band-tailed Pigeon, *Patiogioenas fasciata*. Accidental

Eurasian Collared-Dove, *Streptopelia decaocto*. CoRes, HuTC, HuFR (self-introduced since the 1990s)

White-winged Dove, *Zenaida asiatica*. OcSuRes; HuTC, HuFR

Mourning Dove, *Zenaida macroura*. AbSuRes; HuTC, HuFR, GrAll, WoUp

Passenger Pigeon, *Ectopistes migratorius*. Extinct since the early 1900s

Inca Dove, *Columbina inca*. VRMig; HuTC, HuFR

Common Ground Dove, *Columbina passerina*. Accidental

Order Psittacifomes (Parrots)

FAMILY PSITTACIDAE (PARROTS)

(Monk Parakeet, *Myopsitta monarchus.*) Hypothetical Carolina Parakeet, *Conuropsis carolinensis.* Extinct since the early 1900s

Order Cuculiformes (Cuckoos)

FAMILY CUCULIDAE (CUCKOOS)

Yellow-billed Cuckoo, *Coccyzus americanus.* UnSuRes; RiDe, FoUD (T2-NNLP-18)

Black-billed Cuckoo, *Coccyzus erythropthalmus.* OcSuRes; RiDe, FoUD (T1-NNLP-18)

Groove-billed Ani, *Crotophaga sulcirostris.* Accidental

Order Caprimulgiformes (Goatsuckers)

FAMILY CAPRIMULGIDAE (GOATSUCKERS)

Common Nighthawk, *Chordeiles minor.* UnSuRes; HuTC

Common Poorwill, *Phalaenoptilus nuttallii.* UnSuRes (We, PR, WH); SRUP, WoUp, RoBL

Chuck-will's-widow, *Antrostomus carolinensis.* UnSuRes (Ea); RiDe (T2-NNLP-18)

Eastern Whip-poor-will, *Antrostomus vociferus.* UnSuRes (Ea); FoUD, RiDe (T2-NNLP-18)

Order Apodiformes (Swifts and Hummingbirds)

FAMILY APODIDAE (SWIFTS)

Chimney Swift, *Chaetura pelagica.* CoSuRes (Ea), HuTC

White-throated Swift, *Aeronautes saxatalis.* CoSuRes (PR, WH); RoCl (CC1, T2-NNLP-18)

FAMILY TROCHILIDAE (HUMMINGBIRDS)

Ruby-throated Hummingbird, *Archilochus colubris.* UnSuRes (Ea); RiDe, HuPG (T2-NNLP-18)

Black-chinned Hummingbird, *Archilochus alexandri.* Accidental

Costa's Hummingbird, *Calypte costa.* Accidental

Broad-tailed Hummingbird, *Selasphorus platycercus.* OcFa-Mig; RiDe, HuPG

Rufous Hummingbird, *Selasphorus rufus.* OcFaMig; RiDe, HuPG (CC1)

Calliope Hummingbird, *Selasphorus calliope.* VRMig; RiDe, HuPG (CC1)

Order Gruiformes (Rails and Cranes)

FAMILY RALLIDAE (RAILS, GALLINULES, AND COOTS)

Yellow Rail, *Coturnicops noveboracensis.* RaMig; WeSW

Black Rail, *Laterallus jamaicensis.* VRMig; WeSW

Clapper Rail, *Rallus longirostris.* Accidental

King Rail, *Rallus elegans.* RaSuRes (Ea); WeSW (CC, T2-NNLP-18)

Virginia Rail, *Rallus limicola.* UnSuRes; WeSW

Sora, *Porzana carolina.* CoSuRes; WeSW

Purple Gallinule, *Porphyrio martinica.* Accidental

Common Gallinule, *Gallinula galeata*. RaSuRes; WeSW

American Coot, *Fulica americana.* CoSuRes; WeSW

FAMILY GRUIDAE (CRANES)

Sandhill Crane, *Antigone canadensis.* AbMig; RaSuRes; HuCL, RiWe (T2-NNLP-18)

Common Crane, *Grus grus.* VRMig; HuCL, RiWe

(Hooded Crane, *Grus monacha.*) Hypothetical

Whooping Crane, *Grus americana.* OcMig; RiWe, HuCL (CC, federally endangered, T1-NNLP-18)

Order Charadriiformes (Shorebirds)

FAMILY RECURVIROSTRIDAE (AVOCETS AND STILTS)

Black-necked Stilt, *Himantopus mexicanus.* LoSuRes (W, SH); WeSW (T2-NNLP-18)

American Avocet, *Recurvirostra americana.* LoSuRes (W, SH); WeSW (T2-NNLP-18)

FAMILY CHARADRIIDAE (PLOVERS)

Black-bellied Plover, *Pluvialis squatarola*. UnMig; HuCL, HuFR

American Golden-plover, *Pluvialis dominica*. UnMig; HuCL, HuFR

Killdeer, *Charadrius vociferus*. CoSuRes; HuFR, RiWe, RiHe

Semipalmated Plover, *Charadrius semipalmatus*. UnMig; RiWe

Piping Plover, *Charadrius melodus*. LoSuRes RiWe (CC, federally and state threatened, NT, T1-NNLP-18)

Mountain Plover, *Charadrius montanus*. CoSuRes (Pan); SaSG (CC, state threatened, NT, T1-NNLP-18)

Snowy Plover, *Charadrius nivosus*. RaSuRes, RiWe (T2-NNLP-18)

FAMILY SCOLOPACIDAE (SANDPIPERS AND SNIPES)

Upland Sandpiper, *Bartramia longicauda*. UnSuRes; GrSH, GrMG

Eskimo Curlew, *Numenius borealis*. Probably extinct since the late 1900s

Whimbrel, *Numenius phaeopus*. VRMig; WeSW

Long-billed Curlew, *Numenius americanus*. LoSuRes (Pan); SH; GrSH, WeSW (CC, NT, T1-NNLP-18)

Hudsonian Godwit, *Limosa haemastica*. UnSpMig (Ea); NT; WeSW

Marbled Godwit, *Limosa fedoa*. UnMig; SESW

Ruddy Turnstone, *Arenaria interpres*. OcMig; WeSW

Red Knot, *Calidris canutus*. VRMig (federally and state threatened); WeSW

Sanderling, *Calidris alba*. OcMig (Ea); WeSW

Semipalmated Sandpiper, *Calidris pusilla*. CoMig; WeSW

Western Sandpiper, *Calidris mauri*. UnFaMig (We); WeSW

Baird's Sandpiper, *Calidris bairdii*. CoMig; WeSW

Least Sandpiper, *Calidris minutilla*. CoMig; WeSW

White-rumped Sandpiper, *Calidris fuscicollis*. CoSpMig; WeSW

Pectoral Sandpiper, *Calidris melanotos*. CoMig; WeSW

Sharp-tailed Sandpiper, *Calidris acuminata*. Accidental

Dunlin, *Calidris alpina*. OcMig; Ea

Stilt Sandpiper, *Calidris himantopus*. CoMig; WeSW

Ruff, *Philomachus pugnax*. VRMig

Buff-breasted Sandpiper, *Tryngites subruficollis*. UnMig (Ea)
   (CC, NT, T1-NNLP-18)

Short-billed Dowitcher, *Limnodromus griseus*. OcMig (Ea)

Long-billed Dowitcher, *Limnodromus scolopaceus*. CoMig

Wilson's Snipe, *Gallinago delicata*. LoSuRes; WeSW

Spotted Sandpiper, *Actitis macularia*. CoSuRes; WeSW, RiLo

Solitary Sandpiper, *Tringa solitaria*. UnMig; WeSW

Lesser Yellowlegs, *Tringa flavipes*. CoMig; WeSW

Willet, *Tringa semipalmatus*. CoMig; WeSW

Greater Yellowlegs, *Tringa melanoleuca*. CoMig; WeSW

American Woodcock, *Scolopax minor*. LoSuRes (Ea); FoUD,
   RiWe (T2-NNLP-18)

Wilson's Phalarope, *Phalaropus tricolor*. LoSuRes (SH); WeSW

Red-necked Phalarope, *Phalaropus lobatus*. UnSpMig (We);
   WeSW

Red Phalarope, *Phalaropus fulicarius*. VRMig; WeSW

FAMILY LARIDAE (GULLS AND TERNS)

Black-legged Kittiwake, *Rissa tridactyla*. VRMig; WeDW

Sabine's Gull, *Xema sabini*. VRMig; WeDW, WeSW

Bonaparte's Gull, *Chroicocephalus philadelphia*. UnMig;
   WeDW, WeSW

Black-headed Gull, *Chroicocephalus ridibundus*. Accidental

Little Gull, *Hydrocoloeus minutus*. VRMig; WeDW, WeSW

Ross's Gull, *Rhodostethia rosea*. Accidental

Laughing Gull, *Leucophaeus atricilla*. VRMig; WeDW, WeSW

Franklin's Gull, *Leucophaeus pipixcan*. CoMig; WeDW, WeSW

Mew Gull, *Larus canus*. VRMig; WeDW

Ring-billed Gull, *Larus delawarensis.* CoMig; WeDW, WeSW
California Gull, *Larus californicus.* RaMig; WeDW, WeSW
Herring Gull, *Larus argentatus.* UnMig; WeDW, WeSW
Iceland Gull, *Larus glaucoides.* RaMig; WeDW
Lesser Black-backed Gull, *Larus fuscus.* RaMig; WeDW
(Slaty-backed Gull. *Larus shistasagus.*) Hypothetical
Glaucous Gull, *Larus hyperboreus.* RaWiMig; WeDW
Great Black-backed Gull, *Larus marinus.* RaMig; WeDW
Least Tern, *Sternula antillarum.* LoSuRes RiWe (CC, federally and state threatened, T1-NNLP-18)
Caspian Tern, *Hydroprogne caspia.* UnMig; WeDW
Black Tern, *Chlidonias niger.* LoSuRes (SH); WeSW (T1-NNLP-18)
Common Tern, *Sterna hirundo.* UnMig; WeDW, WeSW
Arctic Tern, *Sterna paradisaea.* Accidental
Forster's Tern, *Sterna forsteri.* LoSuRes (SH); WeDW, WeSW (T2-NNLP-18)
Royal Tern, *Thalasseus maximus.* Accidental

FAMILY STERCORARIIDAE (JAEGERS)

Pomarine Jaeger, *Stercorarius pomarinus.* VRMig; WeDW
Parasitic Jaeger, *Stercorarius parasiticus.* VRMig; WeDW
Long-tailed Jaeger, *Stercorarius longicaudus.* VRMig; WeDW

FAMILY ALCIDAE (AUKS)

Ancient Murrelet, *Synthliboramphus antiquus.* Accidental

Order Phaethoniformes (Tropicbirds)
FAMILY PHAETHONTIDAE (TROPICBIRDS)

(White-tailed Tropicbird, *Phaethon lepturus.*) Hypothetical

Order Gaviiformes (Loons)
FAMILY GAVIIDAE (LOONS)

Red-throated Loon, *Gavia stellata.* VRMig; WeDW

Pacific Loon, *Gavia pacifica.* VRMig; WeDW

Common Loon, *Gavia immer.* UnMig; WeDW

Yellow-billed Loon, *Gavia adamsii.* Accidental

Order Ciconiiformes (Storks)

FAMILY CICONIIDAE (STORKS)

Wood Stork, *Mycteria americana.* Accidental

Order Suliformes (Cormorants and Relatives)

FAMILY FREGATIDAE (FRIGATEBIRDS)

(Magnificent Frigatebird, *Fregata magnificens.*) Hypothetical

FAMILY PHALACROCORACIDAE (CORMORANTS)

Neotropic Cormorant, *Phalacrocorax brasilianus.* VRMig; WeDW, WeSW

Double-crested Cormorant, *Phalacrocorax auritus.* LoSuRes (We); WeDW, WeSW

FAMILY ANHINGIDAE (ANHINGAS)

Anhinga, *Anhinga anhinga.* Accidental

Order Pelecaniformes (Pelicans, Herons, and Ibises)

FAMILY PELECANIDAE (PELICANS)

American White Pelican, *Pelecanus erythroryhnchos.* CoMig; LoSuVis, WeDW, WeSW

Brown Pelican, *Pelecanus occidentalis.* VRMig; WeDW, WeSW

FAMILY ARDEIDAE (HERONS AND EGRETS)

American Bittern, *Botaurus lentiginosus.* LoSuRes (SH); WeSW

Least Bittern, *Ixobrychus exilis.* LoSuRes; WeSW

Great Blue Heron, *Ardea herodias.* CoSuRes; WeSW, RiDe

Great Egret, *Ardea alba.* OcMig; WeSW

Snowy Egret, *Egretta thula.* RaSuVis, WeSW

Little Blue Heron, *Egretta caerulea.* OcMig; WeSW

Tricolored Heron, *Egretta tricolor.* Accidental

Reddish Egret, *Egretta rufescens.* Accidental

Cattle Egret, *Bubulcus ibis.* OcMig; HuFR

Green Heron, *Butorides virescens.* CoSuRes; WeSW

Black-crowned Night-Heron, *Nycticorax nycticorax.* LoSuRes (SH, RB); WeSW (T2-NNLP-18)

Yellow-crowned Night-Heron, *Nyctanassa violacea.* OcSuVis, WeSW

FAMILY THRESKIORNITHIDAE (IBISES AND SPOONBILLS)

White Ibis, *Eudocimus albus.* Accidental

Glossy Ibis, *Plegadis falcinellus.* LoSuRes (SH); WeSW

White-faced Ibis, *Plegadis chihi.* UnSuRes (SH, RB); WeSW

Roseate Spoonbill, *Platalea ajaja.* Accidental

Order Cathartiformes (Vultures)

FAMILY CATHARTIDAE (NEW WORLD VULTURES)

Black Vulture, *Coragyps atratus.* Accidental

Turkey Vulture, *Cathartes aura.* UnSuRes; RoBL, RoCl, HuFR

Order Accipitriiformes (Eagles, Hawks, and Ospreys)

FAMILY PANDIONIDAE (OSPREYS)

Osprey, *Pandion haliaetus.* RaSuRes; RiWe

FAMILY ACCIPITRIDAE (HAWKS, EAGLES, AND KITES)

Swallow-tailed Kite, *Elanoides forficatus.* Extirpated; two recent records (CC1)

White-tailed Kite, *Elanus leucurus.* Accidental

Mississippi Kite, *Ictinia mississippiensis.* RaSuRes (We); HuTC, HuPG (T2-NNLP-18)

Golden Eagle, *Aquila chrysaetos.* UnRes (W, PR); RoBL, SRUP (T2-NNLP-18)

Bald Eagle, *Haliaeetus leucocephalus.* UnRes; RiDe, RiWe (CC, T1-NNLP-18)

Northern Harrier, *Circus hudsonicus.* UnSuRes (Cen); RiWe

Sharp-shinned Hawk, *Accipiter striatus*. UnWiMig; HuTC, HuPG

Cooper's Hawk, *Accipiter cooperii*. UnRes; FoUD, FoWC

Northern Goshawk, *Accipiter gentilis*. RaRes (We); FoWC

Harris's Hawk, *Parabuteo uncinctus*. Accidental

Red-shouldered Hawk, *Buteo lineatus*. RaRes (SE); RiDe (T2-NNLP-18)

Broad-winged Hawk, *Buteo platypterus*. OcSuRes (SE); RiDe

Swainson's Hawk, *Buteo swainsoni*. UnSuRes (We); GrAll (CC1)

Zone-tailed Hawk, *Buteo albonotatus*. Accidental

Red-tailed Hawk, *Buteo jamaicensis*. CoSuRes; GrAll, FoUD, FoWC

Ferruginous Hawk, *Buteo regalis*. UnSuRes (We); SaSG (CC, T1-NNLP-18)

Rough-legged Hawk, *Buteo lagopus*. UnWiMig (We); GrAll

Order Strigiformes (Owls)

FAMILY TYTONIDAE (BARN OWLS)

Barn Owl, *Tyto alba*. UnRes; HuFR, RoUc, GrAll (T1-NNLP-18)

FAMILY STRIGIDAE (TYPICAL OWLS)

(Flammulated Owl, *Otus flammeolus*.) Hypothetical

Eastern Screech-Owl, *Megascops asio*. CoRes; FoUD, HuPG, HuFR

Great Horned Owl, *Bubo virginianus*. CoRes; TeAll

Snowy Owl, *Bubo scandiacus*. UnWiVis, GrAll

Northern Hawk Owl, *Surnia ulula*. Accidental

Burrowing Owl, *Athene cunicularia*. UnSuRes (We, SH); GrSG, SaSG (CC, T1-NNLP-18)

Barred Owl, *Strix varia*. UnSuRes (Ea); FoUD

Great Gray Owl, *Strix nebulosa*. Accidental

Long-eared Owl, *Asio otus*. UnSuRes; FoUD

Short-eared Owl, *Asio flammeus*. UnRes; GrAll (CC, CC1, T1-NNLP-18)

Boreal Owl, *Aegolius funereus*. Accidental

Northern Saw-whet Owl, *Aegolius acadicus*. RaRes (PR, WH, NC); FoWC (T2-NNLP-18)

FAMILY ALCEDINIDAE (KINGFISHERS)

Belted Kingfisher, *Megacerle alcyon*. UnSuRes; RiLo

Order Piciformes (Woodpeckers)

FAMILY PICIDAE (WOODPECKERS)

Lewis's Woodpecker, *Melanerpes lewis*. RaSuRes; FoWC, FoUD (CC, CC1, T2-NNLP-18)

Red-headed Woodpecker, *Melanerpes erythrocephalus*. UnSuRes (Ea); FoUD, RiDe (CC1)

Acorn Woodpecker, *Melanerpes formicivorous*. Accidental

Red-bellied Woodpecker, *Melanerpes carolinus*. CoRes; FoUD, RiDe

Williamson's Sapsucker, *Sphyrapicus thyroideus*. Accidental

Yellow-bellied Sapsucker, *Sphyrapicus varius*. UnMig; RiDe

Red-naped Sapsucker, *Sphyrapicus nuchalis*. RaFaMig; RiDe, FoWC

Downy Woodpecker, *Dryobates pubescens*. CoRes; FoUD, RiDe

Hairy Woodpecker, *Dryobates villosus*. CoRes; FoUD, RiDe

American Three-toed Woodpecker, *Picoides dorsalis*. Accidental

Black-backed Woodpecker, *Picoides arcticus*. Accidental

Northern Flicker, *Colaptes auratus*. UnRes; FoUD, RiDe

Pileated Woodpecker, *Dryocopus pileatus*. LoRes (SE); FoUD (T2-NNLP-18)

Order Falconiformes (Falcons and Relatives)

FAMILY FALCONIDAE (FALCONS AND CARACARAS)

Crested Caracara, *Caracara cheriway*. Accidental

American Kestrel, *Falco sparverius*. UnRes; FoUD, RiDe

Merlin, *Falco columbarius*. RaRes (PR); FoUD

Gyrfalcon, *Falco rusticolus*. RaWM (SH); GrAll

Peregrine Falcon, *Falco peregrinus.* VRSuRes (Ea); HuTC
(CC, T2-NNLP-18)

Prairie Falcon, *Falco mexicanus.* OcSuRes (We); RoCl (T2-NNLP-18)

Order Passeriformes (Passerine Birds)

FAMILY TYRANNIDAE (TYRANT FLYCATCHERS)

Olive-sided Flycatcher, *Contopus cooperi.* UnMig; FoWC
(CC1, NT)

Western Wood-Pewee, *Contopus sordidulus.* CoSuRes (Pan);
FoWC, FoUD

Eastern Wood-Pewee, *Contopus virens.* CoSuRes (Ea); FoUD,
RiDe

Yellow-bellied Flycatcher, *Empidonax flaviventris.* UnMig
(Ea); RiDe

Acadian Flycatcher, *Empidonax virescens.* UnMig (Ea);
FoUD, RiWe (T2-NNLP-18)

Alder Flycatcher, *Empidonax alnorum.* UnMig; RiWe

Willow Flycatcher, *Empidonax traillii.* UnSuRes; RiWe (CC1)

Least Flycatcher, *Empidonax minimus.* CoMig; FoUD, HuPG

Hammond's Flycatcher, *Empidonax hammondii.* VRFaMig
(We); FoWC

Cordilleran Flycatcher, *Empidonax occidentalis.* LoSuRes
(NW); FoWC (T2-NNLP-18)

Gray Flycatcher, *Empidonax wrightii.* Accidental or
extremely rare

Dusky Flycatcher, *Empidonax oberholseri.* VRMig (We); WoUp

Eastern Phoebe, *Sayornis phoebe.* CoSuRes (Ea); FoUD,
RiDe

Say's Phoebe, *Sayornis saya.* CoSuRes (We); WoUp

Vermilion Flycatcher, *Pyrocephalus rubinus.* Accidental

Ash-throated Flycatcher, *Myiarchus cinerascens.* VRSuRes
(We); WoUp

Great Crested Flycatcher, *Myiarchus crinitus.* CoSuRes (Ea);
FoUD

Cassin's Kingbird, *Tyrannus vociferans.* OcSuRes (PR, WH); FoWC (T2-NNLP-18)

Western Kingbird, *Tyrannus verticalis.* CoSuRes; GrAll, HuPG, HuFR

Eastern Kingbird, *Tyrannus tyrannus.* CoSuRes; GrAll, RiDe, HuPG, HuFR

Scissor-tailed Flycatcher, *Tyrannus forficatus.* RaSuRes (Ea); HuFR (T2-NNLP-18)

FAMILY LANIIDAE (SHRIKES)

Loggerhead Shrike, *Lanius ludovicianus.* UnSuRes; GrAll (T1-NNLP-18)

Northern Shrike, *Lanius borealis.* UnWiMig; GrAll

FAMILY VIREONIDAE (VIREOS)

White-eyed Vireo, *Vireo griseus.* UnMig (Ea); FoUD

Bell's Vireo, *Vireo bellii.* UnSuRes, RiWe, SRUP (CC, CC1, NT)

Black-capped Vireo, *Vireo atricapilla.* Extirpated

Yellow-throated Vireo, *Vireo flavifrons.* UnSuRes (Ea); RiDe, FoUD (T2-NNLP-18)

Plumbeous Vireo, *Vireo plumbeus.* UnSuRes (PR); SRUP (T2-NNLP-18)

Cassin's Vireo, *Vireo cassinii.* UnFaMig (Pan); FoUD

Blue-headed Vireo, *Vireo solitarius.* UnMig (Ea); FoUD

Warbling Vireo, *Vireo gilvus.* CoSuRes; RiDe, FoUD

Philadelphia Vireo, *Vireo philadelphicus.* UnMig (Ea); RiDe

Red-eyed Vireo, *Vireo olivaceus.* CoSuRes; FoUD

FAMILY CORVIDAE (CROWS, JAYS, AND MAGPIES)

Canada Jay, *Perisoreus canadensis.* VRWiMig; FoWC

Pinyon Jay, *Gymnorhinus cyanocephalus.* LoRes (PR, WH); WoUp (CC1, Vul, T1-NNLP-18)

Steller's Jay, *Cyanocitta stelleri.* RaWiMig

Blue Jay, *Cyanocitta cristata.* CoRes

(Woodhouse's Scrub-Jay, *Aphelocoma woodhouseii*.) Hypothetical

Clark's Nutcracker, *Nucifraga columbiana*. VRRes (PR); FoWC

Black-billed Magpie, *Pica hudsonia*. CoRes (We); WoUp, SRUP (T1-NNLP-18)

American Crow, *Corvus brachyrhynchos*. CoRes; TeAll

Fish Crow, *Corvus ossifragus*. Accidental

Chihuahuan Raven, *Corvus cryptoleucus*. Extirpated since the mid-twentieth century

Common Raven, *Corvus corax*. Extirpated in the late 1800s; a few recent sight records

FAMILY ALAUDIDAE (LARKS)

Horned Lark, *Eremophila alpestris*. CoRes; GrAll

FAMILY HIRUNDINIDAE (SWALLOWS)

Bank Swallow, *Riparia riparia*. CoSuRes; RiHe, RiWe

Tree Swallow, *Tachycineta bicolor*. CoSuRes; HuFR, HuPG

Violet-green Swallow, *Tachycineta thalassina*. CoSuRes (PR, WH); RoCl, FoWC (T2-NNLP-18)

Northern Rough-winged Swallow, *Stelgidopteryx serripennis*. CoSuRes; RiHe, RiWe

Purple Martin, *Progne subis*. CoSuRes (Ea); HuTC

Barn Swallow, *Hirundo rustica*. CoSuRes; HuFR, HuPG, HuTC

Cliff Swallow, *Petrochelidon pyrrhonota*. CoSuRes; RoCl, RoUc, RiWe

Cave Swallow, *Petrochelidon fulva*. Accidental

FAMILY PARIDAE (CHICKADEES AND TITMICE)

(Carolina Chickadee, *Poecile carolinensis*.) Hypothetical

Black-capped Chickadee, *Poecile atricapillus*. CoRes; HuPG, HuTC, FoUD

Mountain Chickadee, *Poecile gambeli*. RaWiMig (We)

(Boreal Chickadee, *Poecile hudsonicus.*) Hypothetical

Tufted Titmouse, *Baeolophus bicolor.* CoRes (SE); RiDe (T2-NNLP-18)

FAMILY SITTIDAE (NUTHATCHES)

Red-breasted Nuthatch, *Sitta canadensis.* UnRes (We); (PR, WH); FoWC, HuPG

White-breasted Nuthatch, *Sitta carolinensis.* CoRes; FoUD, HuPG

Pygmy Nuthatch, *Sitta pygmaea.* UnRes (PR, WH); FoWC (T2-NNLP-18)

Brown-headed Nuthatch, *Sitta pusilla.* Accidental

FAMILY CERTHIIDAE (CREEPERS)

Brown Creeper, *Certhia americana.* UnRes; FoUD, RiDe (T2-NNLP-18)

FAMILY TROGLODYTIDAE (WRENS)

Rock Wren, *Salpinctes obsoletus.* CoSuRes (Pan, SW); RoCl, RoUc

Canyon Wren, *Catherpes mexicanus.* Accidental

House Wren, *Troglodytes aedon.* CoSuRes; RiDe, FoUD

Winter Wren, *Troglodytes troglodytes.* UnWiMig; RiSS, HuPG

Pacific Wren, *Troglodytes pacificus.* Accidental

Sedge Wren, *Cistothorus platensis.* CoSuRes (Ea); RiHe

Marsh Wren, *Cistothorus palustris.* CoSuRes; WeSW (T2-NNLP-18)

Carolina Wren, *Thryothorus ludovicianus.* UnRes (Ea); HuPG

Bewick's Wren, *Thryomanes bewickii.* VRSuRes (Ea); SRUP, WoUp

FAMILY POLIOPTILIDAE (GNATCATCHERS)

Blue-gray Gnatcatcher, *Polioptila caerulea.* CoSuRes (SE, We); RiDe, SRUP

FAMILY CINCLIDAE (DIPPERS)

American Dipper, *Cinclus mexicanus.* VRMig (We); RiLo

FAMILY REGULIDAE (KINGLETS)

Golden-crowned Kinglet, *Regulus satrapa.* UnMig; HuPG, FoUD

Ruby-crowned Kinglet, *Regulus calendula.* UnMig; HuPG, FoUD

FAMILY TURDIDAE (THRUSHES)

(Northern Wheatear, *Oenanthe oenanthe.*) Hypothetical

Eastern Bluebird, *Sialia sialis.* CoSuRes (Ea); HuFR, FoUD

Western Bluebird, *Sialia mexicana.* Accidental (We)

Mountain Bluebird, *Sialia currucoides.* CoSuRes (PR, WH); FoWC

Townsend's Solitaire, *Myadestes townsendi.* CoWiMig (We); WoJu (T2-NNLP-18)

Veery, *Catharus fuscescens.* UnMig; FoUD, HuPG

Gray-cheeked Thrush, *Catharus minimus.* CoMig; FoUD, HuPG

Swainson's Thrush, *Catharus ustulatus.* CoMig; FoUD, HuPG (CC1)

Hermit Thrush, *Catharus guttatus.* UnMig; FoUD, HuPG

Wood Thrush, *Hylocichla mustelina.* UnSuRes (Ea); FoUD, HuPG (T1-NNLP-18)

American Robin, *Turdus migratorius.* AbRes; FoUD, HuFR, HuPG, HuTC

Varied Thrush, *Ixoreus naevius.* RaMig (We); FoUD, HuPG

FAMILY MIMIDAE (THRASHERS AND MOCKINGBIRDS)

Gray Catbird, *Dumetella carolinensis.* CoSuRes; SRUP

Northern Mockingbird, *Mimus polyglottos.* OcSuRes (Ea); SRUP

Brown Thrasher, *Toxostoma rufum.* CoSuRes; SRUP

Curve-billed Thrasher, *Toxostoma curvirostre.* VRMig (We); SRUP

Sage Thrasher, *Oreoscoptes montanus.* UnSuRes (Pan); SaSG

FAMILY STURNIDAE (STARLINGS)

European Starling, *Sturnus vulgaris.* AbRes (self-introduced, first reported in 1939); HuTC

FAMILY BOMBYCILLIDAE (WAXWINGS)

Bohemian Waxwing, *Bombycilla garrulus.* RaWiMig; WoJu, HuPG

Cedar Waxwing, *Bombycilla cedrorum.* CoSuRes; WoJu, HuPG

FAMILY PTILOGONATIDAE (SILKY FLYCATCHERS)

Phainopepla, *Phainopepla nitens.* Accidental

FAMILY PASSERIDAE (OLD WORLD SPARROWS)

House Sparrow, *Passer domesticus.* AbRes (introduced in the late 1800s); HuTC, HuFR

Eurasian Tree Sparrow, *Passer montanus.* Accidental

FAMILY MOTACILLIDAE (PIPITS AND WAGTAILS)

American Pipit, *Anthus rubescens.* CoMig; WeSW, GrAll

Sprague's Pipit, *Anthus spragueii.* UnMig; RMig, GrTG (Vul, T1-NNLP-18)

FAMILY FRINGILLIDAE (BOREAL FINCHES)

Brambling, *Fringilla montifringilla.* Accidental

Evening Grosbeak, *Coccothraustes vespertinus.* RaWiMig; FoWC, HuPG

Gray-crowned Rosy-Finch, *Leucosticte tephrocotis.* RaWiMig (Pan); GrSG

(Black Rosy-Finch, *Leucosticte atrata.*) Hypothetical

Pine Grosbeak, *Pinicola enucleator.* RaWiMig; FoWC, HuPG

Purple Finch, *Haemorhous purpureus.* UnWiMig; HuPG

Cassin's Finch, *Haemorhous cassinii.* RaWiMig (Pan); FoWC, HuPG

House Finch, *Haemorhous mexicanus.* AbRes; HuPG, HuTC

Common Redpoll, *Acanthis flammeus.* UnWiMig; FoWC, HuPG

Hoary Redpoll, *Acanthis hornemanni.* RaWiMig; FoWC, HuPG

Red Crossbill, *Loxia curvirostra.* UnRes (PR, WH); FoWC

White-winged Crossbill, *Loxia leucoptera.* RaWiMig; FoWC

Pine Siskin, *Spinis pinus.* UnRes (PR, WH); FoWC, HuPG
(T2-NNLP-18)

Lesser Goldfinch, *Spinis psaltria.* LoSuRes (We); SRUP

American Goldfinch, *Spinis tristis.* CoRes; SRUP, HuPG

FAMILY CALCARIIDAE (LONGSPURS)

Lapland Longspur, *Calcarius lapponicus.* CoWiMig; GrAll

Chestnut-collared Longspur, *Calcarius ornatus.* CoSuRes
(Pan); GrSG (NT, T1-NNLP-18)

Smith's Longspur, *Calcarius pictus.* RaWiMig; GrAll

McCown's Longspur, *Rhynchophanes mccownii.* UnSuRes
(Pan); GrSG (CC, CC1, T1-NNLP-18)

Snow Bunting, *Plectrophenax nivalis.* OcWiMig; GrAll

FAMILY PASSERELLIDAE (NEW WORLD SPARROWS AND TOWHEES)

Cassin's Sparrow, *Peucaea cassinii.* RaSuRes (Pan); SaSG,
GrSG (T2-NNLP-18)

Grasshopper Sparrow, *Ammodramus savannarum.* CoSuRes;
GrAll

Black-throated Sparrow, *Amphispiza bilineata.* VRMig; GrAll

Lark Sparrow, *Chondestes grammacus.* CoSuRes; GrAll, SRUP

Lark Bunting, *Calamospiza melanocorys.* CoSuRes (We);
GrSG, GrSH

Chipping Sparrow, *Spizella passerina.* CoSuRes; FoWC,
HuPG

Clay-colored Sparrow, *Spizella pallida.* CoMig; GrMG, GrSG

Field Sparrow, *Spizella pusilla.* CoSuRes; SRUP, GrMG

Brewer's Sparrow, *Spizella breweri.* UnSuRes (Pan); SaSG
(CC, CC1, T1-NNLP-18)

Fox Sparrow, *Passerella iliaca.* UnMig; RiSS

American Tree Sparrow, *Spizelloides arborea.* CoWiMig; GrAll, HuFR

Dark-eyed Junco, *Junco hyemalis.* CoRes (PR); CoWM (Pan, Cen, Ea), SRUP, GrAll, HuPG (includes *J. h. aikeni*, T2-NNLP-18)

White-crowned Sparrow, *Zonotrichia leucophrys.* CoWiMig; HuFR, HuPG, SRUP

Golden-crowned Sparrow, *Zonotrichia atricapilla.* VRMig; SRUP

Harris's Sparrow, *Zonotrichia querula.* CoWM; HuFR, HuPG, SRUP (CC1)

White-throated Sparrow, *Zonotrichia albicollis.* CoWiMig; HuPG, SRUP

Sagebrush Sparrow, *Artemisiospiza nevadensis.* Accidental

Vesper Sparrow, *Pooecetes gramineus.* CoSuRes; GrAll

LeConte's Sparrow, *Ammospiza leconteii.* UnMig; WeSW

Nelson's Sparrow, *Ammospiza nelsoni.* RaMig; WeSW

Baird's Sparrow, *Centronyx bairdii.* UnMig; GrMG (CC, T1-NNLP-18)

Henslow's Sparrow, *Centronyx henslowii.* UnSuRes (SE); GrTG (CC, CC1, NT, T1-NNLP-18)

Savannah Sparrow, *Passerculus sandwichensis.* CoSuRes; WeSW (T2-NNLP-18)

Song Sparrow, *Melospiza melodia.* CoRes; RiWe, WeSW

Lincoln's Sparrow, *Melospiza lincolnii.* CoMig; RiSS, RiHe

Swamp Sparrow, *Melospiza georgiana.* UnSuRes; WeSW (T2-NNLP-18)

Green-tailed Towhee, *Pipilo chlorurus.* UnMig (We); SaSG, SRUP

Spotted Towhee, *Pipilo maculatus.* CoSuRes (We); WoUp, SRUP

Eastern Towhee, *Pipilo erythropthalmus.* CoSuRes (Ea); WoUp, SRUP

(Canyon Towhee, *Melazone fusca.*) Hypothetical

Yellow-breasted Chat, *Icteria virens.* UnSuRes (We); RiSS

FAMILY ICTERIDAE (ORIOLES, BLACKBIRDS, AND COWBIRDS)

Yellow-headed Blackbird, *Xanthocephalus xanthocephalus.*
CoSuRes (SH, RB); WeSW

Bobolink, *Dolichonyx oryzivorus.* CoSuRes; WeSW

Eastern Meadowlark, *Sturnella magna.* CoSuRes (Ea, SH);
GrTG, GrSH (T2-NNLP-18)

Western Meadowlark, *Sturnella neglecta.* CoSuRes; GrAll

Orchard Oriole, *Icterus spurius.* CoSuRes; SRUP, RiDe,
HuPG, HuFR

Hooded Oriole, *Icterus cucullatus.* Accidental

Bullock's Oriole, *Icterus bullockii.* CoSuRes (Pan); RiDe,
HuFR, FoWC (T2-NNLP-18)

Baltimore Oriole, *Icterus galbula.* CoSuRes (Ea, Cen);
FoUD, RiDe, HuFR, FoWC

Scott's Oriole, *Icterus parisorum.* Accidental

Red-winged Blackbird, *Agelaius phoeniceus.* CoSuRes; WeSW,
RiSS, RiHe, SRUP

Brown-headed Cowbird, *Molothrus ater.* AbSuRes; GrAll,
SRUP, HuFR, HuPG

Rusty Blackbird, *Euphagus carolinus.* UnMig (We, SW)
(CC1, Vul)

Brewer's Blackbird, *Euphagus cyanocephalus.* CoMig (We);
SRUP, GrAll

Common Grackle, *Quiscalus quiscula.* CoSuRes; HuFR,
HuPG, HuTC

Great-tailed Grackle, *Quiscalus mexicanus.* UnSuRes; WeSW,
HuFR, HuPG, HuTC

FAMILY PARULIDAE (NEW WORLD WARBLERS)

Ovenbird, *Seiurus aurocapilla.* UnSuRes (Ea); RiDe, FoUD

Worm-eating Warbler, *Helmitheros vermivorum.* VRMig (Ea);
RiDe, WoUp

Louisiana Waterthrush, *Parkesia motacilla.* LoSuRes (Ea);
RiDe (T2-NNLP-18)

Northern Waterthrush, *Parkesia noveboracensis.* UnMig;
RiDe

Golden-winged Warbler, *Vermivora chrysoptera.* VRMig (Ea);
WoUp

Blue-winged Warbler, *Vermivora cyanoptera.* RaMig (Ea);
WoUp (CC1)

Black-and-white Warbler, *Mniotilta varia.* UnSuRes; RiDe,
FoWC (T2-NNLP-18)

Prothonotary Warbler, *Protonotaria citrea.* LoSuRes (SE);
RiDe, WeSW (CC1, T2-NNLP-18)

Swainson's Warbler, *Limnothlypis swainsonii.* Accidental

Tennessee Warbler, *Leiothlypis peregrina.* CoMig; FoUD,
SRUP

Orange-crowned Warbler, *Leiothlypis celata.* CoMig; FoUD,
WoUp

Nashville Warbler, *Leiothlypis ruficapilla.* CoMig; WoUp

Virginia's Warbler, *Leiothlypis virginiae.* VRMig (We); SRUP

Connecticut Warbler, *Oporornis agilis.* UnMig; WoUp

MacGillivray's Warbler, *Geothlypis tolmiei.* RaMig (We); RiSS

Mourning Warbler, *Geothlypis philadelphia.* UnMig (Ea);
RiSS

Kentucky Warbler, *Geothlypis formosa.* LoSuRes (Ea); WoUp
(CC1, T2-NNLP-18)

Common Yellowthroat, *Geothlypis trichas.* CoSuRes; WeSW

Hooded Warbler, *Setophaga citrina.* RaMig (Ea); RiDe

American Redstart, *Setophaga ruticilla.* CoSuRes (Ea); FoUD

Cape May Warbler, *Setophaga tigrina.* UnMig (Ea); FoUD,
FoWC, HuPG

Cerulean Warbler, *Setophaga cerulea.* LoSuRes (SE); FoUD,
RiDe (CC, Vul)

Northern Parula, *Setophaga americana.* LoSuRes (Ea);
WeSW

Magnolia Warbler, *Setophaga magnolia.* UnMig (Ea); FoUD, FoWC

Bay-breasted Warbler, *Setophaga castanea.* UnMig (Ea); FoUD

Blackburnian Warbler, *Setophaga fusca.* UnMig (Ea); FoUD, FoWC

Yellow Warbler, *Setophaga petechia.* CoSuRes; RiSS, WeSW

Chestnut-sided Warbler, *Setophaga pensylvanica.* UnMig (Ea); SRUP

Blackpoll Warbler, *Setophaga striata.* CoMig (Ea); RiDe

Black-throated Blue Warbler, *Setophaga caerulescens.* RaMig (Ea); WoUp, HuPG

Palm Warbler, *Setophaga palmarum.* UnMig (Ea); SRUP

Pine Warbler, *Setophaga pinus.* VRMig (Ea); RMig

Yellow-rumped Warbler, *Setophaga coronata.* CoSuRes (PR, Pan); FoWC

Yellow-throated Warbler, *Setophaga dominica.* SuRes (SE); FoUD (T2-NNLP-18)

Prairie Warbler, *Setophaga discolor.* VRMig (Ea); SRUP

Grace's Warbler, *Setophaga graciae.* Accidental

Black-throated Gray Warbler, *Setophaga nigrescens.* VRMig (We); SRUP

Townsend's Warbler, *Setophaga townsendi.* VRMig (We); WoUp

Hermit Warbler, *Setophaga occidentalis.* Accidental

Black-throated Green Warbler, *Setophaga virens.* UnMig (Ea); WoUp

Canada Warbler, *Cardillina canadensis.* OcMig (Ea); RiSS (CC1)

Wilson's Warbler, *Cardillina pusilla.* CoMig; RiSS

FAMILY CARDINALIDAE
(CARDINALS, TANAGERS, AND GROSBEAKS)

Hepatic Tanager, *Piranga flava.* Accidental

Summer Tanager, *Piranga rubra.* UnSuRes (Ea); FoUD
(T2-NNLP-18)

Scarlet Tanager, *Piranga olivacea.* UnSuRes (Ea); FoUD

Western Tanager, *Piranga ludoviciana.* UnSuRes (PR);
FoWC

Northern Cardinal, *Cardinalis cardinalis.* CoRes (Ea);
HuPG, HuTC

Rose-breasted Grosbeak, *Pheucticus ludovicianus.* CoSuRes
(Ea); FoUD, RiDe

Black-headed Grosbeak, *Pheucticus melanocephalus.* CoSuRes
(Pan, SW); FoWC, FoUD, RiDe

Blue Grosbeak, *Passerina caerulea.* UnSuRes; GrAll, SRUP

Lazuli Bunting, *Passerina amoena.* CoSuRes (PR, WH);
SRUP

Indigo Bunting, *Passerina cyanea.* CoSuRes (Ea); FoUD

Painted Bunting, *Passerina ciris.* Accidental

Dickcissel, *Spiza americana.* CoSuRes; GrTG (CC1)

The taxonomic nomenclature of Ballinger, Lynch, and Smith (2010) is used here, but English vernacular names follow those of Powell, Conant, and Collins (2016) and the 2010 species list of the Society for the Study of Amphibians and Reptiles. Some obsolete or unsettled Latin and English names that have appeared in the literature of Nebraska herpetiles are shown parenthetically. Within families, the sequence of taxa is alphabetic by genera, and (with one exception) within genera the sequence of species is also alphabetic. The abbreviated status code sequence (see the list of abbreviations in the front matter) following each species' name is Range (in-state or county distribution), Abundance; Habitats (conservation status). "State-protected" species have specific possession limits or state prohibitions on collecting, possessing, or selling these species. Counties mentioned as part of a species' range are those having county specimen records (Fogell 2010; Ballinger, Lynch, and Smith 2010) but do not necessarily indicate its current presence. Species with associated narrative profiles are *bold-italic*.

Amphibians (Class Amphibia)

FAMILY AMBYSTOMATIDAE (SALAMANDERS AND NEWTS)

Barred Tiger Salamander, *Ambystoma mavortium*. Ub, Co; GrAll, WeSW, WeDW

Small-mouthed Salamander, *Ambystoma texanum*. Lo (SE), Un; GrAll, WeSW, WeDW (CC, state protected, T2-NNLP-18)

Eastern Tiger Salamander, *Ambystoma tigrinum*. State abundance and distribution uncertain; GrAll, WeSW, WeDW (CC)

FAMILY PELOBATIDAE (SPADEFOOTS)

Plains Spadefoot, *Spea bombifrons*. Wi, Co; GrSH

FAMILY BUFONIDAE (TOADS)

American Toad, *Anaxyrus (Bufo) americanus*. Lo (Washington, Douglas, and Sarpy Counties), Un; GrAll (CC, state protected, T2-NNLP-18)

Great Plains Toad, *Anaxyrus (Bufo) cognatus.* Wi, Co; GrAll

Woodhouse's Toad, ***Anaxyrus (Bufo) woodhousii.*** Ub, Co; GrAll

FAMILY MICROHYLIIDAE (NARROWMOUTH FROGS)

Great Plains Narrowmouth Frog, *Gastrophryne olivacea.* HL (Webster and Gage Counties), Un; RoUc (state protected, T2-NNLP-18)

FAMILY RANIDAE (FROGS)

American Bullfrog, ***Lithobates (Rana) catesbeiana.*** Ub, Co; WeSW

Plains Leopard Frog, ***Lithobates (Rana) blairi.*** Ub, Co; WeSW

Northern Leopard Frog, ***Lithobates (Rana) pipiens.*** Wi (No, Cen), Co; WeSW (T2-NNLP-18)

FAMILY HYLIDAE (CHORUS FROGS)

Blanchard's (Northern) Cricket Frog, *Acris crepitans (blanchardi).* Wi (Ea, Cen), Co; FoDe, WeSW (T2-NNLP-18)

Cope's Gray Treefrog, ***Hyla chrysoscelis.*** SD (Ea, Cen), Co; FoDe, WeSW

Boreal (Western) Chorus Frog, *Pseudacris triseriata.* Ub, Co; WeSW, TeAll

FAMILY CHELYDRIDAE (SNAPPING TURTLES)

Eastern Snapping Turtle, ***Chelydra serpentina.*** Wi, Co; RiWe

FAMILY EMYDIDAE (POND AND BOX TURTLES)

Western Painted Turtle, ***Chrysemys picta.*** Wi, Co; RiWe

Blanding's Turtle, *Emydoidea blandingii.* SD (NC), Un; WeSW, GrSH (CC, state protected, T1-NNLP-18)

False Map Turtle, *Graptemys pseudogeographica.* Lo (NE, Ea, SE), Un; RiLo, WeSW, WeDW (state protected)

Ornate Box Turtle, ***Terrapene ornata.*** Lo (SH), Co; GrSH

Pond Slider, *Trachemys scripta.* HL (Buffalo and Richardson Counties), Un; WeSW (T2-NNLP-18)

Yellow Mud Turtle, *Kinosternon flavescens*. SD (NC, SW, SC), Un; RiWe

FAMILY TRIONYCHIDAE (SOFTSHELL TURTLES)

Midland Smooth Softshell, *Apalone mutica*. SD (NC, EA, SC), Un; RiWe, WeDW, WeSW (T2-NNLP-18)

Eastern Spiny Softshell Turtle, *Apalone spinifera*. Wi, Un; RiWe, WeDW, WeSW

Lizards (Class Sauria)

FAMILY ANGUIDAE (GLASS LIZARDS)

Western Slender Glass Lizard, *Ophisaurus attenuatus*. HL (Franklin and Johnson Counties), VR; GrSG, SanGr (state protected, T2-NNLP-18)

FAMILY SCINCIDAE (SKINKS)

Six-lined Racerunner, *Cnemidophorus sexlineata*. Wi (most common in SH); GrSH, GrSG, SanGr

Common Five-lined Skink, *Plestiodon fasciatus*. HL (Lancaster and Richardson Counties), Ra; WoUP, RoUc, SanGr (state protected)

Many-lined Skink, *Eumeces (Plestiodon) multivirgatus*. Di (Pan, SH, NE), Co; GrSG, GrMG, SanGr

Great Plains Skink, *Plestiodon obsoletus*. Lo (SW, SC, Morrill County), Ra; GrTG, RoUc, SanGr (CC, state protected)

Northern Prairie Skink, *Plestiodon septentrionalis*. Di (Cen, Ea), Un; RoUc, GrTG, SanGr

FAMILY PHRYNOSOMATIDAE

(HORNED, EARLESS, AND SPINY LIZARDS)

Common Lesser Earless Lizard, *Holbrookia maculata*. Wi, Co; GrSH, SanGr

Greater Short-horned Lizard, *Phrynosoma hernandesi*. Lo (NW, SW, Pan), Un; GrSG, SanGr (state protected, T2-NNLP-18)

Prairie Lizard, *Sceloporus (undulatus) consobrinus.* Wi (Pan, Cen, NE), Co; GrSH, RoUc, SanGr

Common Sagebrush Lizard, *Sceloporus graciosus.* HL (Morrill County), Ra; SaSG, SanGr (state protected, T1-NNLP-18)

Snakes (Class Serpentes)

FAMILY COLUBRIDAE (HARMLESS EGG-LAYING SNAKES)

Glossy Snake, *Arizona elegans.* Lo (Dundy, Hitchcock, and Thomas Counties), Ra; GrMG, GrSG (CC, state protected, T1-NNLP-18)

North American Racer, *Coluber constrictor.* Wi, Co; TeAll

Yellow-bellied Kingsnake, *Lampropeltis calligaster.* Lo (SE), Un; GrTG (state protected, T2-NNLP-18)

Speckled Kingsnake, *Lampropeltis holbrooki.* Lo (Morrill County, SE), Un; WeSW (state protected, T2-NNLP-18)

Eastern Milksnake, *Lampropeltis triangulum.* Wi, Un; RoBL, WoUp

Coachwhip, *Masticophis (Coluber) flagellum.* Lo (Morrill, Chase, Dundy, and Red Willow Counties), Un; GrAll

Smooth Greensnake, *Opheodrys (Liochlorophis) vernalis.* Di (Cen, SC, Ea, NE), Un; GrAll (state protected, T2-NNLP-18)

Great Plains Ratsnake, *Pantherophis emoryi.* Lo (Harlan to Richardson County), Un; GrAll, RoBL

Western Ratsnake, *Pantherophis (Scotophis) obsoletus.* Lo (SE: Thayer and Washington to Richardson County), Un; GrAll, RoBL

Western Foxsnake, *Pantherophis (Mintonicus) vulpinus (ramspotti).* SD (NE, EA), Un; TeAll, WeSW (T2-NNLP-18)

Gophersnake (Bullsnake), *Pituophis catenifer.* Wi (We, SC, Ea), Co; GrAll, RoUc

Plains Blackhead Snake, *Tantilla nigriceps.* Lo (SW, seven counties), Un; GrSG (state protected, T2-NNLP-18)

Northern Watersnake, *Nerodia sipedon*. Wi, Un; AqAll

Graham's Crayfish Snake, *Regina grahamii*. Lo (Douglas, Sarpy, Lancaster, Jefferson, Nemaha, and Richardson Counties), Un; AqAll (state protected, T2-NNLP-18)

Dekay's Brownsnake, *Storeria dekayi*. SD (NE, Ea, SC), Un; RiDe, RiSS, WeSW (T2-NNLP-18)

Black Hills Red-bellied Snake, *Storeria occipitomaculata*. Lo (Dawson, Buffalo, Hall, and Phelps Counties), Un; RiDe, RiSS, WeSW (state protected, T1-NNLP-18)

Wandering Gartersnake, *Thamnophis elegans*. HL (Sioux County), Ra; WeSW (state protected, T2-NNLP-18)

Western Ribbonsnake, *Thamnophis proximus*. Lo (Saline, Lancaster, Sarpy, Cass, and Richardson Counties), Un; WeSW (state protected, T2-NNLP-18)

Plains Gartersnake, *Thamnophis radix*. Ub, Co; WeSW

Common Gartersnake, *Thamnophis sirtalis*. Ub, Co; TeAll

Lined Snake, *Tropidiclonion lineatum*. SD (SC, SE), Un; GrTG, GrMG

Western Wormsnake, *Carphosis vermis*. Lo (Sarpy, Cass, Otoe, Pawnee, and Richardson Counties), Un; WoUp (state protected, T2-NNLP-18)

Ring-necked Snake, *Diadophis punctatus*. SD (NC, SC, SE, Ea, NE), Un; WoUp, RoUc

Plains Hog-nosed Snake, *Heterodon nasicus*. Wi (We, Cen), Un; SanGr

Eastern Hog-nosed Snake, *Heterodon platirhinos*. Di (NC, NE, Ea, SE, SW, SC), Un; SanGr (T2-NNLP-18)

Copperhead, *Agkistrodon contortrix*. HL (Gage and Richardson Counties), Oc; WoUp, RoUc (state protected, T2-NNLP-18)

Timber Rattlesnake, *Crotalus horridus.* Lo (Cass, Jefferson, Gage, Pawnee, and Richardson Counties), Ra; FoDe, WoUp (CC, state protected, T1-NNLP-18)

Prairie Rattlesnake, *Crotalus viridis.* Wi (Pan, NC, NE, SW, SC), Un; RoUc, RoBL, GrSG, GrMG

Massasauga, *Sistrurus catenatus.* Lo (EC, SE), Ra; WeSW, GrTG (CC, T1-NNLP-18)

CHAPTER 6

Some Natural Treasures of Nebraska

AGATE FOSSIL BEDS NATIONAL MONUMENT. Area, 11,617 acres. Located in Sioux County, 22 miles south of Harrison, or 34 miles north of Mitchell on State Rd. 29. This famous fossil site preserves vast numbers of mostly early Miocene mammals from about 22 to 20 million years ago (Hunt, Skolnick, and Kaufman 2018; Hunt 1981). The site was first excavated in 1904 and has produced hundreds of museum specimens. They include such remarkable ancestral forms as *Miohippus* (Miocene horse), *Monoceras* (Miocene rhino), *Stenomylus* (Miocene camel), and *Paleocaster* (Oligocene-Miocene land beaver), as well as now long-extinct mammalian lines, including large predators such as the bear-dog *Amphicyon* (*Amphicyonidae*) (Hunt, Skolnick, and Kaufman 2018). There was also *Daeodon* (= *Dinohyas*) (a large pig-like omnivore, Antelodontidae) and *Moropus* (a clawed horse-like herbivore, Chalcotheridae). Graetz, Garrott, and Craven (1995) published a general faunal survey, including 124 bird species. Dibner et al. (2018) summarized the site's physical characteristics and ecology and provided lists of some of the area's vascular plants and fish. Prairie rattlesnakes are common during summer along the sandy trail to the excavation sites, and carrying a supply of water during hot weather is strongly advised. No admission fee; the site is largely undeveloped. The visitor center has reconstructed skeletons and other fossil exhibits of *Moropus*, *Dinohyas*, and other mammals found at Agate, as well as some Lakota artifacts. Address: 301 River Rd., Harrison, Nebraska 69346 (ph. 308-665-4113).

ASHFALL FOSSIL BEDS STATE HISTORIC PARK. Area, 260 acres. Located in Antelope County, 2 miles west and 6 miles north of Royal. This world-famous fossil site includes the exposed in situ fossil remains from dozens of mid-Miocene mammals, which died after a fallout of volcanic ash that originated from eruptions in the Yellowstone region about 12–10 million years ago (Voorhies 1981; Johnsgard 2014b). There are also one- and three-toed horses (five genera), camels (three genera), canids (three genera), saber-toothed deer (one genus), and a few birds, including a crane that is anatomically similar to modern African crowned cranes (*Balearica*). A large "rhino barn" allows for easy viewing of many partially excavated rhinos and camels, as well as painted reconstructions of many of the regional fossils. There is a visitor center with local fossil exhibits and a small bookstore. Operated by the University of Nebraska and the Nebraska Game and Parks Commission. Open daily, Memorial Day to Labor Day; also limited hours in May, September, and October. Admission fee. Address: 86930 517th Ave., Royal, Nebraska 68773 (ph. 402-893-2000).

BOYER CHUTE NATIONAL WILDLIFE REFUGE. Area, ca. 2,000 acres. This relatively new refuge is located about 5 miles east of Fort Calhoun beside the Missouri River. This refuge includes a reconstructed side channel (chute) on the west side of the Missouri River. The refuge is still under development, with additional acres planned for acquisition. There is no local bird checklist, but the list for nearby DeSoto Bend (see below) is applicable. Address: c/o DeSoto National Wildlife Refuge, 1434 316th Ln., Missouri Valley, Iowa 51255 (ph. 712-388-4800).

CENTRAL PLATTE RIVER VALLEY. Although the Platte River's headwaters are in Colorado and Wyoming, a 70-mile stretch of the river, extending from about Lexington to Grand Island, is the center of the Central Flyway. This relatively short stretch (the "Big Bend") of the Platte River in central Nebraska, centered roughly between Lexington and Grand Island, hosts one of the world's great bird spectacles in early spring, when millions of waterfowl and a half-million sandhill cranes descend into the Platte Valley and the adjoining Rainwater Basin immediately to the south (Johnsgard 2013b; see account below). The migration includes nearly 30 species of waterfowl, both species of American cranes, and up to two dozen species of shorebirds (Johnsgard 1979a, 2011b, 2012a, 2012b). Estimates of waterfowl numbers vary greatly and depend on water conditions, but it is commonly estimated that up to ten million birds, including up to an estimated maximum of seven million snow geese, can be here from mid-February to mid-March. There are also up to hundreds of thousands each of Canada geese and greater white-fronted geese. There may be as many as 100,000 cackling geese and 10,000–50,000 Ross's geese, although these species' numbers are still poorly documented because of field identification difficulties. Ducks, mallards, and northern pintails are the most common and, along with common mergansers and common goldeneyes, are the earliest to arrive. By the end of March, the last of a dozen or more duck species will have arrived, and the Arctic-bound geese and sandhill cranes are leaving. By mid-April the sandhill cranes (mostly tundra-breeding lesser sandhills headed for Alaska and Siberia) will have been replaced by family-sized groups of whooping cranes and early shorebird migrants. The approximately two dozen species of shorebirds usually peak by early to mid-May. Much of this section of the Platte River is under protection as a result of efforts by the Nature Conservancy, the Platte River Recovery Implementation Program, the National Audubon Society, the Crane Trust, and the Nebraska Game and Parks Commission. I (Johnsgard 2007a) provided a natural history guide to the region, including species lists of all the major plant and animal taxa. Brown and Johnsgard (2013) documented the nearly 400 species of birds of this 11-county, 10,000-square-mile region.

CRESCENT LAKE NATIONAL WILDLIFE REFUGE. Area, 45,818 acres. Located in Garden County, 28 miles north of Oshkosh (or 20 miles south of Lakeside, Sheridan County), and reached only over mostly sandy and gravel roads. There are about 20 wetland complexes on this enormous Nebraska Sandhills refuge; these total 8,251 acres and comprise almost 20 percent of the refuge. Crescent Lake National Wildlife Refuge is in the central Nebraska Sandhills region, which extends over about 19,000 square miles and is the largest region of (now mostly stabilized) dunes in North America (Johnsgard 1995). At least 32 species of waterfowl have been reported at the refuge, and 14 are known or suspected breeders. Three grebes (western, eared, and pied-billed) are also breeders. Other wetland nesters include the double-crested cormorant, great blue heron, black-crowned night-heron, American bittern, sora, and Virginia rail. The Forster's tern, white-faced ibis, and black-necked stilt breed irregularly. The marshes and shallow lakes in this large and remote Sandhills refuge vary greatly as to their relative alkalinity. At the western edge of the refuge Border Lake marks the eastern boundary of a three-county (Garden, Morrill, and Sheridan) regional area of hypersaline water conditions; the Wilson's phalarope and American avocet are common breeders here. The refuge headquarters has the only source of drinking water and public-access facilities on the refuge. To avoid becoming mired, be careful not to park on bare sand; the nearest gas and car assistance is more than 20 miles away. Gunderson (1973) summarized the refuge's mammals. The refuge bird list includes 273 species and is available from the refuge manager. Free admission. Address: 10630 Rd. 181, Ellsworth, Nebraska 69340 (ph. 308-762-4893). The refuge's website is https://crescentlake.fws.gov/.

DESOTO NATIONAL WILDLIFE REFUGE. Area, 7,823 acres. This federal refuge is mostly located in Iowa but includes some Nebraska acreage owing to river channel movements. The refuge can be reached by driving 5 miles east of Blair, Nebraska. It includes a 750-acre oxbow lake of the Missouri River and encompasses floodplain lands, including river-bottom forest and adjacent grasslands, marshes, and cultivated lands. A visitor center contains panoramic viewing windows and a collection of artifacts recovered from the steamboat *Bertrand*, which sunk when it hit a snag in 1865 DeSoto

National Wildlife Refuge has extensive riparian and bottomland woods that make for fine spring birding; 21 species of warblers are on the spring list, and at least four (yellow, black-and-white, American redstart, and common yellowthroat) have been reported as nesting. This refuge has historically been a major spring and fall staging area for snow geese and other migratory waterfowl. The transient snow goose population sometimes reaches nearly a million birds during spring migration, and other very abundant spring species are the mallard, ring-necked pheasant, mourning dove, and red-winged blackbird. A checklist of 240 bird species is available from the refuge manager, RR 1-B, Missouri Valley, Iowa 51555 (ph. 712-647-4121). The refuge charges a small daily admission fee. Website: https://www.fws.gov/refuge/desoto/.

FONTENELLE FOREST PRESERVE. Area, ca. 1,300 acres. This privately owned large area of mature riverine hardwood forest includes 17 miles of footpaths, a mile-long boardwalk, a combined nature center and museum, and a raptor recovery center that treats over 500 birds annually. An observation blind overlooks a small marsh, and there are organized bird or nature hikes plus many other programs. Summer species of special interest as known or possible breeders include American woodcock, broad-winged and red-shouldered hawks, whip-poor-will, Acadian flycatcher, Carolina wren, yellow-throated vireo, wood thrush, brown creeper, American redstart, parula, prothonotary and Kentucky warblers, and scarlet and summer tanagers. The yellow-throated warbler is at the north edge of its normal breeding range here, and the rare pileated woodpecker regularly breeds. There is a bird checklist of 246 species that have been reported in the past decade, and more than 100 of these are species that potentially breed. Barth and Ratzlaff (2004, 2007) authored and photographically illustrated two field guides to the herbaceous and woody plants of Fontenelle Forest and Neale Woods, a 562-acre forest and prairie preserve in northern Omaha that is also owned and managed by the Fontenelle Forest Nature Association. The preserve and Neale Woods are open 8:00 a.m. to

5:00 p.m. daily; there is an admission fee. The preserve's address is 1111 N. Bellevue Blvd., Omaha, Nebraska 68005 (ph. 402-731-3140).

FORT NIOBRARA NATIONAL WILDLIFE REFUGE. Area, 19,122 acres. Includes about 4,350 acres of mostly riparian woods and 375 acres of wetlands. It is mostly riparian hardwood forest along the Niobrara River and upland sandhills prairie, with some spring-fed ponds. Notable wetland breeding species include the wood duck, upland sandpiper, and long-billed curlew. The most abundant Neotropical migrants nesting in the refuge area are the common yellowthroat, ovenbird, black-and-white warbler, and red-eyed vireo. Located about 5 miles east of Valentine along Nebraska Hwy 12 (ph. 402-376-3789). The most recent refuge bird list includes 230 species, many of which are riparian woodland species with primarily eastern zoogeographic affinities, but some western species also occur, along with hybrids between eastern and western counterpart species (Johnsgard 2007b). There is also a prairie dog colony and a managed population of bison, and elk are regularly present during winter. Bogan and Ramotnik (1995) documented the refuge's mammals. A bird list is available from the refuge manager, Hidden Timber Rte., HC 14, Box 67, Valentine, Nebraska 69201 (ph. 402-376-3789). No admission fee. The refuge website is https://fortniobrara.fws.gov/5.

INDIAN CAVE STATE PARK. Area, 2,381 acres. This is a mostly wooded riverine park about 10 miles east of Shubert on Nebraska Hwy 64E. It consists of mature riverine red oak–basswood–hickory forest, with a good diversity of southern-oriented trees and several breeding species of otherwise rare Nebraska birds, such as the whip-poor-will, chuck-will's-widow, pileated woodpecker, Acadian flycatcher, and Louisiana waterthrush. A shallow Paleozoic limestone cave of Pennsylvanian age contains a few Native American petroglyphs, which are almost hidden among modern graffiti. There are 20 miles of hiking trails. Camping is permitted. A state park entry permit is required. Address: 65296 720 Rd., Shubert, Nebraska 68437 (ph. 402-8834-2575).

LAKE MCCONAUGHY STATE RECREATION AREA. Area, 6,492 acres. The state recreation area occupies much of the north side of this reservoir impounded by Kingsley Dam. The reservoir, the largest body of water in Nebraska, attracts vast numbers of migrant grebes (up to about 40,000 western grebes), waterfowl, gulls, and shorebirds. Hundreds of bald eagles also occur in winter, attracted by dead fish and the wintering duck and goose populations. Extensive marshy wetlands exist at **Clear Creek Wildlife Management Area**, at the lake's western end, with rails, herons, and least bitterns often observed. Over 100 miles of shoreline are present along the lake, with the southern shoreline mostly rocky and steep and the northern shore gradually sloping and sandy. The sandy shorelines support rare shorebirds such as nesting piping plovers and least terns, and a few snowy plovers have also nested in recent years. Both eastern and western wood-pewees reportedly occur, as do east–west species pairs of towhees, orioles, grosbeaks, and buntings. Kingsley Dam offers a good vantage point for birds along both the deep eastern end of Lake McConaughy and the shallower and much smaller Lake Ogallala, located at the eastern base of Kingsley Dam. Lake Ogallala (part of **Lake Ogallala State Recreation Area**) receives the spillway water from Lake McConaughy, and its level fluctuates greatly and frequently, depending on irrigation-based diversion needs. However, it is very attractive to migrant ducks, ospreys, gulls, Caspian terns, cliff swallows, white pelicans, double-crested cormorants, and other bird species, some of which breed locally. The shoreline of Lake Ogallala has some deciduous wooded habitats with a rich array of nesting passerines, but frequent lake fluctuations limit nesting for aquatic-nesting and shoreline species. Lake Ogallala State Recreation Area is classified as a Nebraska Important Bird Area. Some of the many rare gulls that have been seen here repeatedly are the mew, Iceland (including "Thayer's"), glaucous, and lesser black-backed. Rare wintering or migrant waterfowl include the trumpeter swan, greater scaup, all three scoters, long-tailed duck, and Barrow's goldeneye. An eagle-watching cabin at the eastern base of Kingsley Dam is freely available during winter, when 200–300 bald eagles are usually present. The cabin is open from late December through early March, Thursdays and Fridays, 8:00 a.m. to noon, Saturdays and Sundays, 8:00 a.m. to 4:00 p.m. A recent bird checklist for the Lake McConaughy and adjacent North Platte Valley region has 362 species (Brown, Dinsmore, and Brown 2012). The University of Nebraska's Cedar Point Biological Station (CPBS) (nearly 400 acres) is located along the south shoreline of Lake Ogallala and has an active summer research and teaching program, but it is not open to the general public. Information on CPBS classes and representative faunal and floral lists can be found at https://cedarpoint.unl.edu/. For information on Lake McConaughy State Recreation Area, see its website: https://www.lakemcconaughy.com/ngp.html (ph. 866-386-2862). A state park entry permit is required for the state recreation areas.

LILLIAN ANNETTE ROWE SANCTUARY AND IAIN NICOLSON AUDUBON CENTER. Area, ca. 2,400 acres. Located about 4 miles southwest of Gibbon on Elm Island Road (drive south 2.1 miles off I-80 Exit 285, then turn west on Elm Island Road and go 2.1 miles). The sanctuary is situated along the Platte River in the center of the sandhill crane's spring staging area. The sanctuary protects nearly 5 miles of prime crane feeding and roosting habitat, and up to 70,000 roosting cranes can often be seen from its blinds. A "Crane Cam" provides year-round live streaming video of crane roosting areas and river views via the Internet. Least terns and piping plovers often nest on barren sandbars that are also used by roosting cranes. Summer breeding birds include the dickcissel, upland sandpiper, and bobolink, as well as riparian woodland habitat species such as the rose-breasted grosbeak and willow flycatcher. The Iain Nicolson Audubon Center is open year-round and provides guided sunrise and sunset crane-viewing visits to riverside blinds from early March until early April (admission fee; reservations available from January 1 onward annually). There are several viewing blinds, the largest of which overlooks a roost attracting as many as 70,000 cranes during peak migration in late March. Single or two-person blinds can also be rented for overnight use for those photographers who are able to tolerate the cold and cost. As an alternative to the Rowe blinds, a free public viewing platform at the bridge on Gibbon Road (2 miles south of I-80 Exit 305, or a few hundred yards north of the east end of Elm Island Road). The platform provides excellent free sunset and sunrise crane viewing, with crane roosts usually visible both upstream and downstream. A hike-bike trail bridge crosses the Platte at the **Fort Kearney State Recreation Area** and offers another good viewing choice. It is 10 miles west of Rowe, off Nebraska Hwy L-50A (or from Kearney, via Nebraska Hwy 44 south to the Nebraska Hwy L-50A junction about 2 miles south of Kearney, then east 5 miles to the access road). The major biota of Rowe and its nearby surroundings includes about 250 species of birds, nearly 30 mammals, 17 reptiles, 7 amphibians, 70 butterflies, and numerous native forbs and grasses, many of which are listed on the sanctuary's website: https://rowe.audubon.org/conservation/flora-fauna -central-platte. A book on the biology of the Rowe Sanctuary and vicinity is in early stages of publication (Johnsgard forthcoming). Rowe Sanctuary is also located north of the ecologically important seasonal wetlands of the western Rainwater Basin (see the description below). Address: 44450 Elm Island Rd., Gibbon, Nebraska 68840 (ph. 308-468-5282). No admission fee, but a small donation is requested. An Audubon-sponsored celebration of the cranes and Platte Valley natural history has been held in Kearney annually for more than 40 years during the weekend of the first day of spring.

NIOBRARA STATE PARK. Area, 1,632 acres. This state park is located at the confluence of the Niobrara River and the backwaters of the Missouri River that are impounded by Gavins Point Dam and form Lewis and Clark Lake. The park is mostly grassland but also has extensive riparian woodlands. There are more than 12 miles of hiking trails, and a 2-mile hike-bike trail extends along the park's northern boundary. It is a designated Nebraska Important Bird Area, with over 260 bird species reported from the park and surrounding area. Locally breeding bird species associated with the eastern deciduous forest community include the American redstart, black-and-white warbler, and whip-poor-will. East of the town of Niobrara is **Bazile Creek Wildlife Management Area** (4,300 acres). There Bazile Creek flows north into the marshy backwaters of Lewis and Clark Lake, where it becomes part of a vast shallow wetland of several thousand acres that attracts countless wading birds such as American bitterns, great blue herons, great and snowy egrets, and rails. Bald eagles and ospreys are common to abundant fall to spring migrants. Niobrara State Park has an interpretive center, and both modern cabins and primitive camping facilities are present. A state park entry permit is required. Address: 89261 522 Ave., Niobrara, Nebraska 68760 (ph. 402-857-3373).

NIOBRARA VALLEY PRESERVE. Area, ca. 56,000 acres. This Nature Conservancy preserve is located 16 miles north of Johnstown, Brown County. It includes about 25 miles of the Niobrara River in Brown and Keya Paha Counties and is in the heart of the transition zone between western coniferous and eastern deciduous forest types (Kaul, Kantak, and Churchill 1988; Steinauer and Scudder n.d.; Johnsgard 1995). The preserve lies within the **Niobrara National Scenic River District** (ph. 402-376-3241), which extends for 76 miles. It has a remarkably diverse biotic checklist that includes 581 plant species, as well as 44 mammals, 213 birds, 25 fish and 70 butterflies. There are 75 definitely breeding and 30 more possibly breeding bird species (Brogie and Mossman 1983). Several of the breeding birds have east-west counterpart species (Baltimore and Bullock's orioles, lazuli and indigo buntings, and rose-breasted and black-headed grosbeaks) that hybridize in this ecological transition zone (Johnsgard 2007a). The preserve is partly managed for bison, and research on bison foraging behavior and fire ecology is being conducted. A 2012 fire burned 29,000 acres of the preserve, which are slowly recovering. Trails radiate out from the headquarters, passing through several forest types and Sandhills prairie vegetation. No admission fee. Address: 42269 Morel Rd., Johnstown, Nebraska 69214 (ph. 402-722-4440).

NORTH PLATTE NATIONAL WILDLIFE REFUGE. Area, 5,047 acres. Part of the Crescent Lake / North Platte National Wildlife Refuge complex, including **Crescent Lake National Wildlife Refuge**, Lake Alice (1,377 acres when full, but it is usually dry), Lake Minatare (430 acres), and Winters Creek (700 acres). Of the last three, the best wetland habitat is at Winters Creek, which seasonally supports many migrating waterfowl and sandhill cranes. Located 4 miles north and 8 miles east of Scottsbluff, Scotts Bluff County; free admission. The refuge bird list totals 228 species, including 85 wetland species (13 breeders), and is available online or from the refuge manager, 10630 Road 181, Ellsworth, Nebraska 69340 (ph. 308-762-4893). It is also available online: https://www.npwrc.usgs.gov/resource/birds/chekbird/r6/31.htm. No admission fee.

OGLALA NATIONAL GRASSLAND. Area, 94,344 acres. The southern border of this arid grassland is 7 miles north of Crawford on Nebraska Hwy 71. The grasslands in northern Sioux and Dawes Counties consist of sage-steppe, shortgrass steppe, and mixed-grass prairies over eroded clay and Pierre shale badlands. Within the grassland is the Forest Service's **Toadstool Geological Park**, which is about 15 miles north of Crawford. About 50 mammal species have been reported from the Oglala National Grassland, and a bird list of 302 bird species (which also includes the nearby Nebraska National Forest and associated Pine Ridge region of northwestern Nebraska) is available from the U.S. Forest Service, 270 Pine St., Chadron, Nebraska 69337 (ph. 308-432-3367 or 308-432-4475). Website: www.fs.fed.us/r2/nebraska. No admission fee.

PONCA STATE PARK. Area, 2,166 acres. This state park is situated 2 miles north of Ponca, Dixon County, on Nebraska Hwy 12. It consists of mature and old-growth riverine hardwood forest at the downstream end of the **Missouri National Recreational River** system. The nearby Missouri River is still unchanneled here; it is one of the few downstream stretches that still resembles its original state.

Elk Point Bend Wildlife Management Area / State Recreation Area is located about 2 miles north of Ponca State Park and consists of oak savanna and riparian wetlands along the Missouri River. An interpretive center, the **Missouri National Recreational River Resource and Educational Center**, is located in Ponca State Park and focuses on the ecology and history of the Missouri River. The park is mostly forested, with stands of bur oak, walnut, hackberry, and elm; one of the bur oaks is more than 300 years old. There are 17 miles of hiking trails, modern cabins, and a campground. A local bird list includes about 240 species. Whip-poor-wills are common in summer, bald eagles are present from fall to spring, and large numbers of snow geese migrate through the area in spring and fall. Entry permits are required for the state park and state recreation area. For information contact Ponca State Park, 88090 Spur 26 E, Ponca, Nebraska 68770 (ph. 402-755-2284).

RAINWATER BASIN WETLAND MANAGEMENT DISTRICT. This multicounty region south of the central Platte River contains hundreds of seasonal playa wetlands that are geographically divided into eastern and western components. The **Rainwater Basin Joint Venture** involves various government agencies and many landowners and coordinates the Rainwater Basin's wetland sites. These include approximately 50 federally owned waterfowl production areas and about 30 state-owned wildlife management areas, extending from Phelps County east to Butler and Saline Counties. The Rainwater Basin's importance to Great Plains migrating shorebirds during April and early May is probably second only in the Great Plains to Cheyenne Bottoms Wildlife Management Area in Kansas (Jorgensen 2004, 2012). During wet springs the Rainwater Basin also often holds millions of migrating geese (snow, Ross's, greater white-fronted, Canada, and cackling) during March, but during some springs it is nearly dry except for a few wetlands that are maintained by pumping water. Jorgensen (2012) reported 359 bird species from the Rainwater Basin, and Brown and Johnsgard (2013) reported 373 species from the central Platte Valley, including northern parts of the Rainwater Basin. Freeman and Benedict (1993) documented the mammals of the Platte Valley. Nebraska's playa wetlands are included within the multistate **Playa Lakes Joint Venture** program, which extends geographically from northwestern Nebraska south to western Texas; its address is 103 East Simpson St., LaFayette, Colorado 80026 (ph. 303-926-0777). The address for the Rainwater Basin Wetland Management District is P.O. Box 8, Funk, Nebraska 68940 (ph. 308-263-3000). The Rainwater Basin Joint Venture's address is 2550 N. Diers Ave., Suite L, Grand Island, Nebraska 68803 (ph. 308-395-8586). A collective bird list for the Rainwater Basin and adjacent central Platte Valley containing more than 300 species is available from the joint venture office.

SCOTTS BLUFF NATIONAL MONUMENT. Area, ca. 3,000 acres. This historically famous bluff along the Oregon Trail southwest of Scottsbluff consists of shortgrass plains, eroded rock slopes, and cliffs. It is capped by ponderosa pine habitat and has nearly vertical rocky sides that provide nesting sites for white-throated swifts and other cliff-nesting birds. There is a vehicular road to the summit and a 3-mile nature trail leading from the summit parking lot down to the visitor center. The visitor center includes many historic paintings by the famous geologist-explorer William H. Jackson and was expanded and renovated in 2019. The riverine woodland habitat occupies only about 4 percent of the site's total area but supports 57 percent of all the vertebrate species that Cox and Franklin (1988, 1989) reported for the monument. Their lists of its terrestrial vertebrates include 98 birds, 18 mammals, 8 reptiles, and 5 amphibians. Three prairie dog towns totaling about 63 acres are present and may support burrowing owls. The breeding birds include many other western-oriented species such as the prairie falcon, common poorwill, rock wren, Bullock's oriole, and black-headed grosbeak. The western lazuli and eastern indigo buntings are also breeders, and at least three races of dark-eyed juncos overwinter, including the endemic white-winged race. Scotts Bluff National Monument is a short distance west of Gering on the old Oregon Trail. Free admission. For information contact the superintendent at Box 27, Gering, Nebraska 69341 (ph. 308-436-4340).

SPRING CREEK PRAIRIE AUDUBON CENTER. Area, ca. 850 acres. Located about 20 miles south of Lincoln and 3 miles south of Denton. The prairie is owned by the Nebraska section of the National Audubon Society, with a modern visitor/interpretive center and several miles of trails. This mostly virgin prairie is on a hilly glacial moraine (including a hill having the highest elevation in Lancaster County) and is one of the largest such preserved tallgrass prairies in Nebraska. There are also wetlands, riparian areas, and deciduous woods. Of the approximate 200 prairie-adapted (nontree and non-aquatic) plant species, broad-leaved forbs comprise about 70 percent of the flora, native grasses about 20 percent, shrubs and vines about 8 percent, and trees about 1 percent. The riverine woodlands support about 5 percent of the vascular plants. The site's checklists include about 230 species of birds, 30 mammals, 9 herpetiles, and about 50 butterflies. I (Johnsgard 2018b) have described the natural history and ecology of the preserve. Open daily year-round (weekend afternoons only), except for major holidays. No admission fee, but a small donation is requested. Address: 11700 S.W. 100 St., P.O. Box 117, Denton, Nebraska 68239 (ph. 402-797-2301). Website: www .springcreekprairie.audubon.org.

THE CRANE TRUST. Area, ca. 2,000 acres. The Crane Trust (originally known as the Platte River Whooping Crane Critical Habitat Maintenance Trust) was formed as a result of a legal settlement over the environmental costs of building a North Platte River dam in Wyoming and manages several thousand acres of riparian wetlands and wet meadows along the Platte River of critical importance to cranes. Trust biologists perform annual surveys of sandhill crane numbers and habitat surveys and undertake other ornithological and ecological research. A herd of bison on native prairie meadows provides a tourist attraction, along with 8 miles of walking trails that are available for public use when cranes are not present. The trust's headquarters are located 1 mile east of the Alda Road bridge (about a mile south of I-80 Exit 305), then turn east and continue approximately 1.5 miles on Whooping Crane Drive (restricted access, so phone ahead: 308-384-4633). The **Crane Trust Nature and Visitor Center** is located just south of I-80 Exit 305 and is open year-round. It provides information on cranes and other local wildlife to tourists and has a gift shop and an art gallery. Like Rowe Sanctuary, the visitor center offers fee-based sunrise and sunset visits to riverside crane-viewing blinds from late February to early April (reservations needed, available annually from January 1 onward). The Crane Trust's website is https://cranetrust.org. Address of the visitor center: 9325 S. Alda Rd., Wood River, Nebraska 68883 (ph. 308-382-1820).

TOADSTOOL GEOLOGICAL PARK. Area, ca. 300 acres. Located 15 miles north of Crawford. Drive 4 miles north from Crawford on Nebraska Hwy 2 and Nebraska Hwy 71. Turn northwest on Toadstool Road and drive about 10 miles. Turn left at the Toadstool Park sign and drive about 1.5 miles. The topography here consists of eroded sandstone formations eroded from streambed sediments deposited about 45–26 million years ago, as well as more recent volcanic ash deposits. Fossil remains of several strange Oligocene mammalian groups have been found, such as rhino-like brontotheres (also known as titanotheres), oreodonts (pig-like herbivores with long canines), entelodonts (pig-like omnivores), and hyaenodonts (hyena-like carnivores). Fossils of some still-extant or more recently extinct groups are also present, such as rhinos, camels, three-toed horses, and saber-toothed cats. Pronghorns are often seen here, and prairie rattlesnakes are common. If hiking, bring water, as none is available on site. Wear hiking shoes, since much of the trail is over uneven bare rock surfaces. The **Hudson-Meng Education and Research Center** (20 acres) (also known as Hudson-Meng Bison Kill site, lat 42.8266, long -103.6015) is located south of Toadstool Geological Park. It is accessible from Toadstool Park over 2 miles of foot trail or can be reached by car over 12 miles on a gravel Forest Service road off Toadstool Road that is variously identified on maps as Meng Drive, Sand Creek Road, or FS 904 and FS 905. The enclosed site contains about 600 skeletal remains of bison dating from 8,000–10,000 years ago (Alberta Culture period) that were apparently butchered here and whose bones have been preserved in situ (Agenbroad 1978). Address of Toadstool Park: USDA Forest Service, Nebraska National Forest, 125 N. Main St., Chadron, Nebraska 69337 (ph. 308-432-0300). Address of Hudson-Meng Center: 18 Meng Dr., Crawford, Nebraska 69346 (ph. 308-665-3900). Call before visiting to verify hours of visitor access and admission fee.

VALENTINE NATIONAL WILDLIFE REFUGE. Area, 71,516 acres. Located in Cherry County, 22 miles south of Valentine. Nebraska's largest national wildlife refuge consists mostly of Sandhills prairie, with glacial-age sand dunes and intervening depressions that often are shallow wetlands. The refuge contains nearly 40 mostly small lakes that are surrounded by sand dunes up to 200 feet high, connected by sometimes problematic roads. The roughly 500 refuge plants are largely sand-adapted species (including a Nebraska endemic, blow-out penstemon, *Penstemon haydenii*). Many prairie-dependent grassland birds, such as the long-billed curlew and upland sandpiper, are common. Four grebes (eared, western, Clark's, and pied-billed) have nested here, as has the Wilson's phalarope, American avocet, and white-faced ibis. Trumpeter swans, Canada geese, and nine species of ducks also breed here. Free-access observation blinds for spring viewing of prairie grouse display are available on a first-come, first-served basis. Up to 150,000 migrants can be found seasonally on the refuge. A recent refuge checklist of 272 bird species lists 100 wetland species, including 31 shorebirds, 24 waterfowl, 10 gulls and terns, 5 grebes, and 4 rails. Bogan and Ramotnik (1995) documented the refuge's 57 species of mammals; both mule deer and white-tailed deer occur, as well as confined elk and bison. Refuge reptiles include the nationally rare Blanding's turtle. Address of refuge office: 3993 Refuge Rd., Valentine, Nebraska 69201 (ph. 402-376-3789). Many of the roads are of bare sand and gravel; drive and park carefully. Free admission. Website: https://www.fws.gov/refuge/valentine/.

References

GENERAL REFERENCES

Bleed, A., and C. Flowerday, eds. 1998. *An Atlas of the Sand Hills.* 2nd ed. Resource Atlas No. 5b. Lincoln: Conservation and Survey Division, University of Nebraska.

Cook, J. H. 1980. *Fifty Years on the Old Frontier.* Norman: University of Oklahoma Press.

Cox, M. K., and W. L. Franklin. 1988. "Faunal survey of the birds, mammals, and reptiles at Scotts Bluff National Monument." *Iowa Cooperative Fish and Wildlife Research Unit, Annual Report* 53:26–28.

———. 1989. "Terrestrial vertebrates of Scotts Bluff National Monument, Nebraska." *Great Basin Naturalist* 49:597–613.

Dankert, N., D. Brust, H. Nagel, and S. M. Spomer. 2005. *Butterflies of Nebraska.* University of Nebraska at Kearney, 5 April. https://www.lopers.net/student .org/Nebraskinverts/butterflies/homehtm.

Graetz, J. L., R. A. Garrott, and S. R. Craven. 1995. *Faunal Survey of Agate Fossil Beds National Monument.* Omaha: National Park Service, Midwest Region.

Johnsgard, P. A. 1995. *This Fragile Land: A Natural History of the Nebraska Sandhills.* Lincoln: University of Nebraska Press.

———. 2001. *The Nature of Nebraska: Ecology and Biodiversity.* Lincoln: University of Nebraska Press.

———. 2005. *Prairie Dog Empire: A Saga of the Shortgrass Prairie.* Lincoln: University of Nebraska Press.

———. 2007a. *A Guide to the Natural History of the Central Platte Valley of Nebraska.* Lincoln: University of Nebraska Digital Commons. https://digitalcommons .unl.edu/biosciornithology/40/719.

———. 2007b. *The Niobrara: A River Running through Time.* Lincoln: University of Nebraska Press.

———. 2008a. *A Guide to the Prairies of Eastern Nebraska and Adjacent States.* Lincoln: University of Nebraska Digital Commons. https://digitalcommons.unl .edu/biosciornithology/39/.

———. 2008b. *The Platte: Channels in Time.* 2nd ed. Lincoln: University of Nebraska Press.

———. 2009. "Nebraska's eight great natural wonders." *Nebraska Life,* November 2009, 78–84. https://digitalcommons.unl.edu/johnsgard/3.

————. 2012a. *Nebraska's Wetlands: Their Wildlife and Ecology*. Water Survey Paper No. 78. Lincoln: Conservation and Survey Division, University of Nebraska.

————. 2012b. "Spring Creek Prairie Audubon Center: An 800-acre schoolhouse." *Prairie Fire*, October, 18–20, 22. https://www.prairiefirenewspaper.com/2012 /10//spring-creek-prairie-audubon-center-an-800-acre-schoolhouse.

————. 2014a. "The Hutton Niobrara Ranch Audubon Nature Sanctuary." *Prairie Fire*, July, 12–14. https://www.prairiefirenewspaper.com/2014/07/hutton -niobrara-ranch-wildlife-sanctuary.

————. 2014b. *Seasons of the Prairie: A Nebraska Year*. Lincoln: University of Nebraska Press.

————. 2015. "Nebraska: Where the West begins and the East peters out." *Prairie Fire*, May, 8–9. https://www.prairiefirenewspaper.com/./nebraska-where-the -west-begins-and-the-east-peters-out.

————. 2018. *The Ecology of a Tallgrass Treasure: Audubon's Spring Creek Prairie*. Lincoln: Zea E-Books and University of Nebraska Digital Commons. https:// digitalcommons.unl.edu/zeabook.

————. 2019. *Wyoming Wildlife: A Natural History*. Lincoln: Zea E-Books and University of Nebraska Digital Commons. https://digitalcommons.unl.edu /zeabook/73/.

Jones, S. R. 2000. *The Last Prairie: A Sandhills Journal*. New York: McGraw-Hill.

Kaul, R. B., G. E. Kantak, and S. P. Churchill. 1988. "The Niobrara River Valley, a postglacial migration corridor and refugium of forest plants and animals in the grasslands of central North America." *Botanical Review* 54:44–81.

Keech, C. F., and R. Bentall. 1971. *Dunes on the Plains: The Sand Hills Region of Nebraska*. Resources Report 4. Lincoln: Conservation and Survey Division, University of Nebraska.

Knopf, F. L., and F. B. Samson. 1997. "Conservation of grassland vertebrates." In *Ecology and Conservation of Great Plains Vertebrates*, edited by F. L. Knopf and F. B. Samson, 273–89. New York: Springer-Verlag.

Knue, J. 1992. *Nebraska Wildlife Viewing Guide*. Helena MT: Falcon Press.

Krapu, G. L., ed. 1996. *The Platte River Ecology Study: Special Research Report*. Jamestown ND: Northern Prairie Wildlife Research Station.

LaGrange, T. G. 2005. *Guide to Nebraska Wetlands and Their Conservation Needs*. 2nd ed. Lincoln: Nebraska Game and Parks Commission. https://digitalcommons .unl.edu/nebgamepubs/37/.

McCarraher, D. B. 1977. *Nebraska's Sandhills Lakes*. Lincoln: Nebraska Game and Parks Commission.

McMurtry, M. S., R. Craig, and G. Schildmann. 1972. *Nebraska Wetland Survey*. Lincoln: Nebraska Game and Parks Commission.

Novacek, J. M. 1989. "The water and wetland resources of the Nebraska Sandhills." In *Northern Prairie Wetlands*, edited by A. van der Valk, 340–84. Ames:
Iowa State University Press.

Oberholser, H. C., and W. L. McAtee. 1920. *Waterfowl and Their Food Plants in the
Sandhill Region of Nebraska*. Bulletin 794. Washington DC: U.S. Department
of Agriculture.

Panella, M. J. 2010. *Nebraska's At-Risk Wildlife*. Lincoln: Nebraska Game and Parks
Commission.

Schneider, R., M. Fritz, J. Jorgensen, S. Schainost, R. Simpson, G. Steinauer, and
C. Rothe-Groleau. 2018. *Revision of the Tier 1 and 2 Lists of Species of Greatest
Conservation Need: A Supplement to the Nebraska Natural Legacy Project State Wildlife
Action Plan*. Lincoln: Nebraska Game and Parks Commission.

Schneider, R., K. Stoner, G. Steinauer, M. Panella, and M. Humpert. 2011. *The
Nebraska Natural Legacy Project: State Wildlife Action Plan*. 2nd ed. Lincoln:
Nebraska Game and Parks Commission.

Steinauer, E., and E. Scudder. n.d. *The Niobrara Valley Preserve, Crossroads of Nature:
Animal Species List*. Johnstown NE: Nature Conservancy.

Zuerline, G. 1983. "Value of rivers to aquatic life." In "Nebraska Rivers." Special
issue, *Nebraskaland Magazine* 61(1): 132–33.

BOTANY AND PLANT ECOLOGY

Barth, R., and N. Ratzlaff. 2004. *Field Guide to Wildflowers: Fontenelle Forest and Neale
Woods Nature Centers*. Bellevue NE: Fontenelle Nature Association (includes
373 forbs organized by flower color, 3 duckweeds, and 2 horsetails).

———. 2007. *Field Guide to Trees, Shrubs, Woody Vines, Grasses, Sedges and Rushes:
Fontenelle Forest and Neale Woods Nature Centers*. Bellevue NE: Fontenelle Nature
Association (includes 61 trees, 25 shrubs, 10 woody vines, 60 grasses, 31 sedges,
and 2 rushes).

Bessey, C. E. 1887. "A meeting place of two floras." *Bulletin of the Torrey Botanical
Club* 14:189–91.

Farrar, J. 2011. *Field Guide to Wildflowers of Nebraska and the Great Plains*. 2nd ed.
Iowa City: University of Iowa Press (includes 274 species organized by flower
color).

Kaul, R. 1998. "Plants." In *An Atlas of the Sand Hills*, 2nd ed., edited by A. Bleed
and C. Flowerday, 127–42. Resource Atlas No. 5b. Lincoln: Conservation and
Survey Division, University of Nebraska.

Kaul, R., D. M. Sutherland, and S. B. Rolfsmeier. 2012. *The Flora of Nebraska*. 2nd
ed. Lincoln: School of Natural Resources, University of Nebraska.

Nagel, H. G. 1998. "The Loess Hills Prairies of Central Nebraska." Special issue, *Platte Valley Review*. Kearney: University of Nebraska at Kearney.

Nebraska Department of Agriculture. 1979. *Nebraska Weeds*. Lincoln: Nebraska Department of Agriculture (includes 216 forbs, 27 grasses, 2 sedges, and 2 cacti).

Pound, R., and F. C. Clements. 1898. *Phytogeography of Nebraska*. Lincoln NE: Jacob North and County

Rolfsmeier, S. B., and G. Steinhauer. 2010. *Terrestrial Natural Communities of Nebraska*. Lincoln: Nebraska Game and Parks Commission.

Stubbendieck, J., and K. Kottas. 2005. *Common Grasses of Nebraska*. Extension Bulletin EC05-170. Lincoln: Institute of Agriculture and Natural Resources, University of Nebraska (includes 180 species of grasses and 7 sedges).

———. 2007. *Common Forbs and Shrubs of Nebraska*. Extension Bulletin EC-118. Lincoln: Institute of Agriculture and Natural Resources, University of Nebraska (includes 117 species of forbs, 17 shrubs, and 4 cacti).

GEOLOGY AND PALEONTOLOGY

Agenbroad, L. D. 1978. *The Hudson-Meng Site: An Alberta Bison Kill in the Nebraska High Plains*. Caldwell ID: Caxton Printers, Ltd.

Bouc, K., coordinator. 1994. *The Cellars of Time: Paleontology and Archeology in Nebraska*. *Nebraskaland Magazine* 72(1): 1–162 (also published in *Nebraska History* 75[1]).

Carlson, M. P. 1993. *Geology, Geologic Time and Nebraska*. Lincoln: Conservation and Survey Division, University of Nebraska.

Dibner, R. R., N. Korfanta, G. Beauvais, J. Bowler, K. Freedman, K. C. Trujillo, and V. H. Zero. 2018. *Agate Fossil Beds National Monument Natural Resource Condition Assessment Natural Resource Report*. NPS/NGPN/NRR-2018/1676. Fort Collins CO: National Park Service.

Dixon, D. 2008. *The Complete Book of Dinosaurs*. London: Hermes House.

Hunt, R. H., Jr. 1981. "Geology and vertebrate paleontology of the Agate Fossil Beds National Monument and surrounding region, Sioux County, Nebraska (1972–1978)." *National Geographic Society Research Reports* 13:263–85.

Hunt, R. M., Jr., R. Skolnick, and J. Kaufman. 2018. *The Carnivores of Agate Fossil Beds National Monument: Miocene Dens and Waterhole in the Valley of a Dryland Paleoriver*. Zea E-Books 74. https://digitalcommons.unl.edu/zeabook/74.

Kurten, B., and D. C. Anderson. 1983. *Pleistocene Mammals of North America*. New York: Columbia University Press.

Levin, H. L. 1978. *The Earth through Time*. 5th ed. New York: Harcourt Brace.

Lugn, A. L. 1934. *Outline of Pleistocene Geology of Nebraska*. Bulletin B-1-41. Lincoln: University of Nebraska State Museum.

————. 1935. *The Pleistocene Geology of Nebraska.* Nebraska Geological Survey Bulletin Series 2, No. 10.

————. 1939. "Classification of the Tertiary system in Nebraska." *Bulletin of the Geological Society of America* 50:245–76.

Maher, H. D., Jr., G. F. Engelmann, and R. D. Shuster. 2003. *Roadside Geology of Nebraska.* Missoula MT: Mountain Press.

Mengel, R. M. 1970. "The North American Central Plains as an isolating agent in bird speciation." In *Pleistocene and Recent Environments of the Central Great Plains,* edited by W. Dort and J. K. Jones, 280–340. Lawrence: University Press of Kansas.

Palmer, D., ed. 1999. *The Simon and Schuster Encyclopedia of Dinosaurs and Prehistoric Creatures: A Visual Who's Who of Prehistoric Life.* New York: Simon and Schuster.

Plate, R. 1964. *The Dinosaur Hunters: Nathaniel C. Marsh and Edward D. Cope.* New York: McKay County

Prothero, D. R., and R. J. Emry. 2004. "The Chadronian, Orellan, and Whitneyan North American land mammal ages." In *Late Cretaceous and Cenozoic Mammals of North America: Biostratigraphy and Geochronology,* edited by M. O. Woodburne, 156–68. New York: Columbia University Press.

Skinner, M. F., S. M. Skinner, and R. J. Gooris. 1977. "Stratigraphy and biostratigraphy of late Cenozoic deposits in central Sioux County Nebraska." *Bulletin of the American Museum of Natural History* 158:263–370.

Swinehart, J. B., et al. 1985. "Cenozoic paleontology of western Nebraska." In *Rocky Mountain Paleogeography Symposium 3: Cenozoic Paleogeography of West-Central United States,* edited by R. M. Flores and S. S. Kaplan, 209–29. Denver: Rocky Mountain Section, Society of Economic Paleontologists.

Thornberry, W. 1965. *Regional Geomorphology of the United States.* New York: Wiley.

Trimble, D. E. 1980. *The Geologic Story of the Great Plains.* Geologic Survey Bulletin 1493. Washington DC: U.S. Geological Survey, U.S. Department of the Interior.

Voorhies, M. R. 1981. "Ancient ashfall creates a Pompeii of prehistoric animals, dwarfing the St. Helens eruption." *National Geographic* 159:66–75.

———— 1994a. "Camels, rhinos and four-tuskers." In *The Cellars of Time: Paleontology and Archeology in Nebraska,* coordinated by K. Bouc, 43–54. Special issue, *Nebraskaland Magazine* 72(1): 43–54.

————. 1994b. "Mammoths and musk oxen." In *The Cellars of Time: Paleontology and Archeology in Nebraska,* coordinated by K. Bouc. Special issue, *Nebraskaland Magazine* 72(1): 67–73.

————. 1994c. "Our oldest mammals." In *The Cellars of Time: Paleontology and Archeology in Nebraska,* coordinated by K. Bouc. Special issue, *Nebraskaland Magazine* 72(1): 33–41.

———. 1994d. "Sea monsters and dinosaurs." In *The Cellars of Time: Paleontology and Archeology in Nebraska*, coordinated by K. Bouc. Special issue, *Nebraskaland Magazine* 72(1): 19–29.

———. 1994e. "Zebras and giant camels." In *The Cellars of Time: Paleontology and Archeology in Nebraska*, coordinated by K. Bouc. Special issue, *Nebraskaland Magazine* 72(1): 59–65.

Wayne, W. J., et al. 1991. "Quaternary geology of the northern Great Plains." In *Quaternary Nonglacial Geology: Coterminous U.S. The Geology of North America*, edited by R. B. Morrison, K-2:441–76. Boulder CO: Geological Society of North America.

Wright, H. E., Jr., and F. G. Frey, eds. 1965. *The Quaternary of the United States.* Princeton NJ: Princeton University Press.

MAMMALS

Adams, C. E. 1973. "Population dynamics of fox squirrels, *Sciurus niger*, in selected areas in Seward County, Nebraska." PhD dissertation, University of Nebraska–Lincoln.

Amundson, R. 1943. "Rodents in Nebraska." *Outdoor Nebraska* 21(1): 4–7, 20.

Andelt, F. 1992. "Nebraska's threatened and endangered species: River otter." *Nebraskaland Magazine* 70(10): 1–6.

———. 1995a. "Nebraska's threatened and endangered species: Black-footed ferret." *Nebraskaland Magazine* 73(10): 1–6.

———. 1995b. "Swift fox investigations in Nebraska, 1995." In *Swift Fox Conservation Team 1995 Annual Report*, edited by S. H. Allen, J. W. Hoagland, and E. D. Stukel, 81–89. Bismarck: North Dakota Game and Fish Department.

———. 1997. "Swift fox investigations in Nebraska, 1997." In *Swift Fox Conservation Team 1997 Annual Report*, edited by B. Giddings, 77–79. Helena: Montana Department of Fish, Wildlife and Parks.

Anderson, C. R., and F. G. Lindzey. 2005. "Experimental evaluation of population trend and harvest composition of a Wyoming cougar population." *Wildlife Society Bulletin* 33:179–88.

Armstrong, D. M., J. R. Choate, and J. K. Jones Jr. 1986. "Distributional patterns of mammals in the Plains States." *Occasional Papers of the Museum* 105:1–27. Lubbock: Texas Tech University.

Armstrong, D. M., J. P. Fitzpatrick, and C. A. Meaney. 2011. *Mammals of Colorado.* 2nd ed. Boulder: University Press of Colorado.

Atkeson, T. D., R. L. Marchinton, and K. V. Miller. 1988. "Vocalizations of white-tailed deer." *American Midland Naturalist* 120:194–200.

Aubry, K. B., K. S. McKelvey, and J. P. Copeland. 2007. "Distribution and broad-scale habitat relations of the wolverine in the contiguous United States." *Journal of Wildlife Management* 71:2147–58.

Barbour, R. W., and W. H. Davis. 1969. *Bats of America.* Lexington: University of Kentucky Press. 286 pp.

Barclay, R. M. R., M. B. Fenton, and D. W. Thomas. 1979. "Social behavior in the little brown bat, *Myotis lucifugus*: II. Vocal communication." *Behavioral Ecology and Sociobiology* 6(2): 137–46.

Baumann, W. L. 1982. "Microhabitat use in three species of rodents on a Nebraska Sandhills prairie." MS thesis, University of Nebraska–Lincoln.

Beckoff, M., and C. B. Lowe, eds. 2007. *Listening to Cougar.* Boulder: University Press of Colorado.

Bee, J. W., G. E. Glass, R. S. Hoffmann, and R. R. Patterson. 1981. *Mammals in Kansas.* Lawrence: University of Kansas Museum of Natural History.

Beebe, C. 2004. "An ecological study of *Perognathus fasciatus*." MS thesis, University of Colorado, Boulder.

Beed, W. E. 1935. "A preliminary study of the animal ecology of Nebraska short-grass plains grazed only by native animals." MA thesis, University of Nebraska–Lincoln.

———. 1936. "A preliminary study of animal ecology of the Niobrara Game Preserve." *University of Nebraska, Bulletin of the Conservation and Survey Division* 10:1–33.

Belitsky, D. 1981. *Small Mammals of the Salt Wells.* Pilot Butte WY: Bureau of Land Management, Pilot Butte Planning Unit. 104 pp.

Benedict, R. A. 1999a. "Location and characteristics of a hybrid zone between short-tailed shrews (*Blarina*)." *Journal of Mammalogy* 80:135–41.

———. 1999b. "Morphological and mitochondrial DNA variation in a hybrid zone between short-tailed shrews (*Blarina*) in Nebraska." *Journal of Mammalogy* 80:112–34.

———. 2004. "Reproductive activity and distribution of bats in Nebraska." *Western North American Naturalist* 64:231–48.

Benedict, R. A., P. W. Freeman, and H. H. Genoways. 1996. "Prairie legacies—mammals." In *Prairie Conservation: Preserving North America's Most Endangered Ecosystem,* edited by F. B. Samson and F. L. Knopf, 149–66. Washington DC: Island Press.

Benedict, R. A., H. H. Genoways, and P. W. Freeman. 2000. "Shifting distributional patterns of mammals in Nebraska." *Transactions of the Nebraska Academy of Sciences* 26:55–84.

Bischof, R. 2003a. "Indian Cave's night gliders." *Nebraskaland Magazine* 81(10): 20–23 (flying squirrel).

———. 2003b. "Prairie dog survey." *Nebraskaland Magazine* 81(10): 9.

———. 2003c. "Status of the northern river otter in Nebraska." *Prairie Naturalist* 35:117–20.

Boer, A. H., ed. 1992. *Ecology and Management of the Eastern Coyote.* Fredericton: Wildlife Research Unit, New Brunswick University.

Bogan, M. A. 1997. "Historical changes in the landscape and vertebrate diversity of north central Nebraska." In *Ecology and Conservation of Great Plains Vertebrates,* edited by F. L. Knopf and F. B. Samson, 105–30. New York: Springer-Verlag.

Bogan, M. A., K. Geluso, and J. A. White. 2004. *Mammalian Inventories of Two National Wildlife Refuges in Nebraska, 2001–2002.* Final Report to the U.S. Fish and Wildlife Service, Lakewood CO (Fort Niobrara and Valentine National Wildlife Refuges).

Bogan, M. A., M. Jennings, and F. Knopf. 1995. "A portrait of faunal and floral change in the Sandhills of northern Nebraska." In *A Biological Survey of the Fort Niobrara and Valentine National Wildlife Refuges,* edited by M. A. Bogan, 6–24. Fort Collins CO: Midcontinent Ecological Science Center, National Biological Service, U.S. Department of the Interior.

Bogan, M. A., and C. A. Ramotnik. 1995. "The mammals." In *A Biological Survey of the Fort Niobrara and Valentine National Wildlife Refuges,* edited by M. A. Bogan, 140–86. Fort Collins CO: Midcontinent Ecological Science Center, National Biological Service, U.S. Department of the Interior.

Bouc, K. 1998. "White-tailed deer and mule deer." *Nebraskaland Magazine* 76(7): 56–63.

Boyce, M. S., and L. D. Hayden-Wing, eds. 1979. *North American Elk: Ecology, Behavior and Management.* Laramie: University of Wyoming Press.

Bradbury, O. C. 1919. "A study in animal ecology of an eastern Nebraska region." PhD dissertation, University of Nebraska.

Buskirk, S. W. 2016. *Wild Mammals of Wyoming and Yellowstone Park.* Berkeley: University of California Press.

Byers, J. A. 1997. *American Pronghorn: Social Adaptations and the Ghosts of Predators Past.* Chicago: University of Chicago Press. 300 pp.

Camenzind, F. J. 1978. "Behavioral ecology of coyotes (*Canis latrans*) on the National Elk Refuge, Jackson, Wyoming." PhD dissertation, University of Wyoming, Laramie.

Carroll, D. 2007. "The elk population in Nebraska." *Nebraskaland Magazine* 85(7): 23.

Choate, J. R., and J. K. Jones Jr. 1981. "Provisional checklist of South Dakota mammals." *Prairie Naturalist* 13:65–87.

Clark, T. W. 1971. "Ecology of the western jumping mouse in Grand Teton National Park." *Northwest Science* 45:228–38.

———. 1973. "Local distribution and interspecies interactions in microtines, Grand Teton National Park, Wyoming." *Great Basin Naturalist* 33:205–17.

———, ed. 1986. *The Black-footed Ferret.* Great Basin Naturalist Memoirs No. 8:1–308. Brigham University, Provo, UT.

———. 1989. *Conservation Biology of the Black-footed Ferret.* Special Scientific Report No. 3. Wildlife Preservation Trust International, Philadelphia, PA. 175 pp.

Clark, T. W., T. M. Campbell III, D. G. Socha, and D. E. Casey. 1982. "Prairie dog colony attributes and associated vertebrate species." *Great Basin Naturalist* 42:572–82.

Clark, T. W., and M. R. Stromberg. 1987. *Mammals of Wyoming.* Public Education Series No. 10. Lawrence: University of Kansas Museum of Natural History.

Clausen, M. K. 1983. "The ecology of a relict population of woodrats." MS thesis, University of Nebraska–Lincoln (eastern wood rat).

Cockrum, E. L. 1956. "Reproduction in North American bats." *Transactions of the Kansas Academy of Science* 58:487–511.

Cole, G. 1969. *The Elk of Grand Teton and Southern Yellowstone National Parks.* National Park Service, US Department of the Interior Research Report GRTE-N-1. 192 pp.

Cook, H. J. 1931. "A mountain sheep record from Nebraska." *Journal of Mammalogy* 12:170–71.

Corcoran, A. J., J. R. Barber, and W. E. Conner. 2009. "Tiger moth jams bat radar." *Science* 325:325–27.

Corcoran, A. J., and T. J. Weller. 2018. "Inconspicuous echolocation in hoary bats (*Lasiurus cinereus*)." *Proceedings. Biological Sciences* 285:20180441.

Covell, D. F. 1992. "Ecology of the swift fox (*Vulpes velox*) in southeastern Colorado." MS thesis, University of Wisconsin–Madison. 111 pp.

Cover, M. A. 2000. "Ecology of elk in the Pine Ridge region of northwestern Nebraska: Seasonal distribution, characteristics of wintering sites, and herd health." MS thesis, University of Nebraska–Lincoln.

Cowan, I. M. 1940. "Distribution and variation in native sheep of North America." *American Midland Naturalist* 24:505–80.

Czaplewski, N. J. 1976. "Distribution of bats in Nebraska with notes on natural history." MS thesis, Kearney State College, Kearney NE.

Czaplewski, N. J., J. P. Farney, J. K. Jones Jr., and J. D. Druecker. 1979. "Synopsis of bats of Nebraska." *Occasional Papers of the Museum* 61:1–24. Lubbock: Texas Tech University.

Czura, P. 1960. "Notes on Nebraska fauna: Least weasel." *Outdoor Nebraska* 38(8): 26–27.

Danz, P. 1997. *Of Bison and Man*. Boulder: University Press of Colorado. 231 pp.

Darden, S. K., T. Dabelsteen, and S. B. Pedersen. 2003. "A potential tool for swift fox (*Vulpes velox*) conservation: Individuality of long-range barking sequences." *Journal of Mammalogy* 84:1417–27.

Dark-Smiley, D. N., and D. A. Keinath. 2003. "Species assessment for swift fox (*Vulpes velox*) in Wyoming." Wyoming Natural Diversity Database website, http://www.blm.gov/style/medialib/blm/wy/wildlife/animal-assessmnts.Par.72741.File.dat/SwiftFox.pdf.

DeBaca, R. S., and J. R. Choate. 2002. "Biogeography of heteromyid rodents on the Great Plains." *Occasional Papers of the Museum* 212:1–22. Lubbock: Texas Tech University.

Desmond, M. J., J. A. Savidge, and K. M. Eskridge. 2000. "Correlations between burrowing owl and black-tailed prairie dog declines: A 7-year analysis." *Journal of Wildlife Management* 64:1067–75.

Desmond, M. J., J. A. Savidge, and T. F. Seibert. 1995. "Spatial patterns of burrowing owl nests within black-tailed prairie dog towns." *Canadian Journal of Zoology* 73:1375–79.

Dice, L. R. 1941. "Variation of the deer-mouse on the Sand Hills of Nebraska and adjacent areas." *Contributions Laboratory of Vertebrate Genetics, University of Michigan* 15:1–19.

Doby, J. F. 2006. *The Voice of the Coyote*. 2nd ed. Edison NJ: Castle Books. 386 pp.

Dragoo, J. W., J. R. Choate, T. L. Yates, and T. P. O'Farrell. 1990. "Evolutionary and taxonomic relationships among North American arid-land foxes." *Journal of Mammalogy* 71:318–32.

Dubay, S. A. 2000. "Mycophagy as a nutritional strategy for small mammals in the Rocky Mountains." PhD dissertation, University of Wyoming, Laramie.

Eisenberg, J. F. 1963. *The Behavior of Heteromyid Rodents*. University of California Publications in Zoology 69:1–114.

Epperson, C. J. 1978. "The biology of the bobcat in Nebraska." MS thesis, University of Nebraska–Lincoln.

Erickson, L. G. 2006. "The effect of age structure on movement rate, group cohesion, and leadership in female bison (*Bos bison*)." MS thesis, Iowa State University.

Errington, P. L. 1963. *Muskrat Populations*. Ames: Iowa State University Press.

Escherich, P. 1981. "Social behavior of the bushy-tailed woodrat, *Neotoma cinerea*." *University of California Publications in Zoology* 110:1–131.

Fairbanks, W. S. 1985. "Habitat use and foraging behavior of semicaptive bighorn sheep at Fort Robinson State Park, Nebraska." MS thesis, Colorado State University, Fort Collins.

Farney, J. P. 1975. "Natural history and northward dispersal of the hispid cotton rat in Nebraska." *Platte Valley Review* 3:11–16.

Farrar, J. 1974a. "Prairie life: Barking squirrel." *Nebraskaland Magazine* 52(9): 38–41 (prairie dog).

———. 1974b. "Prairie life: Passing of the buffalo." *Nebraskaland Magazine* 52(8): 36–39.

———. 1975b. "Prairie life: Bats." *Nebraskaland Magazine* 53(10): 34–37.

———. 1987. "Notes on Nebraska fauna: Ord's kangaroo rat." *Nebraskaland Magazine* 65(10): 64.

———. 1992. "Musquash: Grazer of the marsh." *Nebraskaland Magazine* 70(5): 14–23 (muskrat).

———. 1996. "Porky in the pigweeds." *Nebraskaland Magazine* 74(7): 32–35 (porcupine).

———. 1999. "Egg-sucking weasels." *Nebraskaland Magazine* 77(6): 30–35.

———. 2007. "Trader rat." *Nebraskaland Magazine* 85(4): 38–45 (woodrat).

Fichter, E. 1950. "Watching coyotes." *Journal of Mammalogy* 31:66–73.

Fichter, E., G. Schildman, and J. H. Sather. 1955. "Some feeding patterns of coyotes in Nebraska." *Ecological Monographs* 25:1–37.

Fichter, E., and J. K. Jones Jr. 1953. "The occurrence of the black-footed ferret in Nebraska." *Journal of Mammalogy* 34:385–88.

Findley, J. S., and C. Jones. 1964. "Seasonal distribution of the hoary bat." *Journal of Mammalogy* 45:461–70.

Forsberg, M. 1997. "Fox squirrels: Acrobats of the trees." *Nebraskaland Magazine* 75(2): 8–15.

———. 2000. "Coyote: Song dog of the High Plains." *Nebraskaland Magazine* 78(1): 10–19.

Forsyth, A. 1999. *Mammals of North America: Temperate and Arctic Regions*. Buffalo NY: Firefly Books. 350 pp.

Foster, N. S. 1990. "A report on black-tailed prairie dogs in Nebraska: Their biology, behavior, ecology, management, and response to a visual barrier fence." MS thesis, University of Nebraska–Lincoln.

Freeman, P. W. 1990. "Mammals." In *An Atlas of the Sand Hills*, edited by A. Bleed and C. Flowerday, 193–200. Resource Atlas No. 5a. Lincoln: Conservation and Survey Division, University of Nebraska (35 species listed).

———. 2005. "Nebraska's endangered species. Part 6: Threatened and endangered mammals." *Museum Notes, University of Nebraska State Museum* 120:1–4.

Freeman, P. W., and R. A. Benedict. 1993a. "Flat water mammals." *Nebraskaland Magazine* 71(6): 24–35 (Platte Valley mammals).

————. 1993b. *Mammals of the Platte River Valley.* Final Report to the U.S. Fish and Wildlife Service, Grand Island NE.

Freeman, P. W., K. N. Geluso, and J. S. Altenbach. 1997. "Nebraska's flying mammals." *Nebraskaland Magazine* 75(6): 38–47 (Nebraska bats).

Freeman, P. W., and H. H. Genoways. 1998. "Recent northern records of the nine-banded armadillo (Dasypodidae) in Nebraska." *Southwestern Naturalist* 43:491–95.

Frey, J. K. 1992. "Response of mammalian faunal element to climatic changes." *Journal of Mammalogy* 73:43–50.

Frost, J. S. 2007. "Small mammal and bird community composition in response to forest habitat structure and composition in the Niobrara River Valley, Nebraska." MS thesis, University of Nebraska–Lincoln.

Fullard, J. H., and M. B. Fenton. 1979. "Jamming bat echolocation: The clicks of arctiid moths." *Canadian Journal of Zoology* 57:647–49.

Garner, H. W. 1974. "Population dynamics, reproduction and activities of the kangaroo rat, *Dipodomys ordii,* in western Texas." *Graduate Studies, Texas Tech University* 7:1–28.

Geist, V. 1971. *Mountain Sheep: A Study in Behavior and Evolution.* Chicago: University of Chicago Press. 383 pp.

Geluso, K. 2006. "Bats in a human-made forest of central Nebraska." *Prairie Naturalist* 38:13–23.

Geluso, K. N., R. A. Benedict, and F. L. Kock. 2004. "Seasonal activity and reproduction in bats of east-central Nebraska." *Transactions of the Nebraska Academy of Sciences* 29:33–44.

Geluso, K., J. P. Damm, and E. W. Valdez. 2008. "Late-seasonal activity and diet of the evening bat in Nebraska." *Western North American Naturalist* 68:21–24.

Geluso, K., J. J. Huebschman, J. A. White, and M. A. Bogan. 2004. "Reproduction and seasonal activity of silver-haired bats in western Nebraska." *Western North American Naturalist* 64:353–58.

Genoways, H. H., J. D. Hoffman, P. W. Freeman, K. Geluso, R. A. Benedict, and J. J. Huebschman. 2008. *Mammals of Nebraska: Checklist, Key, and Bibliography.* Bulletin of the University of Nebraska State Museum 23.

Genoways, H. H., and D. A. Schlitter. 1967. "Northward dispersal of the hispid cotton rat in Nebraska and Missouri." *Transactions of the Kansas Academy of Sciences* 69:356–57.

Gersb, D. 1986. "Nebraska furbearers." *Nebraskaland Magazine* 64:18–34.

Gese, E. M., T. E. Stotts, and S. Grothe. 1996. "Interactions between coyotes and red foxes in Yellowstone National Park, Wyoming." *Journal of Mammalogy* 77:377–82.

Gilbert, F. F. 1969. "Analysis of basic vocalizations of the ranch mink." *Journal of Mammalogy* 50:625–27.

Glass, B. P. 1982. "Seasonal movements of Mexican free-tail bats *Tadarida brasiliensis mexicana* banded in the Great Plains." *Southwestern Naturalist* 27:127–33.

Goldingay, R. L., and J. S. Scheibe, eds. 2000. *Biology of Gliding Squirrels*. Fürth, Germany: Filander Verlag.

Goldman, E. A. 1950. *Raccoons of North and Middle America*. North American Fauna 60.

Gompper, M. E., and H. M. Hackett. 2005. "The long-term, range-wide decline of a once common carnivore: The eastern spotted skunk." *Animal Conservation* 8:195–201.

Goodrich, J. M., and S. W. Buskirk. 1998. "Spacing and ecology of North American badgers (*Taxidea taxus*) in a prairie-dog (*Cynomys leucurus*) complex." *Journal of Mammalogy* 79:171–79.

Gould, E., N. C. Negus, and A. Novicki. 1964. "Evidence for echolocation in shrews." *Journal of Experimental Biology* 156:19–38.

Grier, B. 1985. "Elk in the Pine Ridge." *Nebraskaland Magazine* 63(8): 6–9.

———. 1998a. "Bighorns." *Nebraskaland Magazine* 76(3): 10–21.

———. 1998b. "Pronghorn." *Nebraskaland Magazine* 76(7): 64–67.

———. 1998c. "The smallest and rarest." *Nebraskaland Magazine* 66(4): 18–25 (swift fox).

———. 2002. "Bighorns in the Wildcats." *Nebraskaland Magazine* 80(2): 10–17.

———. 2003. "Little fox of the shortgrass." *Nebraskaland Magazine* 81(4): 28–32 (swift fox).

Gubanyi, J. A. 2001. "Effects of high deer abundance on forests in eastern Nebraska." PhD dissertation, University of Nebraska–Lincoln.

Gunderson, H. L. 1968. "Mammals living in Nebraska." *Museum Notes, University of Nebraska State Museum* 47:1–8.

———. 1973. "Recent mammals of Crescent Lake National Wildlife Refuge, Garden County, Nebraska." *Nebraska Bird Review* 41:71–76.

———. 1979. "The life and times of the buffalo." *Nebraskaland Magazine* 57(6): 2–29, 47.

———. 1983. "Notes on Nebraska fauna: Porcupine." *Nebraskaland Magazine* 61(10): 49–50.

Gunderson, H. L., and B. R. Mahan. 1980. "Analysis of sonograms of American bison (*Bison bison*)." *Journal of Mammalogy* 61:379–81.

Haberman, C. G., and E. D. Fleharty. 1972. "Natural history notes on Franklin's ground squirrel in Boone County, Nebraska." *Transactions of the Kansas Academy of Sciences* 74:76–80.

Hall, E. R. 1951. *American Weasels.* Lawrence: University of Kansas Museum of Natural History.

———. 1955. *Handbook of the Mammals of Kansas.* Lawrence: University of Kansas Museum of Natural History.

———. 1981. *The Mammals of North America.* 2 vols. New York: John Wiley and Sons.

Harbaugh, M. J. 1941. "A study of the animal communities of a streamside forest succession." PhD dissertation, University of Nebraska–Lincoln.

Harder, A. K. W. 1966. "A comparison of habitat and learning behavior for rural and urban populations of fox squirrels (*Sciurus niger* Linnaeus)." MS thesis, University of Nebraska–Lincoln.

Hart, F. M., and J. A. King. 1996. "Distress vocalizations of young in two subspecies of *Peromyscus maniculatus.*" *Journal of Mammalogy* 47:287–93.

Hartnett, D. C., A. A. Steuter, and K. R. Hickman. 1997. "Comparative ecology of native and introduced ungulates." In *Ecology and Conservation of Great Plains Vertebrates*, edited by F. L. Knopf and F. B. Samson, 72–101. New York: Springer-Verlag.

Heffner, H. 2005. "Hearing and sound localization in the kangaroo rat (*Dipodomys merriami*)." *Journal of the Acoustical Society of America* 61:S59.

Henderson, F. R. 1979. "The status of prairie dogs in the Great Plains." *Proceedings of the Great Plains Wildlife Damage Control Conference* 4:101–10.

Henzlik, R. E. 1960. "A study of the animal ecology of a man-made forest in the Nebraska Sandhills." PhD dissertation, University of Nebraska–Lincoln.

———. 1965. "Biogeographic extensions into a coniferous forest plantation in the Nebraska Sandhills." *American Midland Naturalist* 74:87–94.

Higgins, K. E., E. D. Stukel, J. M. Goulet, and D. C. Backlund. 2002. *Wild Mammals of South Dakota.* Pierre: South Dakota Department of Game, Fish and Parks.

Hill, J. E., and J. D. Smith. 1984. *Bats: A Natural History.* Austin: University of Texas Press.

Hines, T. D., and R. M. Case. 1991. "Diet, home range, movements, and activity periods of swift fox in Nebraska." *Prairie Naturalist* 23:131–38.

Hines, T. D., R. M. Case, and R. Lock. 1981. "The swiftest fox." *Nebraskaland Magazine* 59(8): 20–27.

Hoffman, J. D., and H. H. Genoways. 2005. "Recent records of formerly extirpated carnivores in Nebraska." *Prairie Naturalist* 37:225–45.

Hoffman, J. D., H. H. Genoways, and J. R. Choate. 2007. "Long-distance dispersal and population trends of moose in the central United States." *Alces* 42:115–31.

Hoffmann, R. 2002. "Nebraska mountain lions." *Nebraskaland Magazine* 80(5): 46–47.

———. 2005. "Elk on the home place." *Nebraskaland Magazine* 83(8): 20–25.

—————. 2006. "Nebraska's bobcat population sound." *Nebraskaland Magazine* 84(9): 46.

Hoffmann, R., and C. Taylor. 2007. "Pronghorn." *Nebraskaland Magazine* 85(3): 38–45.

Hoffmann, R. S., and J. K. Jones Jr. 1970. "Influence of late-glacial and post-glacial events on the distribution of Recent mammals on the northern Great Plains." In *Pleistocene and Recent Environment of the Central Great Plains*, edited by W. Dort Jr. and J. K. Jones Jr., 355–94. Lawrence: University Press of Kansas.

Holyoak, D. T. 2001. *Nightjars and Their Allies*. New York: Oxford University Press.

Homolka, C. L. 1967. "Notes on Nebraska fauna: Black-footed ferret." *Nebraskaland Magazine* 45(8): 52–53.

Hones, T. 1980. "An ecological study of *Vulpes velox* in Nebraska." MS thesis, University of Nebraska–Lincoln.

Hoogland, J. L. 1995. *The Black-tailed Prairie Dog: Social Life of a Burrowing Mammal*. Chicago: University of Chicago Press.

—————, ed. 2006. *Conservation of the Black-tailed Prairie Dog*. Washington DC: Island Press.

Houston, D. B. 1982. *The Northern Yellowstone Elk: Ecology and Management*. New York: Macmillan.

Huebschman, J. J. 2007. "Distribution, abundance, and habitat associations of Franklin's ground squirrel." *Illinois Natural History Survey Bulletin* 38:1–57.

Hurt, J. J. 1969. "Notes on Nebraska fauna: Northern grasshopper mouse." *Nebraskaland Magazine* 47(6): 46–47.

Huwaldt, B. 2006. "Squirrel of a different color." *Nebraskaland Magazine* 84(3): 5 (fox squirrel).

Hyde, D. O. 1986. *Don Coyote: The Good Times and Bad Times of a Much Maligned American Original*. New York: Arbor House.

Irby, L. R., and J. E. Knight, eds. 1998. *Bison Ecology and Management in North America*. Bozeman: Montana State University Press.

Jochum, E. S. 1980. "An ecological study of red foxes in Nebraska." MS thesis, University of Nebraska–Lincoln.

Johnsgard, P. A. 2005. *Prairie Dog Empire: A Saga of the Shortgrass Prairie*. Lincoln: University of Nebraska Press. 244 pp.

—————. 2014a. "Secrets of the most sincerely dead: Agate Fossil Beds National Monument." *Prairie Fire*, November, 15–17. https://wwwprairiefirenewspaper.com/2014/11/secrets-of-the-most-sincerely-dead-agate-fossil-beds-national-monument.

—————. 2014b. "Secrets of the very long dead: Ashfall Fossil Beds State Historical Park." *Prairie Fire*, October, 1, 3, 4. https://wwwprairiefirenewspaper.com/2014/10/secrets-of-the-very-long-dead.

————. 2014c. "To kill a mountain lion." *Prairie Fire,* January, 18–19. https://www.prairiefirenewspaper.com/2014/02/to-kill-a-mountain-lion.

Jones, J. K., Jr. 1954. "Distribution of some Nebraskan mammals." *University of Kansas Publications, Museum of Natural History* 7:479–87.

————. 1957. "Checklist of mammals of Nebraska." *Transactions of the Kansas Academy of Science* 60:273–82.

————. 1964. *Distribution and Taxonomy of Mammals of Nebraska.* University of Kansas Publications, Museum of Natural History 16.

Jones, J. K., Jr., D. M. Armstrong, and J. R. Choate. 1985. *Mammals of the Northern Great Plains.* Lincoln: University of Nebraska Press (coverage includes North Dakota through Nebraska).

Jones, J. K., Jr., D. M. Armstrong, R. S. Hoffmann, and C. Jones. 1983. *Guide to Mammals of the Plains States.* Lincoln: University of Nebraska Press.

Jones, J. K., Jr., and J. R. Choate. 1980. "Annotated checklist of mammals of Nebraska." *Prairie Naturalist* 12:43–53.

King, J., ed. 1968. "Biology of *Peromyscus* (Rodentia)." *American Society of Mammalogists Special Publication* 2:1–593.

Kitchen, D. W. 1974. "Social behavior and ecology of the pronghorn." *Wildlife Monograph* 38:1–96.

Kjar Hirsch, K. J., J. Stubbendieck, and R. M. Case. 1984. "Relationships between vegetation, soils, and pocket gophers in the Nebraska Sand Hills." *Transactions of the Nebraska Academy of Science* 12:5–11.

Knowles, C. J., J. D. Proctor, and S. C. Forrest. 2002. "Black-tailed prairie dog abundance and distribution in the Great Plains based on historic and contemporary information." *Great Plains Research* 12:219–54.

Kruuk, H. 2006. *Otters, Ecology and Conservation.* Oxford: Oxford University Press. 280 pp.

Kurrus, J. 2007. "Black squirrels: An uncommonly common squirrel in Nebraska." *Nebraskaland Magazine* 85(8): 10–17 (fox squirrel).

Landholt, L. M., and H. Genoways. 2000. "Population trends in furbearers in Nebraska." *Transactions of the Nebraska Academy of Sciences* 26:97–110.

Laydet, F. 1988. *The Coyote.* Norman: University of Oklahoma Press.

Lemen, C. A., and P. W. Freeman. 1986. "Habitat selection and movement patterns in Sandhills rodents." *Prairie Naturalist* 18:129–41.

Lepri, J. J., M. Theodorides, and C. J. Wysocki. 1988. "Ultrasonic vocalizations by adult prairie voles, *Microtus ochrogaster.*" *Cellular and Molecular Life Sciences* 44:271–73.

Lerass, H. J. 1938. "Observations on the growth and behavior of harvest mice." *Journal of Mammalogy* 19:441–44.

Lock, R. A. 1973. "Status of the black-footed ferret and black-tailed prairie dog in Nebraska." In *Proceedings of the Black-footed Ferret and Prairie Dog Workshop*, edited by R. L. Linder and C. N. Hillman, 44–46. Brookings: South Dakota State University.

———. 1978. "Notes on Nebraska fauna: Black-footed ferret." *Nebraskaland Magazine* 56(7): 50.

Lueninghoener, E. W. 1973. "An investigation of the melanistic phase of the western fox squirrel in eastern Nebraska and western Iowa." MA thesis, University of Nebraska at Omaha.

Lund, D. E., and J. P. Farney. 1975. "Life history and chromosomal variation in Ord's kangaroo rat in Nebraska." *Platte Valley Review* 3:1–10.

MacDonald, N. F., and S. E. Hygnstrom. 1991. "Little dogs of the prairie." *Nebraskaland Magazine* 69(5): 24–31.

Mahan, B. R. 1978. "Aspects of American bison (*Bison bison*) social behavior at Fort Niobrara National Wildlife Refuge, Valentine, Nebraska, with special reference to calves." MS thesis, University of Nebraska–Lincoln.

Maher, C. R., and J. A. Byers. 1987. "Age-related changes in reproductive effort of male bison." *Behavioral Ecology and Sociobiology* 21:91–96.

Manning, R. W. 1983. "Habitat utilization of mammals in a man-made forest in the Sandhills region of Nebraska." MA thesis, University of Nebraska at Omaha.

Maxell, M. H., and L. N. Brown. l968. "Ecological distribution of rodents on the high plains of eastern Wyoming." *Southwestern Naturalist* 13:143–58.

McGrew, D. 1958. "Notes on Nebraska fauna: Long-tailed weasel." *Outdoor Nebraska* 36(9): 26–27.

McHugh, T. 1958. "Social behavior of the American buffalo (*Bison bison bison*)." *Zoologica* 43:1–40.

McKelvey, K. S. 2000. "History and distribution of lynx in the contiguous United States." In *Ecology and Conservation of Lynx in the United States*, edited by L. F. Ruggiero et al., 207–64. Niwot: University Press of Colorado.

Menzel, K. 1964. "Notes on Nebraska fauna: The bats." *Nebraskaland Magazine* 42(10): 58–59.

———. 1965. "Notes on Nebraska fauna: Gray squirrel." *Nebraskaland Magazine* 43(10): 58–59.

———. 1968. "Notes on Nebraska fauna: Gray fox." *Nebraskaland Magazine* 46(10): 44–45.

———. 1975. "The deer of Nebraska." *Nebraskaland Magazine* 53(4): 10–41.

Menzel, K., and H. Y. Suetsugu. 1966. "Reintroduction of antelope to the Sandhills of Nebraska." *Proceedings of the Annual Antelope States Workshop* 2:50–54.

Meserve, P. L. 1971. "Population ecology of the prairie vole in the western mixed prairie of Nebraska." *American Midland Naturalist* 86:417–33.

Miller, J. 2010. "Stereotypic vocalizations in harvest mice (Reithrodontomys): Harmonic structure contains prominent and distinctive audible, ultrasonic, and non-linear elements." *Journal of the Acoustical Society of America* 128:1501.

Miller, B., S. Forrest, and R. P. Reading. 1996. *Prairie Night: Black-footed Ferrets and the Recovery of Endangered Species.* Washington DC: Smithsonian Books. 320 pp.

Moehrenschlager, A., B. L. Cypher, K. Ralls, R. List, and M. A. Sovada. 2004. "Swift and kit foxes: Comparative ecology and conservation priorities of swift and kit foxes." In *The Biology and Conservation of Wild Canids*, edited by D. W. MacDonald and C. Sillerozubiri, 185–98. New York: Oxford University Press.

Müller-Schwarze, D. 2011. *The Beaver.* 2nd ed. Ithaca NY: Cornell University Press.

Müller-Schwarze, D., and L. Sun. 2003. *The Beaver: Natural History of a Wetlands Engineer.* Ithaca, NY: Cornell University Press.

Muric, O. J. 1951. *The Elk of North America.* Harrisburg PA: Stackpole Company.

Murrant, M. N., C. J. Bowman, B. Prizen, H. Mayberry, and P. Faure. 2013. "Ultrasonic vocalizations emitted by flying squirrels." *PLoS One* 8(8): e73045.

Neal, E. G. 1996. *The Natural History of Badgers.* New York: Facts on File.

Negus, L. P. 2002. "Small mammal population diversity in a wooded and cleared Platte River habitat." Senior research thesis, Department of Biology, University of Nebraska at Kearney.

Nowak, R. M. 1974. *The Cougar in the United States and Canada.* Washington DC: New York Zoological Society and U.S. Fish and Wildlife Service.

O'Gara, B. W., and J. D. Yoakum. 2004. *Pronghorn: Ecology and Management.* Boulder: University Press of Colorado.

Olson, T. L., and F. G. Lindzey. 2002. "Swift fox (*Vulpes velox*) home-range dispersal patterns in southeastern Wyoming." *Canadian Journal of Zoology* 80:2024–39.

Ostenson, B. T. 1947. "Ecologic and geographic variation in pelage color of the mammals in the Nebraska Sandhills and adjacent areas." PhD dissertation, University of Michigan, Ann Arbor.

Pefaur, J. E., and R. S. Hoffman. 1974. "Notes on the biology of the olive-backed pocket mouse on the northern Great Plains." *Prairie Naturalist* 6:7–15.

Phillips, C. L. 1966. "Ecology of the big brown bat (Chiroptera: Vespertilionidae) in northeastern Kansas." *American Midland Naturalist* 75:168–98.

Plettner, R. G. 1984. "Vital characteristics of the black-tailed jack rabbit in east-central Nebraska." MS thesis, University of Nebraska–Lincoln.

Pomerantz, S. M., and L. G. Clemens. 1981. "Ultrasonic vocalizations in male deer mice (*Peromyscus maniculatus bairdi*): Their role in male sexual behavior." *Physiology and Behavior* 27:869–72.

Powell, D. G., and R. M. Case. 1982. "Food habits of the red fox in Nebraska." *Transactions of the Nebraska Academy of Sciences* 10:13–16.

Pringle, L. 1977. *The Controversial Coyote: Predation, Politics and Ecology.* New York: Harcourt Brace Jovanovich.

Putnam, R. 1988. *The Natural History of Deer.* Ithaca NY: Comstock Press. 191 pp.

Quimby, D. C. 1951. "The life history and ecology of the jumping mouse, *Zapus hudsonicus.*" *Ecological Monographs* 21:61–95.

Reid, F. A. 2006. *A Field Guide to the Mammals of North America.* Boston: Houghton Mifflin.

Rinella, S. 2009. *America Buffalo: In Search of a Lost Icon.* New York: Spiegel and Grau.

Robertson, K. 1965. "Notes on Nebraska fauna: Western harvest mouse." *Nebraskaland Magazine* 4(11): 58–59.

———. 1968. "Notes on Nebraska fauna: Bushy-tailed wood rat." *Nebraskaland Magazine* 4(2): 52–53.

Roehrs, Z. P. 2004. "Biogeography and population dynamics of the prairie dog in Nebraska from 1965 to 2003." MS thesis, University of Nebraska–Lincoln.

Royce, M. S. 1989. *The Jackson Elk Herd: Intensive Wildlife Management in North America.* New York: Cambridge University Press.

Roze, U. 1989. *The North American Porcupine.* Washington DC: Smithsonian Institution Press.

Ruggiero, L. F., ed. 1994. *American Marten, Fisher, Lynx and Wolverine in the Western United States.* Denver CO: USDA Forest Service GTR RM-254. 184 pp.

Ryden, H. 1977. *God's Dog: A Celebration of the North American Coyote.* New York: Coward, McCann and Geohagen.

Sather, J. H. 1953. "The life history, habits and economic status of the Great Plains muskrat." PhD dissertation, University of Nebraska–Lincoln.

———. 1959. *Nebraska Muskrats.* Lincoln: Nebraska Game, Forestation and Parks Commission.

Sawyer, H., F. Lindzey, and D. McWhirter. 2005. "Mule deer and pronghorn migration in western Wyoming." *Wildlife Society Bulletin* 33:1266–73.

Schaffer, D. H., ed. 1960. "Notes on Nebraska fauna: Spotted skunk." *Outdoor Nebraska* 38(4): 26–27.

Schauster, E. R., E. M. Gese, and A. M. Kitchen. 2002. "Population ecology of swift foxes (*Vulpes velox*) in southeastern Colorado." *Canadian Journal of Zoology* 80:307–19.

Schildman, G. 1955. "Notes on Nebraska fauna: Thirteen-striped ground squirrel." *Outdoor Nebraska* 3(3): 26.

Schmidt, C. A. 2003. *Conservation Assessment for the Silver-Haired Bat in the Black Hills National Forest South Dakota and Wyoming.* Custer SD: USDA Forest Service, Black Hills National Forest. 22 pp.

Schmidt, C. A., P. D. Sudman, S. R. Marquardt, and D. S. Licht. 2004. *Inventory of Mammals at Ten National Park Service Units in the Northern Great Plains.* Keystone SD: Northern Great Plains Region, National Park Service.

Schmidt, R. H. 1981. "Prey preferences of red foxes: The small mammal component." MS thesis, University of Nebraska–Lincoln.

Schmidt-French, B., E. Gillam, and M. B. Fenton. 2006. "Vocalizations emitted during mother-young interactions by captive eastern red bats *Lasiurus borealis* (Chiroptera: Vespertilionidae)." *Acta Chiropterologica* 8(2): 477–84.

Schrad, M. C. 1976. "The effects of grazing management on small mammal density and diversity in Nebraska Sandhills." MA thesis, University of Nebraska at Omaha.

Schwartz, C. W., and E. R. Schwartz. 1959. *The Wild Mammals of Missouri.* Kansas City MO: Smith-Grieves Co.

———. 2016. *The Wild Mammals of Missouri.* 3rd ed. Columbia: University of Missouri Press, and Jefferson City: Missouri Department of Conservation.

Seal, U., E. T. Thorne, M. Bogan, and S. Anderson, eds. 1989. *Conservation Biology and the Black-footed Ferret.* New Haven CT: Yale University Press.

Sidle, J. G., D. H. Johnson, and B. R. Euliss. 2001. "Estimated areal extent of colonies of black-tailed prairie dogs in the northern Great Plains." *Journal of Mammalogy* 82:928–36.

Silvia, T. D. 1995. "Riparian habitats of the central Platte as a corridor for dispersal of small mammals in Nebraska." MS thesis, University of Nebraska–Lincoln.

Skryja, D. D. 1970. "Some aspects of the ecology of the least chipmunk (*Eutamias minimus operarius*) in the Laramie Mountains of southeast Wyoming." MS thesis, University of Wyoming, Laramie.

———. 1974. "Reproductive biology of the least chipmunk (*Eutamias minimus*) in southwestern Wyoming." *Journal of Mammalogy* 55:221–24.

Slobodchikoff, C. N., B. S. Perla, and J. L. Verdolin. 2009. *Prairie Dogs: Communication and Community in an Animal Society.* Cambridge MA: Harvard University Press.

Snyder, A. M. 2003. "Food habits of coyotes in central Nebraska." MS thesis, University of Nebraska at Kearney.

Springer, J. T. 1986. "Immediate effects of a spring fire on small mammal populations in a Nebraska mixed-grass prairie." *Proceedings of the North American Prairie Conference* 10:1–5.

Springer, J. T., and A. W. Voigt. 1999. "Small mammal populations in a prairie / riparian forest ecotone." *Proceedings of the North American Prairie Conference* 16:127–32.

Steuter, A. A., and L. Hidinger. 1999. "Comparative ecology of bison and cattle on mixed-grass prairie." *Great Plains Research* 9:329–42.

Steuter, A. A., E. M. Steinauer, G. L. Hill, P. A. Bowers, and L. L. Tieszen. 1995. "Distribution and diet of bison and pocket gophers in a Sandhills prairie." *Ecological Applications* 5:756–66.

Stillings, B. A. 1999. "Ecology of elk in northwestern Nebraska: Demographics, effect of human disturbance, and characteristics of calving habitat." MS thesis, University of Nebraska–Lincoln.

Suetsugu, H. 1962. "Notes on Nebraska fauna: Kangaroo rat." *Outdoor Nebraska* 40(5): 34–35.

———. 1967. "Notes on Nebraska fauna: The bobcat." *Nebraskaland Magazine* 45(5): 50–51.

———. 1969. "Notes on Nebraska fauna: Spotted ground squirrel." *Nebraskaland Magazine* 47(8): 52–53.

———. 1975. "The pronghorn antelope." *Nebraskaland Magazine* 53(9): 18–34.

Sundstrom, C., W. G. Hepworth, and K. L. Diem. 1973. "Abundance, distribution and food habits of the pronghorn." *Wyoming Game and Fish Department Bulletin* 12:1–61.

Svenden, G. F. 1976. "Vocalizations of the long-tailed weasel (*Mustela frenata*)." *Journal of Mammalogy* 57:398–99.

Taulman, J. F., and L. W. Robbins. 1996. "Recent range expansion and distributional limits of the nine-banded armadillo in the United States." *Journal of Biogeography* 23:635–48.

Taylor, W. P. 1956. *The Deer of North America*. Harrisburg PA: Stackpole County, and Washington, DC: Wildlife Management Institute.

Tische, J. 1960. "Notes on Nebraska fauna: Short-tailed shrew." *Outdoor Nebraska* 38(6): 26–27.

Toweill, D. E., and J. W. Thomas, eds. 2002. *North American Elk: Ecology and Management*. Washington DC: Smithsonian Institution Press.

Turbak, G. 1995. *Pronghorn: Portrait of the American Antelope*. New York: Cooper Square Publishing. 138 pp.

Tworek, F. A. 1977. "Effects of mowing and burning on small mammal populations of a restored prairie." MS thesis, University of Nebraska at Omaha.

Vacanti, P. L. 1981. "Effects of controlled burning on small mammal populations of a restored tallgrass prairie." MA thesis, University of Nebraska at Omaha.

Van Wormer, J. 1969. *World of the American Elk*. Philadelphia PA: J. B. Lippincott. 159 pp.

VerCauteren, K. C. 1993. "Home range and movement characteristics of female white-tailed deer at DeSoto National Wildlife Refuge." MS thesis, University of Nebraska–Lincoln.

Verts, B. J. 1967. *The Biology of the Striped Skunk*. Urbana: University of Illinois Press.

Wallmo, O. C., ed. 1981. *Mule and Black-tailed Deer of North America.* Lincoln: University of Nebraska Press.

Waring, G. H. 1970. "Sound communications of black-tailed, white-tailed, and Gunnison's prairie dogs." *American Midland Naturalist* 83:167–85.

Webb, O. L., and J. K. Jones Jr. 1952. "An annotated checklist of Nebraska bats." *University of Kansas Publications, Museum of Natural History* 5:269–79.

Weigand, J. P. 1964. "Notes on Nebraska fauna: The lynx." *Nebraskaland Magazine* 42(11): 58–59.

———. 1965. "Notes on Nebraska fauna: Kit fox." *Nebraskaland Magazine* 43(12): 58–59 (swift fox).

Weihe, J. M., and J. F. Cassel. 1978. "Checklist of North Dakota mammals (revised)." *Prairie Naturalist* 10:81–88.

Wengeler, W. R., D. A. Kelt, and M. L. Johnson. 2010. "Ecological consequences of invasive lake trout on river otters in Yellowstone National Park." *Biological Conservation* 143:1144–53.

White, T., ed. 2001a. "Bighorn sheep roam the Wildcat Hills again." *Nebraskaland Magazine* 79(3): 6.

———, ed. 2001b. "Wandering moose makes stop in Madison County." *Nebraskaland Magazine* 79(5): 7.

———, ed. 2002. "Black bear observed in western Nebraska in May." *Nebraskaland Magazine* 80(6): 8.

Whittaker, J. O., Jr. 1980. *National Audubon Society Field Guide to North American Mammals.* Rev. ed. New York: A. A. Knopf. 935 pp.

Wilson, D., and J. F. Hare. 2004. "Ground squirrel uses ultrasonic alarms." *Nature* 430:523.

Wilson, D. E., and S. Ruff. 1999. *The Smithsonian Book of North American Mammals.* Washington DC: Smithsonian Institution Press. 816 pp.

Wilson, G. M., and J. R. Choate. 1997. "Taxonomic status and biogeography of the southern bog lemming on the central Great Plains." *Journal of Mammalogy* 78:444–58.

Wolff, J. O. 1998. "Breeding strategies, mate choice, and reproduction in American bison." *Oikos* 83:529–44.

Wood, R. 1966. "Notes on Nebraska fauna: Beaver." *Nebraskaland Magazine* 44(11): 42–43.

Woods, S. E., Jr. 1980. *The Squirrels of Canada.* Ottawa, ON: National Museums of Canada.

Zeveloff, S. I. 2002. *Raccoons: A Natural History.* Washington, DC: Smithsonian Institution Press.

Allen, A. A. 1924. "A contribution to the life history and economic status of the screech owl (*Otus asio*)." *Auk* 41:1–16.

American Ornithologists' Union. 1998. *The A.O.U. Checklist of North American Birds*. 7th ed. Washington DC: AOU, plus annual supplements through 2019 (*Auk* 134:751–73). See also Chesser et al. 2018.

Anderson, S. H., and J. R. Squires. 1997. *The Prairie Falcon*. Austin: University of Texas Press.

Angell, T. 1969. "A study of the ferruginous hawk: Adult and brood behavior." *Living Bird* 8:225–41.

Austin, G. R. 1964. *The World of the Red-tailed Hawk*. Philadelphia: Lippincott.

Austin, O. L., ed. 1968. *Life Histories of North American Cardinals, Grosbeaks, Buntings, Towhees, Finches, Sparrows and Allies*. 3 vols. Vol. 1: Cardinalidae, Passerellidae (part). Vol. 2: Passerellidae (part). Vol. 3: Passerellidae (part), Calcariidae. Washington DC: Smithsonian Institution Press.

Baicich, P. J., and C. J. O. Harrison. 1997. *A Guide to the Nests, Eggs and Nestlings of North American Birds*. 2nd ed. New York: Academic Press.

Bailey, A. M., R. J. Niedrach, and A. L. Bailey. 1953. *The Red Crossbills of Colorado*. Denver Museum of Natural History, Museum Pictorial 9:1–64.

Baker, J. A. 1967. *The Peregrine*. New York: Harper and Row.

Baker, M. C., and J. T. Boylan. 1999. "Singing behavior, mating associations and reproductive success in a population of hybridizing lazuli and indigo buntings." *Condor* 181:493–503.

Balda, R. P., and G. C. Bateman. 1973. "The breeding biology of the piñon jay." *Living Bird* 11:5–42.

Balgooyen, T. G. 1976. "Behavior and ecology of the American kestrel (*Falco sparverius* L.) in the Sierra Nevada of California." *University of California Publications in Zoology* 103:1–83.

Banko, W. 1960. *The Trumpeter Swan. North American Fauna* 63. Washington, DC: U.S. Fish and Wildlife Service.

Beason, R. C., and E. C. Franks. 1974. "Breeding behavior of the horned lark." *Auk* 91:65–74.

Benkman, C. W. 1993. "Adaptation to single resources and the evolution of crossbill (*Loxia*) diversity." *Ecological Monographs* 63:305–25.

Benkman, C. W., T. L. Parchman, and E. Mezquida. 2010. "Patterns of coevolution in the adaptive radiation of crossbills." *Annals of the New York Academy of Sciences* 1206:1–16.

Bennett, J. R., and D. A. Keinath. 2001. *Distribution and Status of the Yellow-billed Cuckoo (Coccyzus americanus) in Wyoming*. Sheridan WY: Wolf Creek Charitable Foundation.

Bent, A. C. 1919. *Life Histories of North American Diving Birds*. United States National Museum Bulletin 107:1–245.

———. 1921. *Life Histories of North American Gulls and Terns*. U.S. National Museum Bulletin 113 (Laridae).

———. 1926. *Life Histories of North American Marsh Birds*. U.S. National Museum Bulletin 135 (Ardeidae, Gruidae, Rallidae).

———. 1927. *Life Histories of North American Shorebirds*. I. U.S. National Museum Bulletin 142 (Phalaropodidae, Recurvirostridae, Scolopacidae [part]).

———. 1929. *Life Histories of North American Shorebirds*. II. U.S. National Museum Bulletin 146 (Scolopacidac [part], Charadriidae).

———. 1937. *Life Histories of North American Birds of Prey*. Part 1. U.S. National Museum Bulletin 167 (Falconiformes, except Falconidae).

———. 1938. *Life Histories of North American Birds of Prey*. Part 2. U.S. National Museum Bulletin 170 (Falconidae, Tytonidae, Strigidae).

———. 1939. *Life Histories of North American Woodpeckers*. U.S. National Museum Bulletin 174 (Picidae).

———. 1940. *Life Histories of North American Cuckoos, Goatsuckers, Hummingbirds, and Their Allies*. U.S. National Museum Bulletin 176 (Psittacidae, Cuculidae, Alcedinidae, Caprimulgidae, Apodidae, Trochilidae).

———. 1942. *Life Histories of North American Flycatchers, Larks, Swallows, and Their Allies*. U.S. National Museum Bulletin 179 (Tyrannidae, Alaudidae, Hirundinidae).

———. 1946. *Life Histories of North American Jays, Crows, and Titmice*. U.S. National Museum Bulletin 191 (Corvidae, Paridae).

———. 1948. *Life Histories of North American Nuthatches, Wrens, Thrashers, and Their Allies*. U.S. National Museum Bulletin 195 (Sittidae, Certhiidae, Cinclidae, Troglodytidae, Mimidae).

———. 1949. *Life Histories of North American Thrushes, Kinglets, and Their Allies*. U.S. National Museum Bulletin 196 (Turdidae, Polioptilidae, Regulidae).

———. 1950. *Life Histories of North American Wagtails, Shrikes, Vireos, and Their Allies*. U.S. National Museum Bulletin 197 (Motacillidae, Bombycillidae, Ptilogonatidae, Laniidae, Sturnidae, Vireonidae).

———. 1953. *Life Histories of North American Wood Warblers*. U.S. National Museum Bulletin 203 (Parulidae, Ictariidae).

———. 1958. *Life Histories of North American Blackbirds, Orioles, Tanagers, and Allies*. U.S. National Museum Bulletin 211 (Icteridae, Thraupidae).

Bergen, T. 1987. "A multivariate hierarchical examination of habitat selection in *Tyrannis verticalis.*" MS thesis, University of Nebraska–Lincoln.

Bergin, T. 1987. "A multivariate hierarchical examination of habitat selection in *Tyrannus verticalis.*" MS thesis, University of Nebraska–Lincoln.

Bergman, R. D., P. Swain, and M. W. Weller. 1970. "A comparative study of nesting Forster's and black terns." *Wilson Bulletin* 82: 435–44.

Best, L. B. 1972. "First-year effects of sagebrush control on two sparrows." *Journal of Wildlife Management* 36:534–44.

Bicak, T. K. 1977. "Some eco-ethological aspects of a breeding population of long-billed curlews (*Numenius americanus*) in Nebraska." MA thesis, University of Nebraska at Omaha.

Bly, B., L. Snyder, and T. VerCauteren. 2008. "Migration chronology, nesting ecology, and breeding distribution of mountain plover (*Charadrius montanus*) in Nebraska." *Nebraska Bird Review* 76:120–28.

BNA (*Birds of North America*). Life histories of more than 600 breeding species of North American birds, originally printed from 1992 to 2002 as a series of monographs sponsored by the Academy of Natural Sciences, Philadelphia, and the American Ornithologists' Union, Washington DC. The Cornell University Laboratory of Ornithology has updated many of these as *The Birds of North America Online* and reprinted those that have not been updated. They are cited at the end of each species' profile by their original BNA account numbers, followed parenthetically by the names of the author(s) and most recent publication date.

Bock, C. E. 1970. "The ecology and behavior of the Lewis' woodpecker." *University of California Publications in Zoology* 92:1–100.

Bomberger, M. M. 1982. "Aspects of the breeding biology of Wilson's phalarope in western Nebraska." MS thesis, University of Nebraska–Lincoln.

Braaten, D. J. 1975. "Observations at three brown creeper nests in Itasca State Park." *Loon* 47:110–13.

Brewer, D. 2001. *Wrens, Dippers, and Thrashers: A Guide to the Wrens, Dippers, and Thrashers of the World.* New Haven CT: Yale University Press.

Brogie, M. A. 2017. "The official list of the birds of Nebraska: 2017." *Nebraska Bird Review* 85:179–97.

Brogie, M. A., and M. J. Mossman. 1983. "Spring and summer birds of the Niobrara Valley Preserve area, Nebraska." *Nebraska Bird Review* 51:44–51.

Brown, C. R., and M. B. Brown. 1996. *Coloniality in the Cliff Swallow.* Chicago: University of Chicago Press.

———. 2001. *Birds of the Cedar Point Biological Station.* Occasional Papers of the Cedar Point Biological Station No. 1. Lincoln NE.

Brown, C. R., M. B. Brown, P. A. Johnsgard, J. Kren, and W. C. Scharf. 1996. "Birds of the Cedar Point Biological Station area, Keith and Garden Counties, Nebraska: Seasonal occurrence and breeding data." *Transactions of the Nebraska Academy of Sciences* 29:91–108.

Brown, L., and D. Amadon. 1968. *Eagles, Hawks and Falcons of the World.* 2 vols. New York: McGraw-Hill.

Brown, M. B., L. R. Dinan, and J. G. Jorgensen. 2016. *2016 Interior Least Tern and Piping Plover Monitoring, Research, Management, and Outreach Report for the Lower Platte River, Nebraska.* Joint Report of the Tern and Plover Conservation Partnership and the Nongame Bird Program of the Nebraska Game and Parks Commission. Lincoln NE.

Brown, M. B., S. Dinsmore, and C. R. Brown. 2012. *Birds of Southwestern Nebraska.* Lincoln: Conservation and Survey Division, University of Nebraska.

Brown, M. B., and P. A. Johnsgard. 2013. *Birds of the Central Platte River Valley and Adjacent Counties.* Lincoln: Zea E-Books and University of Nebraska Digital Commons. digitalcommons.unl.edu/zeabook/15. Print copies are available at www.lulu.com/product/paperback/.

Burger, J., and L. M. Miller. 1977. "Colony and nest site selection in white-faced and glossy ibises." *Auk* 94:664–75.

Busby, W. H., and J. L. Zimmerman. 2001. *Kansas Breeding Bird Atlas.* Lawrence: University Press of Kansas.

Butterfield, J. D. 1969. "Nest-site requirements of the lark bunting in Colorado." MS thesis, Colorado State University, Fort Collins.

Byers, C., J. Curson, and U. Olsson. 1995. *Sparrows and Buntings: A Guide to the Sparrows and Buntings of North America and the World.* Boston: Houghton Mifflin.

Canterbury, J. L. 2007. "Songs of the wild: Temporal and geographical distinctions in the acoustic properties of the songs of the yellow-breasted chat." PhD dissertation, University of Nebraska–Lincoln.

Canterbury, J. L., and P. A. Johnsgard. 2000. "A century of breeding birds in Nebraska." *Nebraska Bird Review* 68:89–101. https://digitalcommons.unl.edu /biosciornithology/15/.

Chantlier, P. 1995. *Swifts: A Guide to the Swifts and Treeswifts of the World.* 2nd ed. Fresh Meadows NY: Pica Press.

Chesser, R. T., K. J. Burns, C. Cicero, J. L. Dunn, A. W. Kratter, I. J. Lovette, P. C. Rasmussen, J. V. Remsen Jr., D. F. Stotz, B. M. Winger, and K. Winker. 2018. "Checklist of North and Middle American Birds." American Ornithological Society. http://checklist.aou.org/taxa.

Clark, R. J. 1975. *A Field Study of the Short-eared Owl* Otus flammeus *(Pontoppidan) in North America.* Wildlife Monographs 47.

Clawson, S. D. 1980. "Comparative ecology of the northern oriole and the orchard oriole in western Nebraska." MS thesis, University of Nebraska–Lincoln.

Clement, P. 1993. *Finches & Sparrows: An Identification Guide.* Princeton, NJ: Princeton University Press.

———. 2001. *Thrushes.* Princeton NJ: Princeton University Press.

Combellack, C. R. B. 1954. "A nesting of violet-green swallows." *Auk* 71:435–42.

Cornwell, G. W. 1963. "Observations on the breeding biology and behavior of a nesting population of belted kingfishers." *Condor* 65:426–31.

Creighton, P. D. 1971. *Nesting of the Lark Bunting in North-Central Colorado. Grassland Biome.* U.S. International Biological Program, Technical Report No. 29. Denver.

Currier, P. J., G. R. Lingle, and J. G. VanDerwalker. 1985. *Migratory Bird Habitat on the Platte and North Platte Rivers in Nebraska.* Grand Island NE: Whooping Crane Habitat Maintenance Trust.

Curson, J., D. Quinn, and D. Beadle. 1994. *Warblers of the Americas: An Identification Guide.* Boston: Houghton Mifflin.

Davis, D. E. 1959. "Observations on territorial behavior of least flycatchers." *Wilson Bulletin* 71:73–85.

Davis, J., G. F. Fisher, and B. S. Davis. 1963. "The breeding biology of the western flycatcher." *Condor* 65:337–82.

Desmond, M. J. 1991. "Ecological aspects of burrowing owl nesting strategies in the Nebraska Panhandle." MS thesis, University of Nebraska–Lincoln.

Desmond, M. J., and J. A. Savidge. 1996. "Factors influencing burrowing owl nest densities and numbers in western Nebraska." *American Midland Naturalist* 136:143–48.

Dickson, J. G. 1992. *The Wild Turkey: Biology and Management.* New York: Stackpole Books.

Dilger, W. C. 1956. "Hostile behavior and reproductive isolating mechanisms in the avian genera *Catharus* and *Hylocichla*." *Auk* 73:313–53.

Dinsmore, S. 2001. "Population biology of the mountain plover in southern Phillips County, Montana." PhD dissertation, Colorado State University, Fort Collins.

Dority, B., E. Thompson, S. Kaskie, and L. Tschauner. 2017. *The Economic Impact of the Annual Crane Migration on Central Nebraska.* Kearney: University of Nebraska, Bureau of Business and Technology.

Dorn, J. L., and R. D. Dorn. 1990. *Wyoming Birds.* Cheyenne WY: Mountain West Publishing. 138 pp.

Ducey, J. E. 1988. *Nebraska Birds: Breeding Status and Distribution.* Omaha: Simmons-Boardman Books.

———. 1989. "Birds of the Niobrara River Valley." *Transactions of the Nebraska Academy of Sciences* 17:37–60.

————. 2000. *Birds of the Untamed West: The History of Birdlife in Nebraska 1750–1875.* Omaha: Making History Press.

Dunham, D. W. 1964. "Reproductive displays of the warbling vireo." *Wilson Bulletin* 76:170–73.

————. 1966. "Territorial and sexual behavior in the rose-breasted grosbeak." *Zeitschrift für Tierpsychologie* 23:438–51.

Dunkle, S. W. 1977. "Swainson's hawks on the Laramie Plains, Wyoming." *Auk* 94:65–71.

Eckhardt, R. C. 1976. "Polygyny in the western wood pewee." *Condor* 78:561–62.

Edson, J. M. 1943. "A study of the violet-green swallow." *Auk* 60:396–403.

Ellis, F. H., S. R. Swengel, G. W. Archibald, and C. B. Keple. 1998. "A sociogram for the cranes of the world." *Behavioural Processes* 43:125–51.

Ellison, A. E., and C. M. White. 2001. "Breeding biology of mountain plovers (*Charadrius montanus*) in the Uinta Basin." *Western North American Naturalist* 61:223–28.

Emlen, S. T., J. D. Rising, and W. L. Thompson. 1975. "A behavioral and morphological study of sympatry in the indigo and lazuli buntings of the Great Plains." *Wilson Bulletin* 87:145–79.

Enderson, J. H. 1964. "A study of the prairie falcon in the central Rocky Mountain region." *Auk* 81:332–52.

Errington, P. L., F. Hamerstrom, and F. N. Hamerstrom. 1940. "The great horned owl and its prey in the north-central United States." *Iowa State College Research Bulletin* 277:758–850.

Erwin, W. G. 1935. "Some nesting habits of the brown thrasher." *Journal of the Tennessee Academy of Science* 10:179–204.

Faanes, C. E., and G. R. Lingle. 1995. *Breeding Birds of the Platte Valley of Nebraska.* Jamestown ND: Northern Prairie Wildlife Research Center.

Farrar, J., ed. 1985. "Birds of Nebraska." Special issue, *Nebraskaland* 63(1).

————. 2004. "Birding Nebraska." Special issue, *Nebraskaland Magazine* 82(1).

Faulkner, D. W. 2010. *Birds of Wyoming.* Greenwood Village CO: Roberts and County

Fehon, J. H. 1955. "Life-history of the blue-gray gnatcatcher (*Polioptila caerulea caerulea*)." PhD dissertation, Florida State University.

Ferguson-Lees, J., and D. A. Christie. 2001. *Raptors of the World.* Boston: Houghton Mifflin.

Ficken, M. S. 1962. "Agonistic behavior and territory in the American redstart." *Auk* 79:607–32.

Ficken, M. S., and R. W. Ficken. 1962. "Some aberrant characters of the yellow-breasted chat." *Auk* 79:468–71.

————. 1966. "Behavior of myrtle warblers in captivity." *Bird-Banding* 37:273–79.

Fitzner, J. N. 1978. "The ecology and behavior of the long-billed curlew (*Numenius americanus*) in southeastern Washington." PhD dissertation, Washington State University, Pullman.

Forsythe, D. M. 1972. "Observations on the nesting biology of the long-billed curlew." *Great Basin Naturalist* 32:88–90.

Frederickson, L. H. 1970. "Breeding biology of American coots in Iowa." *Wilson Bulletin* 82:445–57.

———. 1971. "Common gallinule breeding biology and development." *Auk* 88:914–19.

Frydendall, M. J. 1967. "Feeding ecology and territorial behavior of the yellow warbler." PhD dissertation, Utah State University, Logan.

Gessner, D. 2008. *Return of the Osprey: A Season of Flight and Wonder.* New York: Ballantine Books.

Glinski, R. L. 1998. *The Raptors of Arizona.* Tucson: University of Arizona Press.

Goodwin, D. 1967. *Pigeons and Doves of the World.* London: British Museum (Natural History).

Goodwin, R. A. 1960. "A study of the ethology of the black tern, *Chlidonias niger surinamensis*." PhD dissertation, Cornell University, Ithaca NY.

Graul, W. D. 1975. "Breeding biology of the mountain plover." *Wilson Bulletin* 87:6–31.

Greer, R. D., and S. H. Anderson. 1989. "Relationships between population demography of McCown's longspurs and habitat resources." *Condor* 91:609–19.

Griscom, L., and A. Sprunt, eds. 1957. *The Warblers of America.* New York: Devin-Adair.

Hamilton, R. C. 1975. *Comparative Behavior of the American Avocet and the Black-necked Stilt (Recurvirostridae).* A.O.U. Monographs No. 17. American Ornithologists' Union.

Hancock, J., and H. Elliott. 1978. *The Herons of the World.* New York: Harper and Row.

Hancock, J., and J. Kushlan. 1984. *The Heron Handbook.* New York: Harper and Row.

Hanson, H. C., and C. W. Kossack. 1963. *The Mourning Dove in Illinois.* Illinois Department of Conservation Technical Bulletin No. 2. Urbana.

Harrap, S., and D. Quinn. 1996. *Tits, Nuthatches & Treecreepers.* London: Christopher Helm.

Heinrich, B. 1979. *Ravens in Winter.* New York: Summit Books.

Hespenheide, H. A. 1964. "Competition and the genus *Tyrannus*." *Wilson Bulletin* 76:265–81.

Hickey, J. J., ed. 1969. *Peregrine Falcon Populations: Their Biology and Decline.* Madison: University of Wisconsin Press.

Higgins, K. F., and L. M. Kirsch. 1975. "Some aspects of the breeding biology of the upland sandpiper in North Dakota." *Wilson Bulletin* 87:96–102.

Hofslund, P. B. 1959. "A life history of the yellowthroat, *Geothlypis trichas.*" *Proceedings of the Minnesota Academy of Science* 27:144–74.

Holcomb, L. C. 1972. "Traill's flycatcher breeding biology." *Nebraska Bird Review* 40:50–67.

Horn, H. S. 1970. "Social behavior of nesting Brewer's blackbirds." *Condor* 72:15–23.

Hostetter, D. R. 1961. "Life history of the Carolina junco, *Junco hyemalis* Brewster." *Raven* 32:97–170.

Howell, J. C. 1942. "Notes on the nesting habits of the American robin (*Turdus migratorius* L.)." *American Midland Naturalist* 28:529–603.

Hubbard, J. P. 1969. "The relationships and evolution of the *Dendroica coronata* complex." *Auk* 86:393–432.

Hyde, A. S. 1939. *The Life History of the Henslow's Sparrow.* University of Michigan Miscellaneous Publications.

Jaramillo, A., and P. Burke. 1999. *New World Blackbirds: The Icterids.* Princeton, NJ: Princeton University Press.

Jenni, D. A. 1969. "A study of the ecology of four species of herons during the breeding season at Lake Alice, Alachua County, Florida." *Ecological Monographs* 39:245–70.

Jenniges, J. J., and R. G. Plettner. 2008. "Least tern nesting at human created habitats in central Nebraska." *Waterbirds* 31:274–82.

Johnsgard, P. A. 1965. *Handbook of Waterfowl Behavior.* Ithaca NY: Cornell University Press.

———. 1973. *Grouse and Quails of North America.* Lincoln: University of Nebraska Press. https://digitalcommons.unl.edu/bioscigrouse/1.

———. 1975a. *North American Game Birds of Upland and Shoreline.* Lincoln: University of Nebraska Press.

———. 1975b. *Waterfowl of North America.* Bloomington: Indiana University Press.

———. 1979a. *Birds of the Great Plains: Breeding Species and Their Distribution.* Lincoln: University of Nebraska Press. See also 2009 Supplement and Revised Maps: https://digitalcommons.unl.edu/bioscibirdsgreatplains/1/.

———. 1979b. "The breeding birds of Nebraska." *Nebraska Bird Review* 47:3–14. https://digitalcommons.unl.edu/johnsgard/10.

———. 1980a. "Copulatory behavior in the American bittern." *Auk* 97:868–69.

———. 1980b. "Where have all the curlews gone?" *Natural History*, August, 30–34.

———. 1981a. *The Plovers, Sandpipers and Snipes of the World.* Lincoln: University of Nebraska Press.

————. 1981b. *Those of the Gray Wind: The Sandhill Cranes*. New York: St. Martin's Press.

————. 1983. *The Cranes of the World*. Bloomington: Indiana University Press.

————. 1987a. *Diving Birds of North America*. Lincoln: University of Nebraska Press.

————. 1987b. "The ornithogeography of the Great Plains states." *Prairie Naturalist* 10:97–112. https://digitalcommons.unl.edu/johnsgard/8.

————. 1990. *Hawks, Eagles and Falcons of North America: Biology and Natural History*. Washington, DC: Smithsonian Institution Press.

————. 1991. *Crane Music: A Natural History of American Cranes*. Washington DC: Smithsonian Institution Press. Repr., Lincoln: University of Nebraska Press, 1997.

————. 1993. *Cormorants, Darters and Pelicans of the World*. Washington DC: Smithsonian Institution Press.

————. 1994. *Arena Birds: Sexual Selection and Behavior*. Washington DC: Smithsonian Institution Press.

————. 1997a. *The Avian Brood Parasites: Deception at the Nest*. New York: Oxford University Press.

————. 1997b. *The Hummingbirds of North America*. 2nd ed. Washington DC: Smithsonian Institution Press.

————. 1998. "A half century of Christmas Bird Counts at Lincoln and Scottsbluff, Nebraska." *Nebraska Bird Review* 66:74–84. http://digitalcommons.unl.edu/nebirdrev/38.

————. 2001a. "A century of ornithology in Nebraska: A personal view." In *Contributions to the History of North American Ornithology*, vol. 2, edited by W. E. Davis and J. A. Jackson, 329–55. Boston: Nuttall Ornithological Club.

————. 2001b. *Prairie Birds: Fragile Splendor in the Great Plains*. Lawrence: University Press of Kansas.

————. 2002a. *Grassland Grouse and Their Conservation*. Washington DC: Smithsonian Institution Press.

————. 2002b. *North American Owls: Biology and Natural History*. 2nd ed. Washington, DC: Smithsonian Institution Press.

————. 2003a. "Great gathering on the Great Plains." *National Wildlife* 41(3): 20–29. https://digitalcommons.unl.edu/johnsgard/38.

————. 2003b. *Lewis and Clark on the Great Plains: A Natural History*. Lincoln: University of Nebraska Press.

————. 2009. "The wings of March." *Prairie Fire*, March, 1, 17–19. https://www.prairiefirenewspaper.com/2009/03/ nature-notes-wings-of-march.

————. 2010a. "The drums of April." *Prairie Fire*, April, 12–13. https://www.prairiefirenewspaper.com/2010/04/the-drums-of-april.

————. 2010b. "The peregrines of Nebraska." *Prairie Fire*, August, 12–14. http://www.prairiefirenewspaper.com/2010/08/the-peregrine-falcons-of-nebraska.

————. 2010c. "Snow geese of the Great Plains." *Prairie Fire*, February, 12–15. http://www.prairiefirenewspaper.com/2010/3/.

————. 2011a. "The feathers of winter." *Prairie Fire*, December, 17–20. http://www.prairiefirenewspaper.com/2011/12/the-feathers-of-winter.

————. 2011b. *A Nebraska Bird-Finding Guide*. Lincoln: Zea E-Books and University of Nebraska Digital Commons. https://digitalcommons.unl.edu/zeabook/51/.

————. 2011c. *Rocky Mountain Birds: Birds and Birding in the Central and Northern Rocky Mountains*. Lincoln: Zea E-Books and University of Nebraska Digital Commons. https://digitalcommons.unl.edu/zeabook/7/.

————. 2011d. *The Sandhill and Whooping Cranes: Ancient Voices over America's Wetlands*. Lincoln: University of Nebraska Press.

————. 2013a. *The Birds of Nebraska*. Rev. ed. Lincoln: Zea E-Books and University of Nebraska Digital Commons.

————. 2013b. *Wings over the Great Plains: The Central Flyway*. Lincoln: Zea E-Books and University of Nebraska Digital Commons. https://digitalcommons.unl.edu/zeabook/13.

————. 2014. "What are blue Ross's geese?" *Nebraska Bird Review* 82:81–85. https://digitalcommons.unl.edu/cgi/viewcontent.cgi?article=2348.

————. 2015a. *Birding Nebraska's Central Platte Valley and Rainwater Basin*. Lincoln: Zea E-Books and University of Nebraska Digital Commons. https://digitalcommons.unl.edu/zeabook/36.

————. 2015b. *A Chorus of Cranes: The Cranes of North America and the World*. Boulder: University Press of Colorado.

————. 2015c. *Global Warming and Population Responses among Great Plains Birds*. Lincoln: Zea E-Books and University of Nebraska Digital Commons. https://digitalcommons.unl.edu/zeabook/.

————. 2016a. *The North American Geese: Their Biology and Behavior*. Lincoln: Zea E-Books and University of Nebraska Digital Commons. https://digitalcommons.unl.edu/zeabook/44/.

————. 2016b. *The North American Grouse: Biology and Behavior*. Lincoln: Zea E-Books and University of Nebraska Digital Commons. https://digitalcommons.unl.edu/zeabook/.

————. 2016c. *The North American Sea Ducks: Their Biology and Behavior*. Lincoln: Zea E-Books and University of Nebraska Digital Commons. https://digitalcommons.unl.edu/zeabook/50/.

————. 2016d. *Swans: Their Biology and Natural History*. Lincoln: Zea E-Books and University of Nebraska Digital Commons. https://digitalcommons.unl.edu/zeabook/38/.

———. 2017a. *The North American Perching and Dabbling Ducks.* Lincoln: Zea E-Books and University of Nebraska Digital Commons. https://digitalcommons .unl.edu/zeabook/53/.

———. 2017b. *The North American Quails, Partridges and Pheasants.* Lincoln: Zea E-Books and University of Nebraska Digital Commons. https://digitalcommons .unl.edu/zeabook/58/.

———. 2017c. *The North American Whistling-Ducks, Pochards and Stifftails.* Lincoln: Zea E-Books and University of Nebraska Digital Commons. https:// digitalcommons.unl.edu/zeabook/54/.

———. 2018a. *The Birds of Nebraska.* 2nd ed. Lincoln: Zea E-Books and University of Nebraska Digital Commons. https://digitalcommons.unl.edu/zeabook/17/.

———. 2018b. *The Ecology of a Tallgrass Treasure: Audubon's Spring Creek Prairie.* Lincoln: Zea E-Books and University of Nebraska Digital Commons. https:// digitalcommons.unl.edu/zeabook.

———. 2018c. *A Naturalist's Guide to the Great Plains.* Lincoln: Zea E-Books and University of Nebraska Digital Commons. https://digitalcommons.unl.edu /zeabook /63/.

———. 2019. *Wyoming Wildlife: A Natural History.* Lincoln: Zea E-Books and University of Nebraska Digital Commons. https://digitalcommons.unl.edu /zeabook/73/.

———. 2020. "Audubon's Lillian Annette Rowe Sanctuary: A Refuge, a River, and a Half-Million Cranes."

Johnsgard, P. A., and J. Dinan. 2005. "Habitat associations of Nebraska birds." *Nebraska Bird Review* 73:20–25. https://digitalcommons.unl.edu/nebbirdrev /1104/.

Johnsgard, P. A., and R. W. Wood. 1968. "Distributional changes and interactions between prairie chickens and sharp-tailed grouse in the Midwest." *Wilson Bulletin* 80:173–88.

Johnson, D. H., L. D. Igl, J. A. Shaffer, and J. P. DeLong, eds. 2019. *The Effects of Management Practices on Grassland Birds.* U.S. Geological Survey Professional Paper 1842. https://doi.org/10.3133/pp1842.

Jorgensen, J. G. 2004. *An Overview of the Shorebird Migration in the Eastern Rainwater Basin, Nebraska.* Nebraska Ornithological Union Occasional Paper No. 8.

———. 2012. *Birds of the Rainwater Basin, Nebraska.* Lincoln: Nebraska Game and Parks Commission. outdoornebraska.gov/rainwaterbasin/.

Jorgensen, J. G., and L. R. Dinan. 2018. *2017 Nebraska Bald Eagle Nest Report.* Nongame Bird Program of the Nebraska Game and Parks Commission.

Joyner, D. E. 1975. "Nest parasitism and brood-related behavior of the ruddy duck." PhD dissertation, University of Nebraska–Lincoln.

Kangarise, C. M. 1979. "Breeding biology of Wilson's phalarope in North Dakota." *Bird-Banding* 50:12–22.

Kendeigh, S. C. 1941. "Territorial and mating behavior of the house wren." *Illinois Biological Monographs* 18:1–120.

Kilham, L. 1961. "Reproductive behavior of red-bellied woodpeckers." *Wilson Bulletin* 73:237–54.

———. 1966. "Reproductive behavior of hairy woodpeckers. I. Pair formation and courtship." *Wilson Bulletin* 78:251–65.

———. 1968. "Reproductive behavior in white-breasted nuthatches." *Auk* 85:477–92.

———. 1972. "Reproductive behavior in white-breasted nuthatches." *Auk* 89:115–29.

———. 1973. "Reproductive behavior in the red-breasted nuthatch. I. Courtship." *Auk* 90:597–609.

———. 1974. "Early breeding season behavior of downy woodpeckers." *Wilson Bulletin* 84:407–18.

Killpack, M. L. 1970. "Notes on sage thrasher nestlings in Colorado." *Condor* 72:486–88.

Kingery, H., ed. 1998. *Colorado Breeding Bird Atlas*. Denver: Colorado Division of Wildlife.

König, C., F. Weick, and J.-H. Becking. 1999. *Owls: A Guide to the Owls of the World*. New Haven CT: Yale University Press.

Krapu, G. L., D. A. Brandt, P. J. Kinzel, and A. T. Pearse. 2014. *Spring Migration Ecology of the Mid-continent Sandhill Crane Population with an Emphasis on Use of the Central Platte River Valley, Nebraska*. Wildlife Monographs 16.

Kren, J. 1996. "Proximate and ultimate mechanisms of red-winged blackbird (*Agelaius phoeniceus*) responses to interspecific brood parasitism." PhD dissertation, University of Nebraska–Lincoln.

Kroodsma, R. L. 1970. "North Dakota species pairs. I. Hybridization in buntings, grosbeaks and orioles. II. Species' recognition behavior of territorial male rose-breasted and black-headed grosbeaks (*Pheucticus*)." PhD dissertation, North Dakota State University, Fargo.

Labedz, T. 1998. "Birds." In *An Atlas of the Sand Hills*, 2nd ed., edited by A. Bleed and C. Flowerday, 161–80. Resource Atlas No. 5b. Lincoln: Conservation and Survey Division, University of Nebraska (312 species).

Lawrence, L. de K. 1949. "Notes on nesting pigeon hawks at Pimisi Bay, Ontario." *Wilson Bulletin* 61:15–25.

Lederer, R. J. 1977. "Winter feeding territories in the Townsend's solitaire." *Bird-Banding* 48:11–18.

Lewis, J. C. 1973. *The World of the Wild Turkey*. Philadelphia: Lippincott.

Lincer, J. L., and K. Steenhof, eds. 1997. *The Burrowing Owl, Its Biology and Management*. Includes the Proceedings of the First International Burrowing Owl Symposium. Raptor Research Foundation, *Journal of Raptor Research* Report 9.

Lingle, G. R. 1989. "Winter raptor use of the Platte and North Platte River Valleys in south-central Nebraska." *Prairie Naturalist* 21:1–16.

———. 1994. *Birding Crane River: Nebraska's Platte*. Grand Island, NE: Harrier Publishing County

Lingle, G. R., and G. L. Krapu. 1986. "Winter ecology of bald eagles in south-central Nebraska." *Prairie Naturalist* 18:65–78.

Lumsden, H. G. 1965. *Displays of the Sharptail Grouse*. Ontario Department of Lands and Forests Technical Series Research Report No. 66.

Lunk, W. A. 1962. "The rough-winged swallow *Stelgidopteryx ruficollis* (Vieillot): A study based on its breeding biology in Michigan." *Publications of the Nuttall Ornithological Club* 4:1–155.

Lynch, W. 2007. *Owls of the United States and Canada*. Baltimore MD: Johns Hopkins University Press.

Madge, S. 1993. *Crows and Jays: A Guide to the Crows, Jays and Magpies of the World*. Boston: Houghton Mifflin.

Maxwell, G. R., II. 1970. "Pair formation, nest-building and egg-laying of the common grackle in northern Ohio." *Ohio Journal of Science* 70:284–91.

Maxwell, G. R., II, and L. S. Putnam. 1972. "Incubation, care of young, and nest success of the common grackle (*Quiscalus quiscala*) in northern Ohio." *Auk* 89:349–59.

McAllister, N. M. 1958. "Courtship, hostile behavior, nest-establishment and egg-laying in the eared grebe." *Auk* 75:290–311.

McIntyre, J. 1988. *The Common Loon: Spirit of the Northern Lakes*. Minneapolis: University of Minnesota Press.

Meng, H. 1951. "The Cooper's hawk." PhD dissertation, Cornell University, Ithaca NY.

Meyeriecks, A. J. 1960. "Comparative behavior of four species of North American herons." *Publications of the Nuttall Ornithological Club* 2:1–158.

Mickey, F. W. 1943. "Breeding habits of McCown's longspur." *Auk* 60:181–209.

Mikkola, H. 2012. *Owls of the World*. Buffalo, NY: Firefly Books.

Mock, D. M. 1976. "Pair-formation displays of the great blue heron." *Wilson Bulletin* 88:185–230.

Mollhoff, W. J. 2001. *The Nebraska Breeding Bird Atlas, 1984–1989*. Lincoln: Nebraska Game and Parks Commission.

———. 2016. *The Second Nebraska Breeding Bird Atlas*. Bulletin of the University of Nebraska State Museum 29.

Moriarty, L. J. 1965. "A study of the breeding biology of the chestnut-collared longspur (*Calcarius ornatus*) in northeastern South Dakota." *South Dakota Bird Notes* 17:76–79.

Morse, D. H. 1989. *American Warblers: An Ecological and Behavioral Perspective.* Cambridge MA: Harvard University Press.

Nethersole-Thompson, D. 1975. *Pine Crossbills: A Scottish Contribution.* Berkhamstead, UK: T. and A. D. Poyser.

Newman, O. A. 1970. "Cowbird parasitism and nesting success of lark sparrows in southern Oklahoma." *Wilson Bulletin* 82:304–9.

Nice, M. M. 1943. "Studies in the life history of the song sparrow. II. The behavior of the song sparrow and other passerines." *Transactions of the Linnaean Society of New York* 6:1–238.

Nice, M. M., and N. E. Collias. 1961. "A nesting of the least flycatcher." *Auk* 78:145–49.

Nickell, W. P. 1965. "Habitats, territory and nesting of the catbird." *American Midland Naturalist* 73:433–78.

Niemeier, M. M. 1979. "Structural and functional aspects of vocal ontogeny in *Grus canadensis.*" PhD dissertation, University of Nebraska–Lincoln.

Norris, R. A. 1958. "Comparative biosystematics and life history of the nuthatches *Sitta pygmaea* and *Sitta pusilla.*" *University of California Publications in Zoology* 56:119–300.

Nuechterlein, G. 1981. "Courtship behavior and reproductive isolation between western grebe morphs." *Auk* 98:335–49.

Oberholser, H. C., and W. L. McAtee. 1920. *Waterfowl and Their Food Plants in the Sandhill Region of Nebraska.* Washington DC: U.S. Department of Agriculture Bulletin 794.

Odum, E. P. 1941. "Annual cycle of the black-capped chickadee." *Auk* 58:314–33, 518–35.

———. 1942. "Annual cycle of the black-capped chickadee." *Auk* 59:499–531.

Ohlendorf, H. M. 1976. "Comparative breeding ecology of phoebes in trans-Pecos Texas." *Wilson Bulletin* 88:255–71.

Ohlendorf, R. R. 1975. *Golden Eagle Country.* New York: A. A. Knopf.

Olson, S. T., and W. H. Marshall. 1952. *The Common Loon in Minnesota.* Occasional Papers of the Minnesota Museum of Natural History No. 5.

Orabona, A., S. Patla, L. Van Fleet, M. Grenier, B. Oakleaf, and Z. Walker. 2015. *Atlas of Birds, Mammals, Amphibians and Reptiles in Wyoming.* Lander: Wyoming Game and Fish Department. https://wgfd.wyo.gov/web2011/Departments /Wildlife/pdfs/wildlife_animalatlas0000328.pdf.

Orians, G. H., and G. M. Christman. 1968. "A comparative study of the behavior of red-winged, tricolored, and yellow-headed blackbirds." *University of California Publications in Zoology* 84:1–85.

Oring, L. W., and M. L. Knudson. 1973. "Monogamy and polyandry in the spotted sandpiper." *Living Bird* 11:59–73.

Ortega, C. P. 1998. *Cowbirds and Other Brood Parasites.* Tucson: University of Arizona Press.

Palmer, R. S., ed. 1962. *Loons through Flamingos.* Vol. 1 of *Handbook of North American Birds.* New Haven CT: Yale University Press.

———, ed. 1976. *Waterfowl.* Vols. 2 and 3 of *Handbook of North American Birds.* New Haven CT: Yale University Press.

———, ed. 1988. *Diurnal Raptors.* Vols. 4 and 5 of *Handbook of North American Birds.* New Haven CT: Yale University Press.

Panella, M. J. 2010. *Nebraska's At-Risk Wildlife.* Lincoln: Nebraska Game and Parks Commission.

Peer, D., S. K. Robinson, and J. R. Herkert. 2000. "Egg rejection by cowbird hosts in grasslands." *Auk* 117:892–901.

Penny, M. 2001. *Golden Eagle: Habitats, Life Cycles, Food Chains, Threats.* Chicago: Heinemann Raintree.

Peterjohn, B. G., and J. R. Sauer. 1999. "Population status of North American grassland birds from the North American Breeding Bird Survey, 1966–1996." In *Ecology and Conservation of Grassland Birds of the Western Hemisphere*, edited by P. D. Vickery and J. R. Herkert, 27–44. Studies in Avian Biology 19. Camarillo CA: Cooper Ornithological Society.

Peterson, A. J. 1955. "The breeding cycle in the bank swallow." *Wilson Bulletin* 67:235–86.

Peterson, R. A. 1995. *The South Dakota Breeding Bird Atlas.* Aberdeen: South Dakota Ornithologists' Union.

Platt, J. 1976. "Sharp-shinned hawk nesting and nest site selection in Utah." *Condor* 78:102–3.

Poling, T. D., and S. E. Hayslette. 2006. "Dietary overlap and foraging competition between mourning doves and Eurasian collared-doves." *Journal of Wildlife Management* 70:998–1004.

Porter, D. K., S. Strong, J. B. M. Giezentanner, and R. A. Ryder. 1975. "Nest ecology, productivity and growth of the loggerhead shrike on the shortgrass prairie." *Southwestern Naturalist* 19:429–36.

Poulin, R. G., G. L. Krapu, D. A. Brandt, and P. J. Kinzel. 2010. "Changes in agriculture and abundance of snow geese affect carrying capacity of sandhill cranes in Nebraska." *Journal of Wildlife Management* 74:479–88.

Power, H. W., III. 1966. "Biology of the mountain bluebird in Montana." *Condor* 68:351–71.

Prescott, K. W. 1965. *The Scarlet Tanager*. Trenton: New Jersey State Museum.

Rashid, S. 2010. *Small Mountain Owls*. Atglen PA: Schiffler.

Reller, A. W. 1972. "Aspects of behavioral ecology of red-headed and red-bellied woodpeckers." *American Midland Naturalist* 88:207–90.

Rich, T. C., C. J. Beardmore, H. Berlanga, P. J. Blancher, M. S. W. Bradstreet, G. S. Butcher, D. W. Demerest, E. H. Dunn, W. C. Hunter, E. E. Inig-Elias, J. A. Kennedy, A. M. Martell, A. O. Punjabi, D. N. Pashley, K. V. Rosenburg, C. M. Rusta, J. S. Wendt, and T. C. Will. 2004. *North American Landbird Conservation Plan*. Ithaca NY: Partners in Flight and Cornell University Laboratory of Ornithology.

Rising, J. D. 1970. "Morphological variation and evolution in some North American orioles." *Systematic Zoology* 19:315–51.

———. 1983. "The Great Plains hybrid zones." *Current Ornithology* 1:131–57.

———. 1996. *A Guide to the Identification and Natural History of the Sparrows of the United States and Canada*. New York: Academic Press.

Robins, J. D. 1971. "A study of Henslow's sparrow in Michigan." *Wilson Bulletin* 83:39–48.

Root, R. B. 1969. "The behavior and reproductive success of the blue-gray gnatcatcher." *Condor* 71:16–31.

Rosche, R. C. 1982. *Birds of Northwestern Nebraska and Southwestern South Dakota*. Chadron NE: Published by the author.

———. 1994. *Birds of the Lake McConaughy Area and the North Platte Valley, Nebraska*. Chadron NE: Published by the author.

Rosche, R. C., and P. A. Johnsgard. 1984. "Birds of Lake McConaughy and the North Platte Valley, Oshkosh to Keystone." *Nebraska Bird Review* 52:26–35.

Samuel, D. E. 1971. "The breeding biology of barn and cliff swallows in West Virginia." *Wilson Bulletin* 83:284–301.

Santee, R., and W. Granfield. 1939. "Behavior of the saw-whet owl on its nesting grounds." *Condor* 41:3–9.

Sauer, J. R., D. K. Niven, J. E. Hines, D. J. Ziolkowski Jr., K. L. Pardieck, J. E. Fallon, and W. A. Link. 2017. *The North American Breeding Bird Survey, Results and Analysis 1966–2015*. Version 2.07. Laurel MD: U.S. Geological Survey, Patuxent Wildlife Research Center.

Savage, C. 1997. *Bird Brains: The Intelligence of Crows, Ravens, Magpies and Jays*. Vancouver BC: Greystone Books.

Schaller, G. B. 1964. "Breeding behavior of the white pelican at Yellowstone Lake, Wyoming." *Condor* 66:3–23.

Scharf, W., J. Kren, P. A. Johnsgard, and L. R. Brown. 2008. *Body Weights and Distributions of Birds in Nebraska's Central and Western Platte Valley.* Digital Commons Papers in Ornithology. http://digitalcommons.unl.edu/biosciornithology /43/.

Schnell, J. H. 1958. "Nesting behavior and food habits of goshawks in the Sierra Nevada of California." *Condor* 60:377–403.

Schrantz, F. G. 1943. "Nest life of the yellow warbler." *Auk* 60:367–87.

Schukman, J. N. 1974. "Comparative nesting ecology of the eastern phoebe (*Sayornis phoebe*) and Say's phoebe (*Sayornis saya*) in west-central Kansas." MS thesis, Fort Hays State College, Fort Hays KS.

Scott, O. K. 1993. *A Birder's Guide to Wyoming.* Colorado Springs CO: American Birding Association.

Sharpe, R. S., W. R. Silcock, and J. G. Jorgensen. 2001. *The Birds of Nebraska.* Lincoln: University of Nebraska Press.

Short, L. L., Jr. 1965. "Hybridization in the flickers (*Colaptes*) of North America." *Bulletin of the American Museum of Natural History* 129:309–428.

———. 1982. *Woodpeckers of the World.* Greenville: Delaware Museum of Natural History.

Shunk, S. 2016. *Peterson Reference Guide to Woodpeckers of North America.* Boston: Houghton Mifflin.

Shurleff, L. L., and C. Savage. 1996. *The Wood Duck and the Mandarin: The Northern Wood Ducks.* Berkeley: University of California Press.

Sibley, D. A. 2000. *The Sibley Guide to Birds.* New York: A. A. Knopf.

Sibley, C. G., and L. L. Short Jr. 1959. "Hybridization in the buntings (*Passerina*) of the Great Plains." *Auk* 76:443–63.

———. 1964. "Hybridization in the orioles of the Great Plains." *Condor* 66:130–50.

Sibley, C. G., and D. A. West. 1959. "Hybridization in the rufous-sided towhees of the Great Plains." *Auk* 76:326–38.

Silcock, W. R., and J. G. Jorgensen. 2007. "Henslow's sparrow status in Nebraska." *Nebraska Bird Review* 75:13–16.

Smith, J. W., and C. W. Benkman. 2007. "A coevolutionary arms race causes ecological speciation in crossbills." *American Naturalist* 169:450–65.

Smith, R. L. 1963. "Some ecological notes on the grasshopper sparrow." *Wilson Bulletin* 75:159–65.

Snow, C. 1973a. *Golden Eagle* (Aquila chrysaetos). Habitat Management Series for Unique or Endangered Species. U.S. Department of the Interior, Bureau of Land Management Technical Note T-N-239.

———. 1973b. *Southern Bald Eagle* (Haliaeetus leucocephalus leucocephalus) *and Northern Bald Eagle* (Haliaeetus leucocephalus alascanus). Habitat Management

Series for Endangered Species. U.S. Department of the Interior, Bureau of Land Management Technical Note T-N-171.

———. 1974a. *Ferruginous Hawk (*Buteo regalis*)*. Habitat Management Series for Unique or Endangered Species. Report No. 13. U.S. Department of the Interior, Bureau of Land Management Technical Note T-N-255.

———. 1974b. *Prairie Falcon (*Falco mexicanus*)*. Habitat Management Series for Endangered Species. U.S. Department of the Interior, Bureau of Land Management Technical Note T-N-240.

Snyder, N. F. R., and H. Snyder. 2006. *Raptors of North America: Natural History and Conservation*. Minneapolis MN: Voyageur Press.

Stewart, R. E. 1953. "A life history study of the yellow-throat." *Wilson Bulletin* 65:99–115.

Stocek, R. F. 1970. "Observations on the breeding biology of the tree swallow." *Cassinia* 52:3–20.

Stokes, A. W. 1950. "Behavior of the goldfinch." *Wilson Bulletin* 62:107–27.

Stout, G. D., ed. 1967. *The Shorebirds of North America*. New York: Viking Press.

Sutherland, C. A. 1963. "Notes on the behavior of common nighthawks in Florida." *Living Bird* 2:31–39.

Sutton, G. M. 1949. *Studies of the Nesting Birds of the Edwin S. George Reserve. Part I. The Vireos*. University of Michigan Museum of Zoology Miscellaneous Publication 74.

Tacha, T. C. 1998. *Social Organization of Sandhill Cranes from Midcontinental North America*. Wildlife Monographs 99.

Tallman, D. A., D. L. Swanson, and J. S. Palmer. 2002. *Birds of South Dakota*. Aberdeen: South Dakota Ornithologists' Union.

Taylor, S. V., and V. M. Ashe. 1976. "The flight display and other behavior of male lark buntings (*Calamospiza melanochorys*)." *Bulletin of the Psychonomic Society* 7:527–29.

Thomas, R. H. 1946. "A study of eastern bluebirds in Arkansas." *Wilson Bulletin* 58:143–83.

Thompson, M. C., C. Ely, B. Gress, C. Otte, S. T. Patti, D. Seibel, and E. A. Young. 2011. *Birds of Kansas*. Lawrence: University Press of Kansas.

Tramontano, J. P. 1964. "Comparative studies of the rock wren and the canyon wren." MS thesis, University of Arizona, Tucson.

Turner, A. K., and C. Rose. 1989. *A Handbook to the Swallows and Martins of the World*. London: A. and C. Black.

Twedt, C. M. 1974. "Characteristics of sharp-tailed grouse display grounds in the Nebraska Sandhills." PhD dissertation, University of Nebraska–Lincoln.

U.S. Fish and Wildlife Service. 2008. *Birds of Conservation Concern 2008*. Arlington VA: U.S. Fish and Wildlife Service, Division of Migratory Bird Management.

Verbeek, N. A. M. 1967. "Breeding biology and ecology of the horned lark in alpine tundra." *Wilson Bulletin* 79:208–18.

Walkinshaw, L. H. 1944. "The eastern chipping sparrow in Michigan." *Wilson Bulletin* 56:193–205.

Watson, A. T. 1977. *The Hen Harrier*. Berkhamstead UK: T. and A. D. Poyser.

Weaver, R. L., and H. L. West. 1943. "Notes on the breeding of the pine siskin." *Auk* 60:492–503.

Weidensaul, S. 2015. *Peterson Reference Guide to the Owls of North America*. Boston: Houghton Mifflin.

West, D. A. 1962. "Hybridization in grosbeaks (*Pheucticus*) of the Great Plains." *Auk* 79:399–424.

White, H. C. 1953. "The eastern belted kingfisher in the Maritime Provinces." *Fisheries Research Board of Canada Bulletin* 97:1–44.

Wilcox, L. R. 1959. "A twenty year banding study of the piping plover." *Auk* 76:129–52.

Williams, L. 1952. "Breeding behavior of the Brewer blackbird." *Condor* 54:3–47.

Willson, M. F. 1964. "Breeding ecology of the yellow-headed blackbird." *Ecological Monographs* 36:51–77.

Winkler, H., D. A. Christie, and D. Nurney. 1995. *Woodpeckers: A Guide to the Woodpeckers of the World*. Boston: Houghton Mifflin.

Wright, R. 2019. *Peterson Reference Guide to the Sparrows of North America*. Boston: Houghton Mifflin.

Zelenak, J. R., J. J. Rotella, and A. R. Harmata. 1997. "Survival of fledgling ferruginous hawks in northern Montana." *Canadian Journal of Zoology* 97:152–56.

Zimmerman, J. 1983. "Cowbird parasitism of dickcissels in differed habitats and different nest densities." *Wilson Bulletin* 95:7–22.

———. 1993. *The Birds of Konza*. Lawrence: University Press of Kansas.

REPTILES AND AMPHIBIANS

Ballinger, R. E., J. D. Lynch, and P. H. Cole. 1979. "Distribution and natural history of amphibians and reptiles in western Nebraska, with ecological notes on the herptiles of Arapaho Prairie." *Prairie Naturalist* 22:65–74.

Ballinger, R. E., J. D. Lynch, and G. R. Smith. 2010. *Amphibians and Reptiles of Nebraska*. Oro Valley AZ: Rusty Lizard Press, and Lincoln: University of Nebraska Press (includes 13 amphibians and 48 reptiles).

Ballinger, R. E., J. W. Meeker, and M. Theis. 2000. "A checklist and distribution maps of the amphibians and reptiles of South Dakota." *Transactions of the Nebraska Academy of Sciences* 26:29–46.

Baxter, G. T., and M. D. Stone. 2011. *Amphibians and Reptiles of Wyoming*. 2nd ed. (1st ed. 1980). Cheyenne: *Wyoming Naturalist* (42 species).

Behler, J. L., and F. W. King. 1996. *The Audubon Society Field Guide to North American Reptiles and Amphibians.* New York: A. A. Knopf (includes 283 reptiles and 194 amphibians).

Benedict, R. 1996. "Snappers, soft-shells, and stinkpots: The turtles of Nebraska." University of Nebraska State Museum, *Museum Notes* 96:1–4 (nine species).

Breckenridge, W. J. 1943. "The life history of the black-banded skink." *American Midland Naturalist* 29:591–601.

Caldwell, J. P., and J. T. Collins. 1981. *Turtles in Kansas.* Lawrence KS: AMS Publishing.

Clark, D. R., and R. J. Hall. 1970. "Function of the blue tail coloration of the five-lined skink." *Herpetologica* 26: 271–74.

Collins, J. T. 1993. *Amphibians and Reptiles of Kansas.* University of Kansas Museum of Natural History Public Education Series 13 (includes color photographs and descriptions of all 97 Kansas herptiles).

Duellman, W. E., and L. Trueb. 1986. *Biology of Amphibians.* New York: McGraw-Hill.

Ernst, C. H., J. E. Lovich, and R. W. Barbour. 1994. *Turtles of the United States and Canada.* Washington, DC: Smithsonian Institution Press (includes all 56 turtle species of the United States and Canada).

Farrar, J. 1998. "Box turtles: Life in the fast lane." *Nebraskaland* 76(5): 24–33.

Ferraro, D. N.d. *A Guide to Snakes, Turtles, Frogs, Lizards, and Salamanders.* https://herpneb.unl.edu (online identification guide to Nebraska species).

Fischer, T. D., D. C. Backlund, K. F. Higgins, and D. E. Naugle. 1999. *Field Guide to South Dakota Amphibians.* Brookings: South Dakota State University Agricultural Extension Service Bulletin 733.

Fitch, H. S. 1954. "Life history and ecology of the five-lined skink, *Eumeces fasciatus.*" *University of Kansas Museum of Natural History Publications* 8:2–156.

————. 1968. "Natural history of the six-lined racerunner (*Cnemidophorus sexlineatus*)." *University of Kansas Museum of Natural History Publications* 11:11–62.

————. 1970. "Reproductive cycles in lizards and snakes." *University of Kansas Museum of Natural History, Miscellaneous Publications* 52:1–247.

————. 1975. "A demographic study of the ringneck snake (*Diadophis punctatus*) in Kansas." *University of Kansas Museum of Natural History, Miscellaneous Publications* 62:1–53.

————. 1985. "Variation in clutch and litter size in New World reptiles." *University of Kansas Museum of Natural History, Miscellaneous Publications* 76:1–76.

Fogell, D. D. 2010. *A Field Guide to the Amphibians and Reptiles of Nebraska.* Lincoln: Conservation and Survey Division, University of Nebraska (includes 14 amphibians and 48 reptiles).

Freeman, P. 1989. "Amphibians and Reptiles." In *An Atlas of the Sand Hills*, 2nd ed., edited by A. Bleed and C. Flowerday, 157–60. Resource Atlas No. 5b. Lincoln: Conservation and Survey Division, University of Nebraska (28 species).

Gibilisco, C. 1975. "Natural hybridization of *Bufo americanus* Holbrook and *Bufo woodhousei* Girard." MA thesis, University of Nebraska at Omaha.

Hammerson, G. A. 1999. *Amphibians and Reptiles in Colorado*. 2nd ed. Boulder: University Press of Colorado.

Holycross, A. T. 1995. "Movements and natural history of the prairie rattlesnake (*Crotalus viridis viridis*) in the Sandhills of Nebraska." MS thesis, University of Nebraska at Omaha.

Hudson, G. E. 1942. *The Amphibians and Reptiles of Nebraska*. Bulletin 24. Lincoln: Conservation and Survey Division, University of Nebraska.

Iverson, J. B. 1975. "Notes on Nebraska reptiles." *Transactions of the Kansas Academy of Science* 78:51–62.

———. 1977. "Notes on Nebraska reptiles." *Transactions of the Kansas Academy of Science* 80:55–59.

Iverson, J. B., and G. R. Smith. 1993. "Reproductive ecology of the painted turtle (*Chrysemys picta*) in the Nebraska Sandhills and across its range." *Copeia* 1993(1): 1–21.

Johnson, S. 2015. *Reptiles and Amphibians of North Dakota*. Bismarck: North Dakota Game and Fish Department.

Johnson, T. R. 2000. *The Amphibians and Reptiles of Missouri*. 2nd ed. Jefferson City: Missouri Department of Conservation.

Jones, S. M., and D. L. Droge. 1980. "Home range size and special distribution of two sympatric lizard species (*Sceloperus undulatus* and *Holbrookia maculata*) in the Sandhills of Nebraska." *Herpetologica* 36:127–32.

Joy, J. E., and D. Crews. 1985. "Social dynamics of group courtship behavior in male red-sided garter snake (*Thamnophis sirtalis parietalis*)." *Journal of Comparative Psychology* 99:145–49.

Kardong, K. V. 1980. "Gopher snakes and rattlesnakes: Presumptive Batesian mimicry." *Northwest Science* 54:1–4.

Kiesoe, A. M. 2006. *Field Guide to Amphibians and Reptiles of South Dakota*. Pierre: South Dakota Department of Game, Fish and Parks.

Klauber, L. M. 1972. *Rattlesnakes: Their Habits, Life Histories, and Influence on Mankind*. 2nd ed. 2 vols. Berkeley: University of California Press.

Krupa, J. 1994. "Breeding biology of the Great Plains toad in Oklahoma." *Journal of Herpetology* 28:217–24.

Kruse, K. C. 1978. "Causal factors limiting the distribution of leopard frogs in eastern Nebraska." PhD dissertation, University of Nebraska–Lincoln.

————. 1981. "Phonotactic responses of female northern leopard frogs (*Rana pipiens*) to *Rana blairi*, a presumed hybrid, and conspecific mating trills." *Journal of Herpetology* 13:145–50.

Legler, J. M. 1960. "Natural history of the ornate box turtle, *Terrapene ornata.*" *University of Kansas Museum of Natural History Publication* 11:527–660.

Littlejohn, M. J., and R. S. Oldham. 1968. "*Rana pipiens* complex mating call structure and taxonomy." *Science* 162:1993–95.

Lynch, J. D. 1978. "The distribution of leopard frogs (*Rana blairi* and *Rana pipiens*) (Amphibia, Anura, Ranidae) in Nebraska." *Journal of Herpetology* 12:157–62.

————. 1985. "Annotated checklist of the amphibians and reptiles of Nebraska." *Transactions of the Nebraska Academy of Sciences* 13:33–57.

Oliver, J. A. 1955. *The Natural History of North American Amphibians and Reptiles.* Princeton NJ: Van Nostrand.

Powell, R., R. Conant, and J. T. Collins. 2016. *Field Guide to Reptiles and Amphibians of Eastern and Central North America.* 4th ed. Boston: Houghton Mifflin (includes 501 species; coverage extends west to Wyoming border).

Rossman, D. A., N. B. Ford, and R. A. Seigel. 1996. *The Garter Snakes: Evolution and Ecology.* Norman: University of Oklahoma Press.

Sherbrooke, W. C., and G. A. Middendorf III. 2001. "Blood-squirting variability in horned lizards (*Phrynosoma*)." *Copeia* 2001:1114–22.

Sievert, G., and L. Sievert. 2018. *A Field Guide to Oklahoma's Amphibians and Reptiles.* Oklahoma City: Oklahoma Department of Wildlife Conservation.

Smith, H. M. 1946. *Handbook of Lizards: Lizards of the United States and Canada.* Ithaca NY: Cornell University Press (includes 136 species).

————. 1956. *Handbook of Amphibians and Reptiles of Kansas. University of Kansas Museum of Natural History, Miscellaneous Publication* 9:1–36 (see also Collins 1993).

Smith, H. M., and E. D. Brodie Jr. 1983. *Reptiles of North America: A Guide to Field Identification.* New York: St. Martin's Press (includes 278 species).

Somma, L. A. 1990. "Observations on the nesting ecology of the prairie skink (*Eumeces septentrionalis*) in Nebraska." *Bulletin of the Chicago Herpetological Society* 25:77–80.

Stebbins, R. C. 1985. *A Field Guide to Western Reptiles and Amphibians.* 2nd ed. Boston: Houghton Mifflin (includes 244 species).

Stebbins, R. C., and N. W. Cohen. 1995. *A Natural History of Amphibians.* Princeton NJ: Princeton University Press.

Werner, J. K., B. A. Maxwell, P. Hendricks, and D. L. Flath. 2004. *Amphibians and Reptiles of Montana.* Missoula MT: Mountain Press (includes 32 species).

Wheeler, G. C., and J. Wheeler. 1966. *The Amphibians and Reptiles of North Dakota.* Grand Forks: University of North Dakota Press (includes 28 species).

Wright, A. H., and A. A. Wright. 1949. *Handbook of the Frogs and Toads of the United States and Canada.* Ithaca NY: Cornell University Press.

———. 1957. *Handbook of the Snakes of the United States and Canada.* Ithaca NY: Cornell University Press.

Zim, H. S., and H. M. Smith. 2001. *Reptiles and Amphibians.* New York: Macmillan (includes 212 species).